Atomic and Nanometer-Scale Modification of Materials: Fundamentals and Applications

NATO ASI Series

Advanced Science Institutes Series

A Series presenting the results of activities sponsored by the NATO Science Committee, which aims at the dissemination of advanced scientific and technological knowledge, with a view to strengthening links between scientific communities.

The Series is published by an international board of publishers in conjunction with the NATO Scientific Affairs Division

A Life Sciences	Plenum Publishing Corporation
B Physics	London and New York
C Mathematical	Kluwer Academic Publishers
** and Physical Sciences**	Dordrecht, Boston and London
D Behavioural and Social Sciences	
E Applied Sciences	
F Computer and Systems Sciences	Springer-Verlag
G Ecological Sciences	Berlin, Heidelberg, New York, London,
H Cell Biology	Paris and Tokyo
I Global Environmental Change	

NATO-PCO-DATA BASE

The electronic index to the NATO ASI Series provides full bibliographical references (with keywords and/or abstracts) to more than 30000 contributions from international scientists published in all sections of the NATO ASI Series.
Access to the NATO-PCO-DATA BASE is possible in two ways:

– via online FILE 128 (NATO-PCO-DATA BASE) hosted by ESRIN,
Via Galileo Galilei, I-00044 Frascati, Italy.

– via CD-ROM "NATO-PCO-DATA BASE" with user-friendly retrieval software in English, French and German (© WTV GmbH and DATAWARE Technologies Inc. 1989).

The CD-ROM can be ordered through any member of the Board of Publishers or through NATO-PCO, Overijse, Belgium.

Series E: Applied Sciences - Vol. 239

Atomic and Nanometer-Scale Modification of Materials: Fundamentals and Applications

edited by

Phaedon Avouris

IBM Research Division,
T.J. Watson Research Center,
Yorktown Heights, New York, U.S.A.

Springer Science+Business Media, B.V.

Proceedings of the NATO Advanced Research Workshop on
Atomic and Nanometer-Scale Modification of Materials:
Fundamentals and Applications
Ventura, California, U.S.A.
August 16-21, 1992

Library of Congress Cataloging-in-Publication Data

Atomic and nanometer-scale modification of materials : fundamentals
  and applications / edited by Phaedon Avouris.
      p.   cm. -- (NATO ASI series. Series E, Applied sciences ; vol.
  239)
    Includes index.
    ISBN 978-94-010-4895-8   ISBN 978-94-011-2024-1 (eBook)
    DOI 10.1007/978-94-011-2024-1
    1. Nanostructure materials--Congresses. 2. Microstructure-
  -Congresses.   I. Avouris, Phaedon, 1945-   . II. Series: NATO ASI
  series. Series E, Applied sciences ; no. 239.
  TA418.9.N35A86   1993
  620.1'1299--dc20   93-1725

ISBN 978-94-010-4895-8

Printed on acid-free paper

This book contains the proceedings of a NATO Advanced Research Workshop held within the programme of activities of the NATO Special Programme on Nanoscale Science as part of the activities of the NATO Science Committee.

Other books previously published as a result of the activities of the Special Programme are:

NASTASI, M., PARKING, D.M. and GLEITER, H. (eds.), *Mechanical Properties and Deformation Behavior of Materials Having Ultra-Fine Microstructures.* (ASIE 233) 1993 ISBN 0-7923-2195-2

VU THIEN BINH, GARCIA, N. and DRANSFELD, K. (eds), *Nanosources and manipulation of Atoms under High Fields and Temperatures: Applications.* (E235) 1993 ISBN 0-7923-2266-5

LEBURTON, J.-P., PASCUAL J. and SOTOMAYOR TORRES, C. (eds.), *Phonons in Semiconductor Nanostructures.* (E236) 1993 ISBN 0-7923-2277-0

AVOURIS P. (ed.), *Atomic and Nanometer-Scale Modification of Materials: Fundamentals and Applications.* (E239) 1993 ISBN 0-7923-2334-3

BLÖCHL, P. E., JOACHIM, C. and FISHER, A. J. (eds.), *Computations for the Nano-Scale.* (E240) 1993 ISBN 0-7923-2360-2

CONTENTS

PREFACE

This volume contains the proceedings of the conference on "Atomic and Nanometer Scale Modification of Materials: Fundamentals and Applications" which was co-sponsored by NATO and the Engineering Foundation, and took place in Ventura, California in August 1992. The goal of the organizers was to bring together and facilitate the exchange of information and ideas between researchers involved in the development of techniques for nanometer-scale modification and manipulation, theorists investigating the fundamental mechanisms of the processes involved in modification, and scientists studying the properties and applications of nanostructures. About seventy scientists from all over the world participated in the conference.

It has been more than 30 years since Richard Feynman wrote his prophetic article: "There is Plenty of Room at the Bottom" (Science and Engineering, **23**, 22, 1960). In it he predicted that some day we should be able to store bits of information in structures composed of only 100 atoms or so, and thus be able to write all the information accumulated in all the books in the world in a cube of material one two-hundredths of an inch high. He went on to say, "the principles of physics, as far as I can see, do not speak against the possibility of maneuvering things atom by atom." Since that time there has been significant progress towards the realization of Feynman's dreams. Among the key advances was the development of the scanning tunneling microscope (STM), the atomic force microscope (AFM) and the rest of the scanned proximal-probe microscopes, and the realization that these probes can be used not only for the atomic and nanometer-scale imaging of surfaces, but as tools for the modification and manipulation of materials on the same scale. To affect these modifications one can utilize a variety of tip-sample interactions, including attractive and repulsive forces, electric-fields, and the effect of the electron current. As the papers in these proceedings show, the feasibility of a number of modifications involving small numbers of atoms or even single atoms or molecules has been demonstrated. These manipulations include: sliding and positioning of atoms or molecules on a surface, transferring atoms from a sample to the tip and redepositing them at another location on the sample, dissociating individual molecules, desorbing atoms, and inducing local heating and phase transitions. Moreover, tip-induced modifications can be performed under a wide range of conditions, i.e. in air, in ultra-high vacuum, at a solid-liquid interface, and even at a solid-solid interface (using the ballistic electron emission microscope, BEEM). At the same time our knowledge of the microscopic interaction mechanisms between tip and sample has improved significantly.

While the feasibility of atomic and molecular-scale modification has been proven, it may be some time before the required control and speed with which such manipulations are performed is achieved and a genuine atomic-scale technology is developed. However, atomic manipulation can and is currently being used as a powerful experimental tool for basic science to, for example, isolate or arrange atoms and molecules in precise configurations, probe their electronic structure and interactions with scanning tunneling microscopy (STS), subject them to intense electric-fields or high current densities and thus obtain novel information on inter-atomic and inter-molecular interactions, field-effects on chemistry, etc. Analytical applications, in which the structure of features in STM images is elucidated by taking these structures apart in a step-wise manner and imaging the resulting products, are already appearing. A particularly powerful combination involves low temperatures and atomic manipulation. At very low temperatures, the mobility of adsorbates is eliminated and thus inert substrates can be used as work benches on which, for example, molecular systems can be deposited and operated on with the tip or the tunneling electrons.

In the nanometer (tens of nanometers) regime, the technological implications are likely to be more immediate. Already the STM is used to provide proximity-focused electron beams for lithography. Unlike conventional high-energy electron lithography, the low-energy STM beam reduces the problems associated with electron backscattering and the generation of secondary electrons. A resolution of ~10nm, and exposure rates comparable to those of conventional electron lithography have been achieved. Significant progress has been made in the fabrication of nanometer-scale tip-arrays. The use of such arrays will undoubtedly enhance the lithographic exposure rates and will, in general, facilitate the integration of proximal-probe tools in nano-technology. The STM electron beam is also being used to locally deposit ultra-small features, such as thin metal lines or nanometer-scale magnets, through the electron-induced decomposition of the appropriate organo-metallic precursors.

Conventional lithography has been advancing, too. Optical lithography, the workhorse of the field, has greatly benefited from the introduction of new, very high contrast photoresists, phase-shifting masks and other optical technologies. It is, however, unlikely that features below ~100nm can be produced. High energy electron lithography using inorganic resists can lead to a resolution as high as 2nm! However, its throughput, as in the case of the STM, is very low. Currently, the best approach that optimizes resolution and throughput appears to be X-ray proximity printing capable of a 50nm resolution at a 1cm^2/s throughput. In addition to these technologies, a number of new ideas have been introduced and are currently being explored. One such idea involves a "natural nanolithography" in which very fine (nm) spheres (polymers, globular proteins) are rafted together in a Langmuir-Blodgett trough or through self-assembly to form in situ a periodic resist pattern. This pattern is then transferred by implanting or alloying through the interstices.

New ideas have also been introduced and significant advances have taken place in the technology for the growth of low-dimensional structures such as quantum wires (1D) and dots (OD). As 2D heterostructures are easily formed by molecular beam epitaxy (MBE) and chemical vapor deposition (CVD), an obvious way to obtain 1D and OD structures has been through patterning of the 2D structures. However, the etching processes used in the pattern transfer introduce damage that adversely affects the electrical properties. As a result, a number of "softer" fabrication approaches has been introduced. Of particular interest are techniques which utilize MBE and CVD deposition on vicinal (i.e. slightly misoriented) surfaces that form quantum wire structures (tilted and serpentine superlattices) through nucleation at well-organized steps. A very different approach utilizes light-pressure forces to manipulate the density profile of an atomic beam as it impinges onto a substrate. For example, if a 1D standing-wave is generated parallel to the substrate and perpendicular to an atomic beam, the light can, under the appropriate conditions, cause the atoms to become localized in a series of parallel stripes which would then lead to the growth of a series of lines on the substrate. This technique is very new and its potential remains to be evaluated.

Concurrent with the advances in fabrication, there have been advances in the understanding of the basic physics of nanostructures: in particular, of resonant tunneling phenomena and Coulomb blockade. The utilization of these phenomena has made it possible to manipulate single electrons and to build devices such as single-electron turnstiles, pumps and the single-electron transistor. So far, operation is limited to temperatures below 1K; however, with further reductions in dimensions and capacitances, operation at higher temperatures, even room temperature, may be possible. The ability of the STM to probe local electrical properties by establishing point-contacts with the tip has been utilized to study the behavior of conventional electronic devices such as the Schottky-diode at an extreme level of miniaturization. Atomic manipulation at low temperatures with the STM has proven the feasibility of switches involving atomic rather than electronic motion.

A very different approach to nano-fabrication that was discussed in the conference attempts to imitate nature and relies on self-assembly of molecules into supramolecular structures. There are several important advantages in this approach. One is its simplicity and speed, since once the appropriate molecules have been synthesized, self-assembly is spontaneous and does not require the intervention of the technologist. Another important characteristic of self-assembled structures is their extremely low defect density. This is due to the fact that the generated structures are at equilibrium and are thus inherently self-repairing. Finally, when organic molecules are used, the existing extensive knowledge on chemical synthesis can be used to modify and functionalize them, and customize their properties. So far, a number of uses of self-assembled layers including passivation layers, dielectric barriers, and chemical etch resists, have appeared. An exciting new application of self-assembly involves biologically-active proteins (enzymes) on electrode surfaces, where the enzymatic activity is controlled by the electrode potential. These developments raise hopes for the prospect of bio-electronic devices.

As structures and devices decrease in size, the need to precisely control their properties and composition becomes more and more demanding. Specific probes are required and several new proximal probes were discussed. Notable examples include the magnetic-STM which uses ferromagnetic tips and can provide magnetic contrast images with atomic resolution, and the photon-STM, which, by detecting light emission induced by the tunneling electrons, provides valuable insight into the chemical composition of nanometer-scale surface structures.

It appears that Feynman's speculations in the late '50's are rapidly becoming a reality. It is likely that in the next 10 years we will see further dramatic advances that will provide us with a precise control of matter on the atomic and nanometer scales. Nano-devices and nanotechnologies, some imitating natural systems, some based on totally new concepts, could then be realized which would profoundly affect our lives.

The meeting on Atomic and Nanometer Scale Modification of Materials was made possible through the effort of many individuals. Among them, I want to acknowledge and thank James Murday, who co-chaired the Conference, and Calvin Quate, Heinrich Rohrer and Gordon Fisher for their valuable advise and support. I also want to thank the Scientific Affairs Committee of NATO and the Engineering Foundation for their generous financial and administrative support. Additional support was provided by the National Research Foundation, the Office of Naval Research, the Army Research Office and the IBM Corporation. Finally, I would like to take the opportunity to express my gratitude to all participants for contributing to the success of the Conference.

Phaedon Avouris
Yorktown Heights, Sept. 1st, 1992.

ATOM MANIPULATION WITH THE SCANNING TUNNELING MICROSCOPE

D. M. EIGLER
IBM Research Division
Almaden Research Center
Dept. K34
650 Harry Road
San Jose, CA 95120 USA

ABSTRACT. Two processes for manipulating atoms and molecules with the scanning tunneling microscope are discussed along with their underlying physical mechanisms. It is shown how the reversible transfer of an atom between the tip and surface can be used to create a bistable switch. Several applications of atomic manipulation as a laboratory tool are presented.

Introduction

In an article titled *There's Plenty of Room at the Bottom*, Richard Feynman considered the prospect of being able to manipulate matter on the atomic scale.[1] Advances in scanning tunneling microscopy have made this prospect a reality; in certain circumstances we can use the scanning tunneling microscope (STM) to place single atoms at selected positions and build structures of our own design... *atom-by-atom*. In a sense we have learned to use the STM as both our eyes and hands in the world of atoms on surfaces. Here we review the processes that we have developed for manipulating atoms and molecules with the STM and describe how we are using this new capability as a laboratory tool.

Discussion

Our everyday experience leads us to tacitly assume that the process of imaging an object does not perturb the object (quantum mechanics teaches otherwise, and for Schrödinger's cat this is a matter of grave consequence). Similarly, the STM is conventionally thought of as a non-perturbative probe; that is, the act of imaging does not significantly perturb the imaged surface. While this is a reasonable approximation for most circumstances, there exists a regime where this approximation breaks down, namely, when the forces between the STM tip and an adsorbate are comparable in magnitude to the forces between the adsorbate and the surface.

The tip of an STM always exerts a force on the atoms of a surface. At large tip-surface separation this force comes from the Van der Waals interaction between the outermost tip atoms and the atoms of the surface. At closer distance the forces associated with chemical bonding become dominant. The application of a voltage between the tip and the surface will also result in electrostatic contributions to the force between the tip and surface. These forces are usually negligible because the tip is maintained at large enough separation from the surface, the voltages applied are

1

P. Avouris (ed.), Atomic and Nanometer-Scale Modification of Materials: Fundamentals and Applications, 1–10.

low enough, and the atoms on the surface are bound strong enough. What happens when the force between the tip and the surface is no longer negligible?

We came to be confronted with just this problem while studying the adsorption of xenon on a platinum surface. As a rule of thumb, the energy barrier to diffusion of an atom across a surface will scale in proportion to the adsorption energy. Similarly, the minimum force necessary to move an atom across a surface will scale with the minimum force necessary to pull an atom away from a surface. The bond between the xenon atom and the underlying platinum surface is comparatively weak, just 300 mV (7 KCal), and as a consequence it takes relatively little force to move a xenon atom laterally across a surface. In the process of studying the nucleation and growth of islands of xenon atoms on the platinum surface we were compelled to evaluate whether the presence of the tip was causing the islands to distort and thus invalidate our observations. To answer this question we had to consider the nature of the forces between the tip and the adsorbed xenon. In doing so, it became apparent that we could tune both the magnitude and the direction of the Van der Waals force on the adsorbed xenon due to the tip by controlling the location of the tip. If this force could be made attractive enough then it would be possible to overcome the lateral forces between the surface and the xenon, resulting in the ability to drag a xenon atom over the surface from one location to another without actually lifting the atom off the surface.

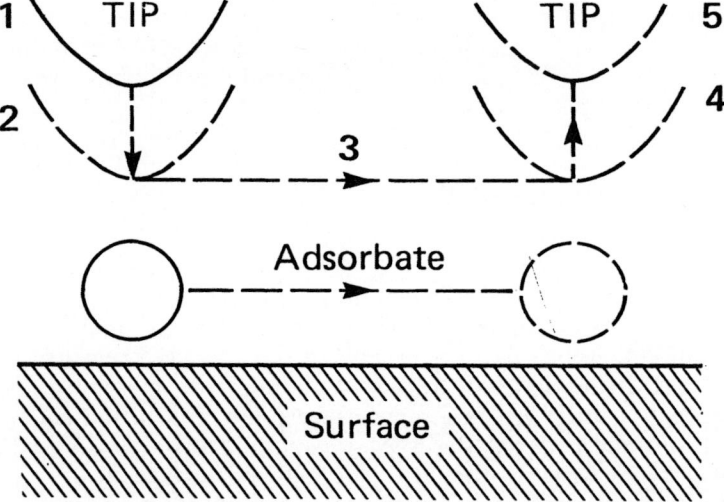

FIG. 1. A schematic illustration of the process for sliding an atom across a surface. The atom is located and the tip is placed directly over it (1). The tip is lowered to position (2), where the atom-tip attractive force is sufficient to keep the atom located beneath the tip when the tip is subsequently moved across the surface (3) to the desired destination (4). Finally, the tip is withdrawn to a position (5) where the atom-tip interaction is negligible, leaving the atom bound to the surface at a new location.

The process for sliding an atom across the surface is shown in figure 1.[2] Here we point out that the essence of this process is to be able to switch between imaging and manipulation modes simply by adjusting the height of the tip above an atom adsorbed on the surface. This process has also been successfully applied to manipulating platinum adatoms on a platinum surface and carbon monoxide molecules on a platinum surface. The sliding process allows atomic-scale precision in so far as we may choose which of any possible stable binding positions we want to place the atom in, and then do so. There are two reasons for this. First, we can control the position of the tip to well within a 1/100th of an atomic diameter. Next, consider that the tip must ultimately exert a force on the atom which overcomes the forces between the atom and the underlying surface. In order to do so the tip must be in near-atomic proximity to the atom. The force between the tip and the atom will thus vary strongly on the length scale of an atom. It is the atomic-scale range of this force that allows us to position atoms with atomic-scale precision.

As mentioned, the force that the tip exerts on an adsorbed atom can have contributions from the Van der Waals interaction, from chemical bonding due to overlap of the electron density of the atom with that of the tip, electrostatic contributions which will depend upon the magnitude of the electric field between the tip and the surface, and possibly from a direct interaction with the electrical current that flows through the atom. We find that the ability to slide a xenon atom over a nickel surface is independent of both the sign and the magnitude of the electric field, the voltage and the current. It does however critically depend upon the separation between the tip and the atom. From this we deduce that it is the Van der Waals interaction, perhaps augmented by chemical binding, that is the dominant interaction between the tip and the xenon atom.

One of the goals of atomic manipulation is to build structures. A limitation of the sliding process is that, so far, all of our attempts to slide an atom over a mono-atomic step on the surface of a sample or to build a three dimensional structure have failed. In order to build three dimensional structures it is desirous to be able to transfer an atom from the surface to the tip of the STM, carry it to a desired location, and then transfer the atom back to the surface, (by now the reader should be discerning a certain correlation between our efforts at atomic manipulation and those of a very young child confronted for the first time with a set of toy building blocks).

Becker, Golovchenko and Swartzentruber[3] found that by applying a voltage pulse between the STM tip and a germanium surface that they could leave behind on the surface an atomic scale perturbation which they suggested might be an atom which transferred from the tip to the surface during the pulse.[4] Following on the ideas of Becker et al, we found that by applying voltage pulses we could reliably pick up xenon atoms from a surface, carry them to a new location, and redeposit them on the surface.[5]

To pick up a xenon atom from the surface we first place the tip of the STM directly above the atom. We find that if the tip is held close enough to the xenon atom, then when we apply a +1 V voltage pulse to the tip we cause the xenon atom to jump from the surface to the tip. We may then use the tip to carry the xenon atom to any desired location. If we then reverse the sign of the voltage pulse we can cause the xenon atom to jump back to the surface at any location we choose.

We have studied the delay between the onset of the voltage pulse and the change in junction conductance due the motion of a xenon atom from a bare nickel (110) terrace to a particular STM tip. For a fixed tip height and pulse voltage, the distribution of delays is a decaying exponential, which indicates a fixed probability of transferring per unit time, i.e. there is a characteristic transfer rate. Figure 2 shows that the transfer rate for this particular 906 KΩ junction has a power law dependence on the pulse voltage, and consequently on the tunnel current, according to $I^{4.9 \pm 0.2}$. We were unable to measure the dependence of the transfer rate on the tip-sample separation because at smaller separations (with a junction resistance R < 700 KΩ) the xenon atom moved spontaneously to the tip without having to apply a positive voltage pulse. At larger separations (R > 1.5 MΩ) the xenon atom tended to hop among several nearby sites on the nickel surface before transferring to the tip, and sometimes escaped from the junction region entirely.

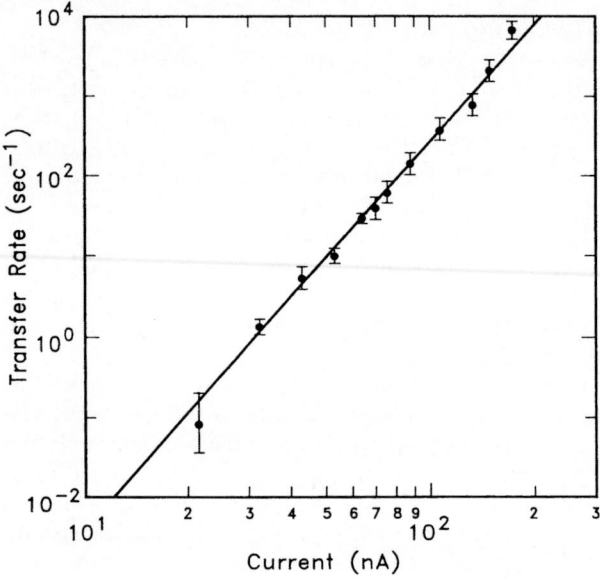

Fig. 2. The measured transfer rate of a xenon atom from the nickel (110) surface to the STM tip as a function of the current during the applied voltage pulse. The transfer rate varies as $I^{4.9 \pm 0.2}$ for this tunnel junction. (For this junction V/I = 906 KΩ plus or minus 2% while the xenon atom is on the surface for the full range of voltages employed.) A heating-assisted electromigration model is consistent with this behavior.

We now turn to the question of what physical mechanism causes the motion of the atom. Any candidate mechanism must be odd in the applied voltage, have the correct sign, and be consistent with the observed transfer delay statistics.

Lyo and Avouris[6] have suggested that ionization followed by field evaporation is the mechanism responsible for the reversible transfer of Si atoms between the tip and Si surface of an STM. This mechanism does not explain the xenon transfer since we find that the motion of the xenon atom is always towards the positively biased electrode.

Mamin et al[9] have argued that negative ion formation and subsequent field evaporation can explain the transfer of gold atoms from a negatively-biased gold tip to a substrate. Haberland et. al.[12] have recently presented evidence for the existence of a negative xenon ion with a lifetime greater than 1×10^{-4} seconds. Negative ion formation is unlikely to be the mechanism, however, because we do not observe the expected threshold electric field (Fig. 2) and because the existence of a ground-state negative xenon ion would require occupation of the xenon 6s resonance. This resonance lies between 4 and 5 volts above the Fermi level[13] and thus is expected to remain largely empty for the range of voltages employed in this experiment.

Ralls et al[14] have studied the electromigration[15] of impurities in metal nanobridges. They found that the dominant contribution to the electromigration-force term comes from the scattering of electrons from the impurity and that the electromigration process is enhanced by heating of the impurity above the lattice temperature by inelastic electron scattering at the impurity. Indeed, they suggested that the mechanisms and forces responsible for electromigration of impurities in solids might be responsible for the motion of an atom across the tunnel junction of an STM. Electromigration is odd in the applied field and can often result in motion of the impurity in the direction of electron flow. The competition between electron heating and the relaxation of vibrational energy to the lattice may result in the observed power law dependence on the current, particularly if multiple inelastic scattering events are required. The observed sideways motion of the xenon atom at greater tip-sample separation is consistent with the xenon atom being vibrationally excited. The absence of sideways motion at closer tip-sample separation may be due to the increased Van der Waals attraction to the tip as the tip is brought closer to the surface. Heating-assisted electromigration is the only mechanism which we are aware of that is consistent with the all the observed phenomena.

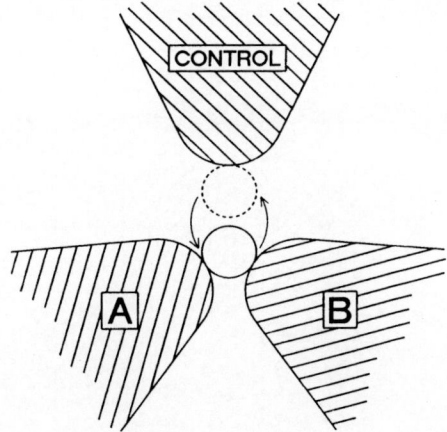

FIG. 3. A three-terminal atom switch. The two states of the switch are represented by the two stable positions of the atom. This results in different conductances between terminals A and B. The atom may be moved by application of a signal to the control terminal.

The ability to cause and atom to reversibly move between two or more locations by the application of an electric signal offers some intriguing possibilities. One of the first that came to mind can be stated as follows: if you can move matter, you can make a switch. Consider three wires coming very close to one another as shown in figure 3. In the central region is located an atom which can be made to reversibly move between the two indicated locations. In one location the atom serves as an

6

electrical bridge between two of the three terminals creating a high conductance condition between these two terminals. In the other location the electrical conductance between these two terminals is reduced. We shall call this a three-terminal *atom switch*.

Now switches are tremendously useful things. They are used as the fundamental logic element of a computers, (transistors are used as switches in a computer), they may also be used to store information. The problem that confronted us was that we had as yet no easy way to bring three wires together so that they were all in simultaneous near-atomic proximity to one another. The solution to this problem came in two parts.

The first part was to realize that one does not need three terminals to make a switch. Two terminal switches are in fact quite common. We can consider a light bulb a two-terminal two-state work-once switch. Application of a high enough voltage to the terminals causes the filament to fail, irretrievably toggling the light bulb into its low-conductance state. Neon bulbs are two-terminal two-state switches which, by virtue of their hysteresis, can be used to store information, and can be toggled as many times as you like.

The second part was to realize that when an atom is transferred between the surface and the tip of an STM the conductance of the tunnel junction should change. This occurs because the conductance of the tunnel junction depends upon the overlap of surface and the tip's wave functions. This overlap is exponentially sensitive to the geometry of the tunnel junction and will also depend upon differences in bonding of the atom to the surface or tip. Transferring an atom from the surface to the tip will thus almost always result in an *identifiably different* conductance according which side of the junction the atom is on. It is this property, the ability to be toggled to an identifiably different state which is the essential property of a switch.

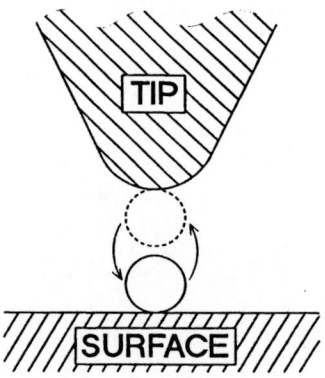

FIG. 4. A two-terminal atom switch. The atom is reversibly transferred between the surface and the tip by application of voltage pulses. The conductance of the tunnel junction switches state according whether the atom is located on the surface or the tip.

Once understood in this framework, it is easy to see how the structure shown in figure 4 would function as a two-terminal switch. The problem of demonstrating an atom switch was thus reduced to that of simply monitoring the current through the tunnel junction while we reversibly transferred a xenon atom back and forth between the tip and the surface through the application of voltage pulses.

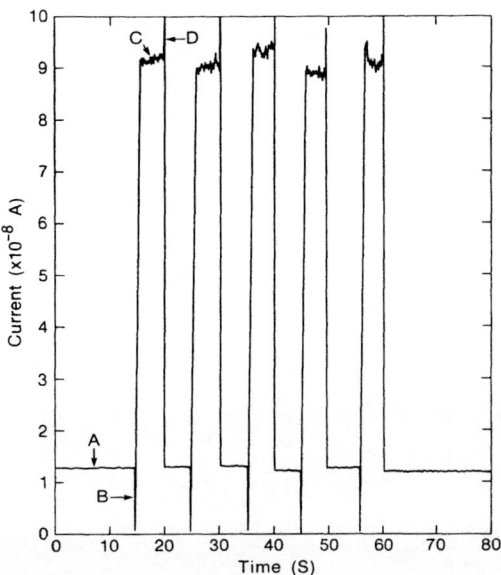

FIG. 5. The time dependence of the current through a two-terminal atom switch during operation. We begin (A) in the low-conductance state with the tip biased at -0.02 volts and the xenon atom bound to the surface. The transient current spike (B) is due to application of a +0.8 volt pulse to the tip for a duration of 64 msec after which the -0.02 volt bias to the tip was reestablished. This results in the transfer of the xenon to the tip and the establishment of the high-conductance state (C). Applying a reverse polarity -0.8 volt pulse for 64 msec gives rise to the transient current spike (D) followed by the reestablishment of the low-conductance state.

The operation of a two-terminal atom switch is shown in figure 5 which is a plot of the current through a two-terminal atom switch as the atom is made to jump back and forth from the surface to the tip of an STM.[5] The changes in the current are due to the atom switch changing from a high conductance to a low conductance condition when the atom moves.

Note that the atom switches we have constructed are *macroscopic* devices in the sense that most of the power dissipation occurs not in the atomic-scale volume of the active element but in a volume comparable to the cube of the inelastic mean free path of electrons in the terminals of the switch. Furthermore, our macroscopic terminals do not exhibit the quantum size effects that atomic-scale leads would exhibit. Whether a truly atomic-scale atom switch (one with nanometer scale leads) can be made to operate remains to be determined.

What good can we expect from being able to manipulate atoms with an STM (or any other atomic manipulation technique for that matter)? Foremost in our minds is that we may use atomic manipulation as a tool for science. We have before us the opportunity to create and study the properties of structures on a scale that heretofore was inaccessible.

As an example, one of the first uses of the sliding process was to *"hand make"* a linear chain of xenon atoms on a nickel surface as shown in figure 6. Besides demonstrating that such structures can be built we also learn about the xenon-nickel system. First we learn that this particular structure is stable. Next we observe that the apparent spacing between the xenon atoms is 5.0 ± 0.2 Å. This corresponds to just twice the length of a unit cell of the underlying nickel (110) surface from which we deduce that such linear chains of xenon atoms order commensurately with the underlying nickel lattice. This indicates that the in-plane xenon-nickel interaction dominates over the in-plane xenon-xenon interaction. In solid xenon the spacing between atoms is just 4.4 Å. Accordingly, attempts to make a more compact linear array of xenon atoms along the rows of nickel atoms met with failure. It was found that in order to pull a xenon atom off the end of the chain the tip had to be lowered closer to the atom than what was found to be necessary to move a lone xenon atom. Thus we learn that the xenon-xenon interaction along the chain is attractive.

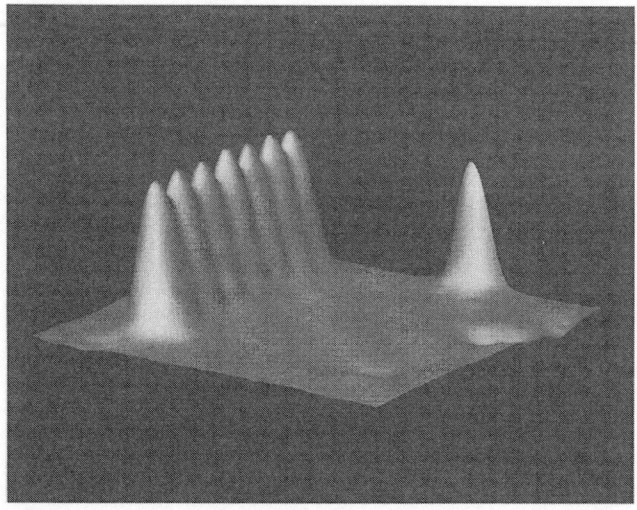

FIG. 6. The first *"hand built"* atomic structure: a linear chain of seven xenon atoms bound to a nickel (110) surface. The xenon atoms are found to order commensurate with the nickel surface indicating that the in-plane xenon-nickel interaction dominates over the in-plane xenon-xenon interaction.

Another example of how we can use atomic manipulation as a tool for science comes from moving carbon monoxide molecules across a platinum surface. Carbon monoxide adsorbed on platinum is one of the most thoroughly studied systems in modern surface science. Carbon monoxide is known to bind to the on-top and bridge sites of the platinum (111) surface and forms a variety of ordered overlayers according to conditions of temperature and coverage. While we can image carbon monoxide on the platinum surface we can not do so and simultaneously observe the positions of the platinum atoms of the surface. Thus, we can not directly discern the binding site from our STM images. Carbon monoxide appears in one of two forms in our images. The appearance of a carbon monoxide molecule may be changed between these two forms by sliding it to a new location on the surface. We refer to these forms as the "bump" and the "sombrero." Both are found to be stable in time, although the sombrero is somewhat delicate and will readily convert to a bump (located just one half of one platinum nearest neighbor distance away) if the tip is brought too close. We thus infer that the bump is the more energetically preferred state. We find that there is just one binding site per surface unit cell in which the carbon monoxide appears as a bump, whereas there are multiple sites per surface unit cell in which the carbon monoxide molecule can appear in the sombrero state. We have been able to assemble an island of bumps in a pattern in which it is known from other experiments that the carbon monoxide binds to the on-top site. Finally, we find (in contrast to the case of xenon on nickel) that there is no range at which the adsorbate-adsorbate interaction appears to be attractive, that is, we see no evidence of bump-bump, or sombrero-sombrero bonding, however a sombrero may be stabilized against conversion to a bump if that sombrero is itself located adjacent to one or more bumps. All of this evidence is consistent with what we already now about carbon monoxide on platinum (111) from other techniques such as infrared spectroscopy, EELS and LEED if we assign the bump state to on-top carbon monoxide and the sombrero state to bridge-bonded carbon monoxide. In this way, we have used our manipulative abilities to help assign bonding sites to the different states in which carbon monoxide appears on this surface.

Conclusion

We have discussed two processes for manipulating atoms and molecules with atomic-scale precision. The mechanism of the sliding process is due to the chemical binding force between the adsorbate and the tip (although this does not preclude the possibility that for some systems the electrostatic forces on the adsorbate can be dominant). Heating assisted electromigration is a mechanism which is consistent with the observed transfer delay statistics found for the motion of xenon between the surface and the tip of the STM, however, no conclusive statement about the mechanism of the carrying process can be made at this moment. We have demonstrated the operation of a switch whose function is derived from the motion of a single atom. Finally, we have applied our ability to manipulate matter on an atomic scale as a tool which allows us to carry out a class of experiments that were heretofore impossible.

As a final observation we point out that one of the challenges that faces us in applying and developing our ability to build structures *from the bottom up* is simply to overcome the mental limitations that result from decades of trying to build things smaller and smaller. When individual atoms are the building blocks, the challenge will be how to build things larger and larger.

Acknowledgement

I am greatly indebted to C. P. Lutz, P. Zeppenfeld and E. K. Schweizer for their contributions to this work.

References

1. Feynman, Richard P. (1960) Engineering and Science.

2. Eigler, Don M., and Schweizer, Erhard K. (1990), Nature, 344, 524.

3. Becker, R. S., Golovchenko, J. A. and Swartzentruber. B. A. (1987) Nature, 325, 419.

4. Becker, R. S., et al were careful to point out that they could not unequivocally demonstrate that the atomic scale perturbation which they had made to the germanium surface was indeed due to an atom which had transferred from the tip to the surface. In retrospect, it seems almost certain that this in fact was the case.

5. Eigler, D. M., Lutz, C. P., and Rudge, W. E., (1991) Nature, 352, 600.

6. Lyo, I. -W., and Avouris, P., (1991) Science 253, 173.

7. Mamin, H. J., Guethner, P. H., and Rugar, D. (1990) Phys. Rev. Lett 65, 2418-2421.

8. Haberland, H. Kolar, T. and Reiners, T. (1989)
   Phys. Rev. Lett. 63, 1219-1222.

9. Eigler, D. M., Weiss, D. M., Schweizer, E. K., and Lang, N. D. (1991) Phys. Rev. Lett. 66, 1189-1192.

10. Ralls, K. S., Ralph, D. C., and Buhrman, R. A. (1989) Phys. Rev. B 40, 11561-11570.

11. Verbruggen, A. H., (1988) IBM J. Res. Dev. 32, 93-98.

STM-INDUCED MODIFICATION AND ELECTRICAL PROPERTIES OF SURFACES ON THE ATOMIC AND NANOMETER SCALES

PH. AVOURIS, I.-W. LYO, and Y. HASEGAWA
IBM Research Division, T.J. Watson Research Center,
Yorktown Heights, New York, 10598.

ABSTRACT. We first discuss the atomic and nanometer scale modification of strongly-bonded materials. Our approach involves short-range tip-sample chemical interactions and the effects of the electrostatic field produced by a voltage pulse. This process is demonstrated by the controlled removal and redeposition of individual atoms or clusters of Si atoms from Si(111). We then show that due to the elastic coupling of surface atoms, there is a material-dependent limit on how local an STM-induced modification can be. We use the modification of the reconstruction of an Au(111) surface as an example. The electrical properties of nanometer-size contacts to metal and silicon surfaces and nanostructures are also explored via tip-sample point-contacts. The metal (tip)-semiconductor contacts are models of Schottky-diodes at an extreme level of miniaturization. Analysis of the resulting I-V curves reveals important differences between the nano-diodes and the corresponding macroscopic devices. Nevertheless, 10 nm diameter diodes are functional electronic devices. Contacts to nanostructures show the importance of carrier scattering at boundaries (steps). We demonstrate that scattering at steps and the resulting interference between incident and reflected waves can be directly imaged in (dI/dV)/(I/V) maps.

1. Introduction

The scanning tunneling microscope (STM) and the other related scanning probe techniques are not just surface structure probes. The multiplicity of interactions that can exist between an STM tip and a sample can be controlled and utilized to modify materials and fabricate structures at the nanometer and, for the first time, at the atomic scale (1). Coupled with the ability of these probes to image and perform local spectroscopy, the above capabilities are helping usher in the era of nanotechnology.

Here we consider two types of applications of the STM in nanometer science and technology: modifying the surface structure of strongly-bonded materials and probing their electrical properties on the nanometer scale. First we discuss an experimental approach that combines chemical STM tip-sample interactions and the effects of the electrostatic field produced by a voltage pulse, to modify surfaces (2). We demonstrate these capabilities by the controlled removal of individual atoms or clusters of atoms from Si(111) and their redeposition.

11

P. Avouris (ed.), Atomic and Nanometer-Scale Modification of Materials: Fundamentals and Applications, 11–24.

Next we explore the factors that determine the spatial extent of an STM-induced modification. We show that the ability to perform very localized modifications with the STM is a function of the nature of the solid involved. Surfaces such as the the adatom surface layer of Si(111)-(7x7), where the surface atoms are weakly coupled to each other, can be modified very locally. In other cases, strong coupling between surface atoms delocalizes the perturbation and large-scale atomic rearrangements may result. We demonstrate this behavior by the modification of the Au(111) surface (3).

In addition to being able to fabricate or modify nanostructures with the STM, we show that one can use STM tip-sample point-contacts to electrically address nanostructures, and employ spectroscopic maps to study their electronic properties. We investigate the properties of nanometer size (~10 nm) contacts to Si(111)-(7x7) and Si(100)-(2x1) surfaces and to nm Si epitaxial islands deposited on top of the surface (4). These contacts provide models for Schottky-diodes at an extreme level of miniaturization. We show that the behavior of the nano-diodes is a function of the Si surface reconstruction in UHV and of the exposure to gases, both being manifestations of a surface electrical transport channel. The Schottky-barriers of the nano-diodes are quite different from the barriers obtained from the corresponding macroscopic diodes. Despite these differences, the nano-diodes have reasonably good electrical characteristics. Point-contacts to epitaxial islands give higher contact resistances than flat terraces, despite the strong coupling of the islands to the substrate; this observation suggests carrier-scattering at the perimeter of the islands. To obtain insight into such local scattering processes, we study electron scattering at steps using an Au(111) surface as a model. We find that $(dI/dV)/(I/V)$ maps of the surface can be used to directly image the oscillations produced by the interference between incident and reflected electron waves at steps. These spectroscopic images can be further analyzed to obtain a quantitative description of the scattering process.

2. Atomic and nanoscale surface modification

2.1. CHEMICALLY-ASSISTED FIELD-EVAPORATION/DESORPTION

A particularly valuable process for the modification of materials and the fabrication of nanostructures involves the ability to zoom on a particular location, break strong chemical bonds, transfer one or a cluster of atoms to the STM tip, and then perhaps redeposit them at another location. We have recently proposed a general scheme involving chemical and electrostatic interactions to accomplish this goal and which can be described as a chemically-assisted field evaporation (CAFE) process (2). The basic principles of the approach can be explained by reference to the diagram shown in Fig. 1. The left side of the diagram illustrates the field-desorption of an adsorbate A from a substrate S, under conditions pertaining to the field-ion microscope (FIM) (5). The energy required to desorb a neutral A atom is Q. Above the neutral ground state are ionic states, such as the S^-A^+ state (at a large distance from the surface its energy will be I-ϕ above the ground state), which upon application of an electric field F is lowered in energy qFz and crosses the neutral curve so that the barrier for field-desorption of A^+ is only Q' (6). In terms of the electronic structure of the system (see bottom picture), an

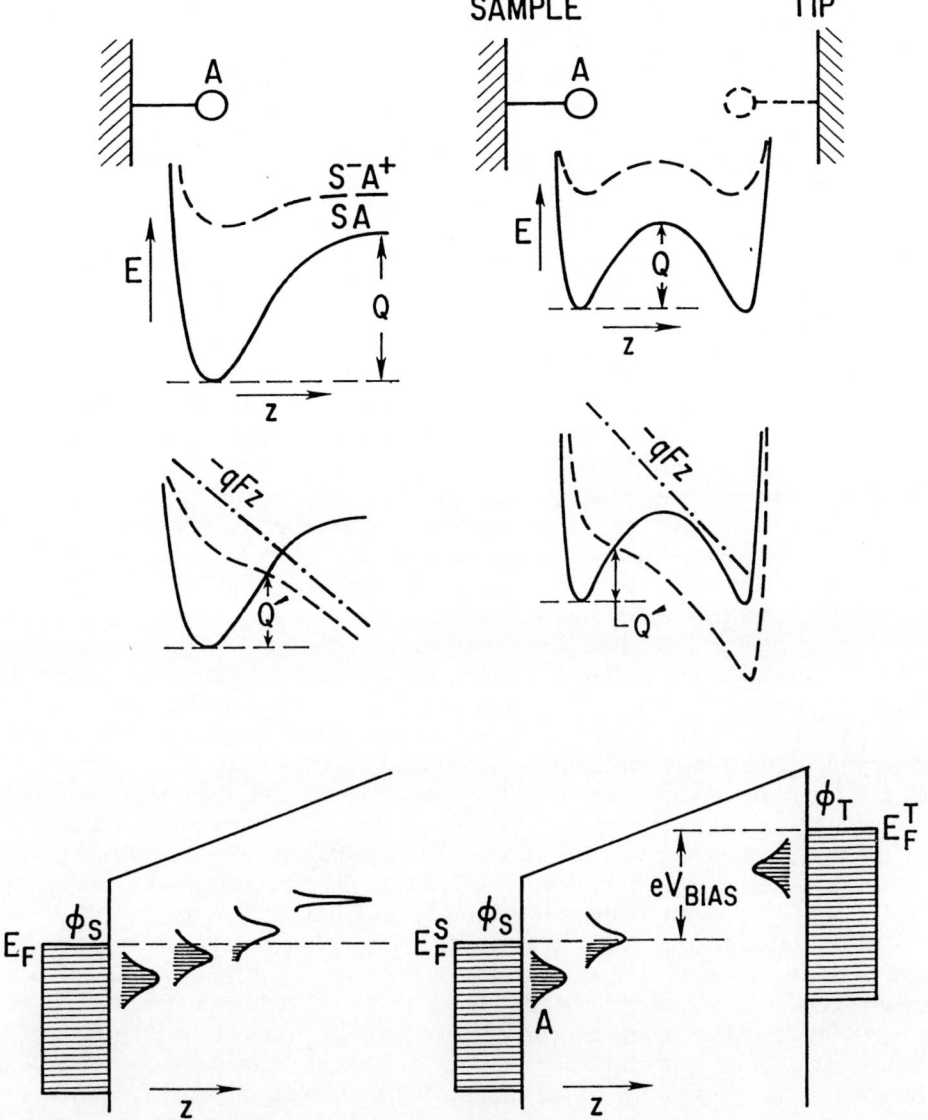

Fig. 1. Schematic diagram illustrating the mechanism and energetics of the field-evaporation/desorption process in the field-ion microscope (left column), and of the chemically-assisted field-evaporation/desorption process in the STM (right column).

occupied adsorbate level lying originally below E_F is gradually emptied as atom A moves in the field.

The main difference between field-evaporation/desorption in the STM compared to that in FIM, is the introduction of a second potential energy well due to the presence of the STM tip (7,8). The barrier Q for transfer of A from sample to tip is a function of the sample-tip distance. As this distance decreases, the barrier decreases drastically. The application of an electric-field F further reduces the barrier and can introduce directionality in the atom-transfer process. Because the electric field varies rather slowly over many atomic sites, the chemical tip-sample interaction is needed to selectively weaken the bonding to the substrate of the chosen atom(s) which is then transferred by the action of the electric field. The presence of the STM tip also has important implications regarding the charge-state of the transferring atom. As shown at the bottom of Fig. 1 (right), the adsorbate level can interact and hybridize with both sample and tip levels. There are two Fermi levels to consider. As the atom moves away from the sample, the occupation of the adsorbate level will start decreasing, but it soon will feel the influence of the tip and upon adsorption on the tip it will be filled again. Thus, the occupation of the level and therefore the charge of the atom A will vary with the distance in a complex manner. In general, however, one should expect that the charge on the transferring atom will be smaller than the charge detected in an FIM experiment and may vary with distance in a non-monotonic manner. This is exactly what is found by the first-principles calculations of Lang (9).

Experimentally, to perform an atom-transfer, we first move the tip from its original tunneling distance closer over the atom we wish to transfer, to establish a chemical interaction. The tip-sample distance is bounded by the positions where the effective tunneling barrier height, ϕ_{eff}, starts decreasing and where it vanishes (ϕ_{eff} is determined independently). Atom-transfer is then accomplished by applying a voltage pulse of the appropriate magnitude and polarity. For the manipulation of Si, the tip-sample distance is typically 1-3 Å away from the $\phi_{eff} = 0$ point and the voltage pulse is adjusted to give a field-strength of ~1 V/Å. Under these conditions, we find that by applying a positive voltage to the sample, the direction of atom transfer is predominantly from sample to tip. Correspondingly, transfer of Si from the tip back to the surface is accomplished by applying a negative pulse to the sample. This behavior is in accord with the conclusions of electronic structure calculations by Lang (9) performed for tip-sample distance and bias conditions corresponding to those in our experiments. In Fig. 2 (top) we show an example of a nanometer-scale modification of Si(111). Panel A shows the modification of the Si(111) surface when a W-tip is brought to 3 Å from the surface and a 3 V pulse is applied to the sample. This leads to a modification with a characteristic morphology composed of a hillock (bright center) surrounded by a depression (dark ring). In a second step, the cluster of atoms forming the hillock is transferred to the tip by placing the tip over the hillock and applying a second 3 V pulse. The Si cluster is then redeposited to the left of the hole by applying a negative (-3 V) voltage pulse.

The characteristic morphology of the modified surface seen in panel A points to yet another interaction involved in the modification, in addition to field-evaporation and the initial tip-sample chemical interaction. When the voltage pulse is applied, Si atoms move towards the apex of the tip, thus bridging the tunnel-gap and connecting the tip to the sample. When the tip is retracted after the voltage pulse, this bridge breaks, leaving behind a central hill such as that seen in panel A. That bridge formation occurs is verified by measurements of the differential conductivity vs. distance (2). An example is shown in Fig. 3 involving an aluminum tip and a Si(111) surface. We see that as the tip approaches the surface, the conductivity increases sharply. At the point indicated by the

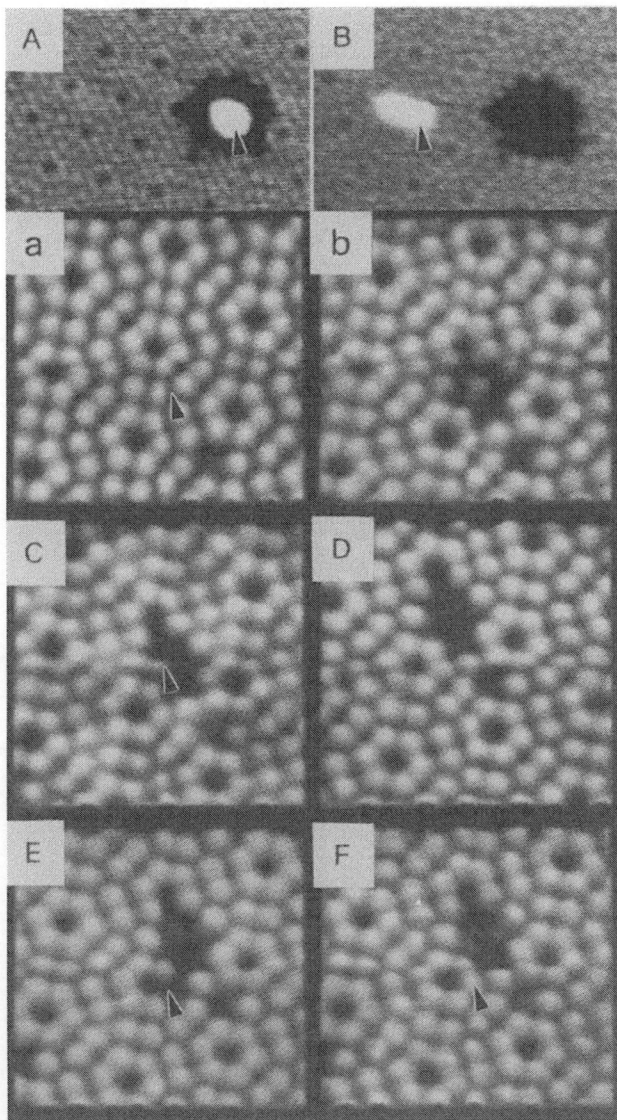

Fig. 2. Top: Nanoscale modification of Si(111). (A) A +3 V pulse applied at ~3 Å from electronic contact has led to the formation of a structure involving a central hill surrounded by a depression. (B) A second +3 V pulse applied over the hill leads to the transfer of the hill to the tip. The tip is then moved to the left of the hole and the cluster is redeposited by applying a -3 V pulse. Bottom: A series of atomic scale manipulations. (a) The tip is placed at ~1 Å from electronic contact over the site indicated by the arrow. (b) A 1 V pulse removes 3 atoms leaving the fourth under the tip. (c) The first attempt to remove this atom leads to its migrating to the left (see arrow). (d) A second pulse removes this fourth atom. (e) A new corner-adatom is removed and in (f) it is placed back to its original position.

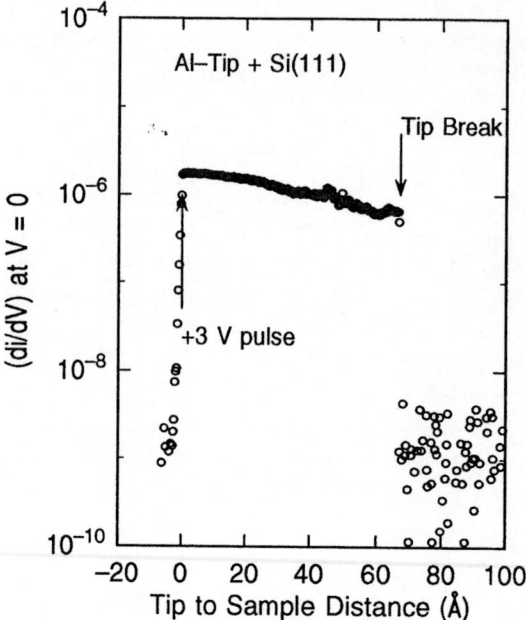

Fig. 3. Tunneling current versus tip displacement curve for an Al tip over a Si(111)-7x7 surface. At the point indicated by the arrow a voltage-pulse is applied and then the tip is retracted.

arrow, a voltage pulse is applied, and subsequently the tip is retracted. We observe that a contact with very strong adhesion has been generated and a nominal z-retraction of ~60 Å is needed to break it. This procedure, which can be described as "nano-welding", provides an excellent way of forming stable electrical contacts to nanostructures.

Let us now turn to atomic scale modification. To remove a particular surface atom, its bonding to the substrate is preferentially weakened by interaction with the tip, and its transfer completed by the application of a near threshold voltage pulse. The STM tip need not only be sharp, but also clean. Tips with oxides or other surface impurities may be adequate for imaging purposes, but not for manipulation. Foreign atoms at the apex of the tip may not interact chemically with the sample, or may have a low field-desorption threshold and thus be transferred to the sample. In Fig. 2 (bottom) we show a series of atomic scale modifications (the voltage pulse is 1 V and the tip-sample distance is ~1 Å). Panel (a) shows a section of a Si(111)-7x7 surface containing a defect in the lower right (dark site) which is used as a marker. In (b) the voltage pulse was applied while the tip was centered over the site indicated by the arrow and three Si atoms were removed, leaving a fourth under the apex of the tip. This fourth atom is the atomic-scale analogue of the hillock (cluster) left behind in Fig. 2A. This Si atom was in an unstable configuration, and it migrated to the left to occupy a center-adatom site of the 7x7 lattice (panel c). This atom was then removed with a second pulse (panel d). Finally, with another pulse a corner-adatom was removed (panel e). One question that arises is: can the STM undo a modification, e.g., can it put back a removed atom or cluster of atoms and reform the original structure? We find that for Si in general and especially for clusters, the answer is no. The redeposited atoms must overcome a size-able activation barrier to occupy their original sites. However, as panel (f) shows, in the case of single atom removal, incorporation occasionally can be achieved by bringing the tip over the vacancy site and applying a negative pulse. We conclude that despite the strong bonding of the surface Si atoms to the substrate, one can achieve considerable control over their atomic-scale manipulation using the CAFE approach.

2.2. HOW LOCAL IS THE STM-INDUCED SURFACE MODIFICATION?

In the above discussion we assumed implicitly that the area of the surface affected is determined by the area of direct interaction with the tip and the high electric field, and this view appears to be supported by the experiment. The 7x7 surface, however, is in some sense unique, in that it is a very open structure with little direct interaction be-tween the top layer atoms. In general, however, surface atoms are elastically coupled to each other so that a perturbation at a certain location on the surface is felt by atoms away from that point. The strength of this interaction depends on the structure of the surface. Thus, in the case of the Si(100)-2x1 surface our CAFE approach tends to re-move individual dimers or multiples of dimers instead of single atoms. The situation can be even more extreme in the case of metal surfaces. To illustrate this point we consider the modification of the Au(111) surface (3).

The Au(111) surface is known to reconstruct to give a structure with a $22x\sqrt{3}$ unit cell (10-12). The surface structure consists of alternating domains where the surface Au atoms occupy hollow sites with an fcc stacking (wider domains), and domains where the stacking is hcp (narrower domains). The fcc and hcp domains are separated by bounda-ries (dislocations) where the Au atoms occupy bridge sites. The dislocations appear in STM images as bright ridges ~0.15 Å higher than the fcc and hcp domains (See Fig. 4(a)). Thus, through STM images one can readily ascertain the coordination of surface Au atoms in different areas of the surface. Let us now induce a surface modification similar to those demonstrated above on a silicon surface, i.e. use a voltage pulse to re-move a number of surface Au atoms and follow the structural changes that this pertur-bation induces. First, in Fig. 4(b), we see that the formation of the hole has led not only to local changes, such as the slight displacement of the two halves of the dislocation line at the hole, but also to a distant rearrangement involving the formation of an edge-dislocation (see arrow) ~80 Å away. As time increases, the altered stress-distribution at the surface leads to more structural rearrangements. In Fig. 4(c), we see that the dislo-cation lines have been split, forming U-loops, and the size of the hole has decreased as diffusing atoms enter the hole. Finally, in Fig. 4(d), the hole has been filled again, but the surface reconstruction has now changed over a sizeable area and a new forked structure has been formed. This new structure appears to be stable at room temperature. Thus, we see that a "local" modification may affect the structure and/or stress-distribution of a surface over large distances due to the long-range elastic interactions present (13). These interactions place a natural limit to the size of modifications that can be performed on a given surface.

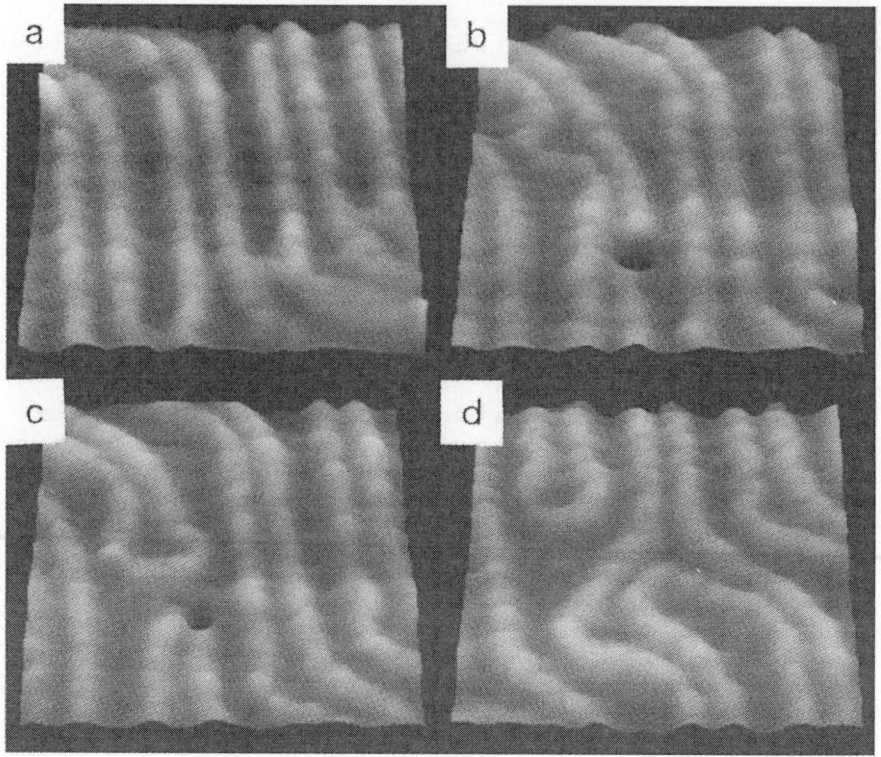

Fig. 4. Large-scale atomic rearrangements on a gold surface induced by a "local" STM tip-induced perturbation. (a) A 200 Åx200 Å area of an Au(111)-(22x√ 3) surface. (b) A voltage pulse is used to remove a number of surface atoms (see dark hole). (c) Same area after ∼9 min. As time progresses the size of the hole decreases and the surface structure rearranges. (d) Same after ∼15 min. The hole has been filled and a new surface structure emerges.

3. Nanometer metal-semiconductor contacts and Schottky-diodes

As the technology for building nanostructures progresses, the issue of being able to electrically address such structures becomes important. There are many factors that can make the properties of nanoscale contacts very different from those of the corresponding macroscopic contacts. At the nanoscale, current-densities and electric fields are much stronger, contact dimensions become comparable with the distance between dopants and, of course, quantum effects may become dominant. Therefore, it may not be possible to predict the electrical properties of nanometer size contacts simply by an extrapolation of what is known about macroscopic contacts.

Here we will briefly consider nanometer metal-semiconductor contacts formed between a metal STM tip and either a flat silicon surface, or model nanostructures composed of Si epitaxial islands grown on top of the flat Si surface by molecular beam epitaxy.

In Fig. 5 we show the differential resistance at zero bias for three systems, Au, Si(111)-7x7 and Si(100)-2x1, as a function of the tip-sample distance. For gold, the

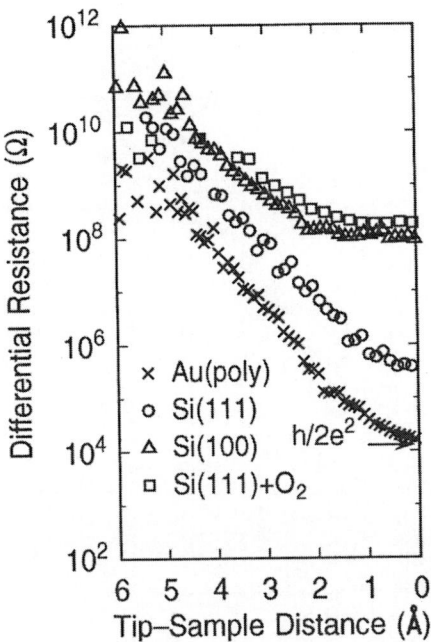

Fig. 5. Differential resistance versus tip-sample distance for a W-tip and three samples: Au, Si(111)-7x7 and Si(100)-2x1.

contact resistance value is very close to the ideal conduction channel "constriction" resistance of $h/2e^2$ = 12.9 kΩ (14,15). A similar result has been reported earlier by Gimzewski and Möller (16) for the contact of an Ir tip and a Ag sample. The resistance of the two Si surfaces, on the other hand, show large deviations from the $h/2e^2$ value. Moreover, while both crystals have similar bulk resistivities, the contact resistance of the Si(111)-7x7 surface is consistently much lower than that of the Si(100)-2x1 surface. The exact values of the contact resistances of both surfaces are also found to be sensitive to the structural perfection of the surfaces and to exposure to gases. As shown in Fig. 7 (squares) after exposure of the Si(111) sample to ~2 L of O_2, its electrical properties become very similar to those of the Si(100) sample.

The fact that the contact resistance is dependent on the reconstruction and condition of the semiconductor surface indicates the presence of a "surface" channel for electrical transport. There are two possible candidates for this channel. One possibility is that electrical transport proceeds via surface states. In this respect, we note that the adatom dangling-bond surface states of the 7x7 surface give it a metallic-like electronic structure (17,18), while the Si(100)-2x1 surface electronic structure shows an energy gap (19). Another possibility involves transport through a surface space-charge layer (20,21). Such a space-charge layer should be present at the Si(111)-7x7 surface because of the pinning of the Fermi level by the dangling-bond surface states.

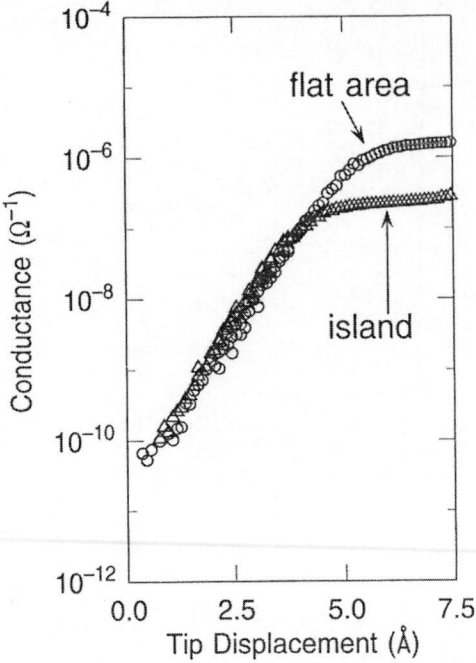

Fig. 6. Differential conductance versus tip-sample distance obtained over a flat Si(111) terrace, and over a 100 nm² epitaxial Si island.

In addition to the "surface" channel there should be transport over and through the Schottky-barrier that should form at the metal-silicon interface. In fact, we are interested to study this Schottky-barrier because this nanometer tip-sample contact provides an opportunity to study the properties of a "simple" solid-state device, the Schottky-diode, at an extreme level of miniaturization. To determine the properties of the contact we analyse the current-voltage (I-V) curves. We find that the surface channel contributes a linear (Ohmic) I-V component which is quenched by exposure to O_2. We thus subtract this linear component and obtain a very good fit of the I-V data to the Schottky equation: $J = J_s(e^{qV/nkT} - 1)$, where n is the diode ideality factor, to obtain J_s and from it the Schottky barrier height, ϕ_{SB} (22).

We have applied this procedure to a diode formed by the contact of a W tip and a B-doped (1 Ωcm) Si(111)-7x7 surface. The diameter of the contact was only ~10 nm (determined by observing the surface area affected after breaking the contact), which corresponds to ~15 Si adatoms of the 7x7 structure. (In contrast, the average distance between dopant atoms in the Si crystal is ~100 nm.) Because of the small area of the diode the current densities produced during its operation are enormous, ~10⁶A/cm². These conditions are perhaps as extreme as one may expect any future miniaturization to lead to. In table 1 we have collected some of the characteristics of this device. Specifically, we give the resistance of the surface channel, R_S, the Schottky barrier height, ϕ_{SB}, and the ideality factor of the diode, n, as a function of the exposure of the surface to different amounts of O_2. We see that the influence of the surface channel decreases (R_S increases) with increasing exposure to O_2 while the Schottky barrier remains relatively constant, ~ 0.3 eV. The diode ideality factor changes from 1.7 in the absence of O_2 to 1.5 after only 2 L O_2 exposure. Our first conclusion is that although not as effective as macroscopic diodes, whose ideality factors are larger than unity by a few percent, the 10 nm diode is still a functional electronic device.

TABLE 1. Characteristics of the W-Si(111) nano-diode as a function of oxygen dose.

O_2 (L)	n	$R_S(\Omega)$	$\phi_{SB}(V)$
0.0	1.69	9.1×10^6	0.28
0.7	1.52	8.6×10^7	0.30
1.4	1.49	7.3×10^8	0.31
2.1	1.47	5.6×10^8	0.31

One major difference between the nano and macro-diodes is the magnitude of the Schottky barriers. For the nano-diodes the Schottky barriers for n- and p-doped Si(111) are 0.2 and 0.3 eV, respectively. For the corresponding macroscopic diodes, the barriers are much higher, 0.67 and 0.45 eV, respectively (22). Moreover, the fact that the sum of the n- and p- Schottky barriers does not add up to the band-gap energy suggests that the Fermi level, which is pinned at macroscopic W-Si(111) contacts, is not pinned in the nano-contacts. This possibility is also supported by our study of the STS spectra as a function of the tip-sample separation which shows that the E_F-pinning surface states disappear as the contact is formed. It remains to be determined to what extent the interesting differences observed reflect the size of the contacts and the associated changes in fields and current-densities, or the mode of formation of the interface.

We now briefly consider the properties of point contacts to nanostructures. We want these nanostructures to be strongly-coupled to the substrate and for this reason use as models Si epitaxial islands on top of flat Si(111) terraces. The ability of the STM to electrically address by point-contact these islands allows us to study the dependence of the contact resistance on variables such as the size and height (number of layers) of the islands. An example is shown in Fig. 6. The resistance at $V = 0$ is measured on top of a 100 nm^2 7x7 island and on the flat 7x7 terrace. We see that the resistance on the island is almost an order of magnitude higher than on the terrace. This is quite remarkable given the strong coupling to the substrate and suggests that perhaps scattering at the boundaries (steps) of the island is responsible for the extra resistance.

4. Direct imaging of electron scattering at steps

It is clear from experiments such as the contacts to the epitaxial islands described above, that we need to understand the role of local scatterers, such as steps, on electrical transport. This subject has been addressed theoretically early on in the 1950's, particularly by Landauer (23) who considered issues such as the resistance of an individual step. An electron wave incident on the step is partially transmitted and partially reflected. If R is the reflection coefficient, the resistance of the step is proportional to R/(1-R) (24).

Here we will describe some initial results which demonstrate that such interactions can in fact be directly imaged with the STM. As a model we again use $Au(111)-22x\sqrt{3}$. This surface has a well-characterized surface state (25,26). We use the STM to populate un-

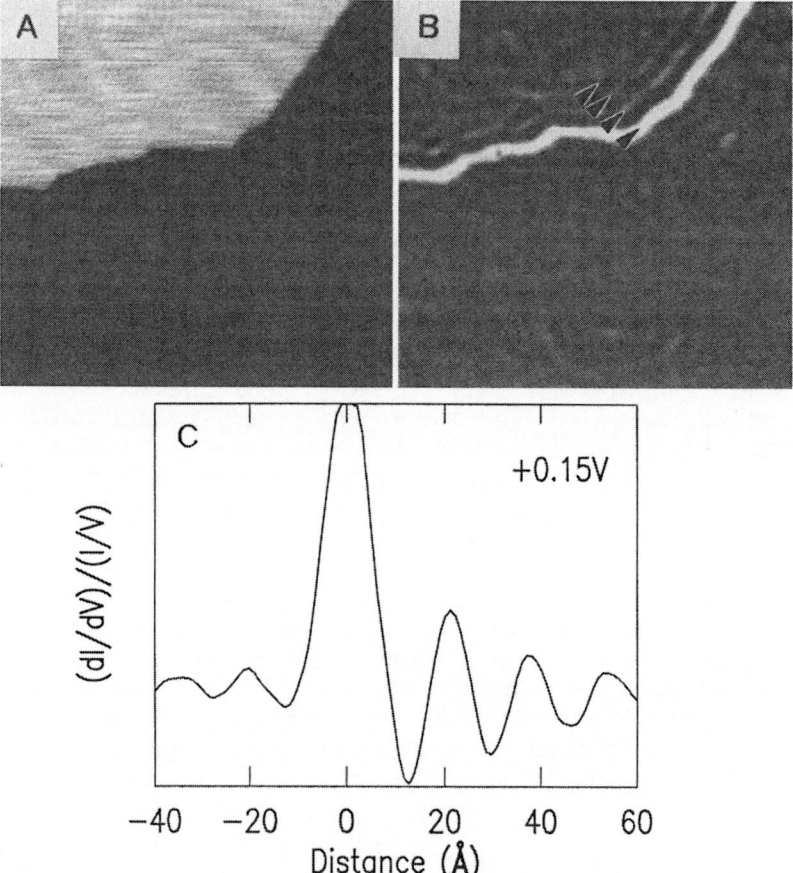

Fig. 7. (A) An STM topograph of an area of a Au(111) surface showing a monoatomic step. (B) A (dI/dV)/(I/V) map of the same area of the Au(111) surface (V = +0.15 V). Note the ripple-like structures on the upper terrace side of the step. (C) A laterally-averaged STM line-scan across the surface step showing oscillations in the charge density. The scan is from the lower terrace (left) to the upper terrace. Sample bias = 0.15 V.

occupied levels of the surface and focus on the interaction of the electron waves with a surface step. In Fig. 7A we show an STM topograph of an area of the Au(111) surface showing two terraces; the upper terrace (top) appears brighter than the bottom terrace. Very faintly one can see the boundaries (dislocation lines) separating the fcc and hcp surface domains. In Fig. 7B we show a map of the quantity $(dI/dV)/(I/V)$ at a sample bias of $+0.15$ V. The quantity $(dI/dV)/(I/V)$ provides a measure of the local density of states (LDOS) (27). From this LDOS map we see that the surface step appears quite prominently and, in addition, a ripple-type structure can be seen on both sides of the step, being strongest at the upper terrace side. As the energy of the electrons is increased by increasing the bias voltage, the wavelength of the oscillation at the step changes. In this way we can obtain the dispersion relation which we find to agree reasonably well with the dispersion of the surface state electrons of Au(111) (25).

The above results indicate that the oscillations observed near the step are the result of the interference between incident and reflected electron waves at the step. Laterally-averaged line scans, such as the one shown in Fig. 7C, obtained at different energies (bias) can be analyzed to obtain the corresponding scattering phase-shifts and coherence lengths.

5. Summary

In conclusion, we have demonstrated that with the STM one can break strong chemical-bonds at surfaces and transfer atoms from sample to tip and vice versa. Electric-field effects, tip-sample attractive forces, and mechanical effects (formation and destruction of a tip-sample bridge) are involved in affecting the transfer of atoms and the morphology of the surface modification. The extent of the localization of an STM-induced surface modification is also material-dependent. Strong elastic-coupling between atoms tends to delocalize the STM-induced perturbation and may lead to large-scale atomic rearrangements.

The STM can not only fabricate or modify nanostructures, but also electrically address them via tip-sample point-contacts. We find that the characteristics of nanometer metal (tip)-semiconductor (sample) contacts show important differences from the characteristics of corresponding macroscopic contacts. Nevertheless, contacts 10-20 atoms in diameter act as functional Schottky-diodes. The study of the electrical properties of model nanostructures suggests the importance of carrier scattering at boundaries. We find that electron scattering at surface steps and the resulting interference between incident and reflected electron waves can be directly imaged in spectroscopic STM maps. A detailed characterization of the scattering process can thus be obtained.

REFERENCES

1. C.F. Quate, in "Highlights in Condensed Matter Physics and Future Prospects", L. Esaski, Editor (Plenum Press, New York, 1991); J.A. Stroscio and D.M. Eigler, Science **254**, 319 (1991).

2. I.-W. Lyo and Ph. Avouris, Science **253**, 173 (1991); Ph. Avouris and I.-W. Lyo, Appl. Surf. Sci. **60/61**, 426 (1992).

3. Y. Hasegawa and Ph. Avouris, Science **258**, 1763 (1992).

4. Ph. Avouris, I.-W. Lyo and Y. Hasegawa, J. Vac. Sci. Technol. A, (1993), to be published.

5. T.T. Tsong, "Atom-Probe Field Ion Microscopy" (Cambridge University Press, Cambridge, 1990).

6. R. Gomer and L.W. Swanson, J. Chem. Phys., **38**, 1613 (1963).

7. H.M. Mamin, P.H. Guether and D. Rugar, Phys. Rev. Lett. **65**, 2418 (1990).

8. I.-W. Lyo and Ph. Avouris, J. Chem. Phys. **93**, 4479 (1990).

9. N.D. Lang, Phys. Rev. B **45**, 13599 (1992).

10. U. Harten, A.M. Lahee, J.P. Toennies and Ch. Wöll, Phys. Rev. Lett. **54**, 2619 (1985).

11. Ch. Wöll, S. Chiang, R.J. Wilson and P.H. Lippel, Phys. Rev. B **39**, 7988 (1989).

12. J.V. Barth, H. Brune, G. Ertl and R.J. Behm, Phys. Rev. B **42**, 9307 (1990).

13. S. Narasinhan and D. Vanderbilt, Phys. Rev. Lett., **69**, 1564 (1992).

14. R. Landauer, Z. Phys. B **68**, 217 (1987).

15. Y. Imry, in "Directions in Condensed Matter Physics", G. Grinstein and G. Mazenko (World Scientific, Singapore, 1986), p. 101.

16. J.K. Gimzewski and R. Möller, Phys. Rev. B **36**, 1284 (1987).

17. V. Backes and H. Ibach, Solid State Commun. **40**, 574 (1981).

18. J.E. Demuth and B.N.J. Persson, Phys. Rev. Lett. **52**, 2214 (1983).

19. F.J. Himpsel and Th. Fauster, J. Vac. Sci. Technol. A, **2**, 815 (1984).

20. A. Many, Y. Goldstein and N.B. Grover, "Semiconductor Surfaces" (North-Holland Publishing Co., Amsterdam, 1965),

21. H.J.W. Zandvliet and A. van Silfhout, Surf. Sci. **195**, 138 (1988).

22. S.M. Sze, "Physics of Semiconductor Devices", 2nd Edition, J. Wiley Publishing Co., New York (1981).

23. R. Landauer, IBM J. Res. Develop. **1**, 223 (1957).

24. R. Landauer, Z. Physik B **21**, 247 (1975).

25. S.D. Kevan and R.H. Gaylord, Phys. Rev. B **36**, 5809 (1987).

26. W.J. Kaiser and R.C. Jaklevic, IBM J. Res. Develop. **30**, 411 (1986).

27. R.M. Feenstra, J.A. Stroscio and A.P. Fein, Surf. Sci. **181**, 295 (1984).

ALKALI METALS ON III-V (110) SEMICONDUCTOR SURFACES: OVERLAYER PROPERTIES AND MANIPULATION VIA STM

L. J. WHITMAN
Naval Research Laboratory
Washington, DC 20375 USA

JOSEPH A. STROSCIO, R. A. DRAGOSET, and R. J. CELOTTA
National Institute of Standards and Technology,
Gaithersburg, MD 20899 USA

ABSTRACT: Field ion microscopists demonstrated more than twenty years ago that polarizable atoms adsorbed on a stepped surface can be induced to diffuse by an electric field due to the field gradients associated with step edges. We have exploited a similar phenomenon, the large field gradients in the vicinity of the STM tip, to induce the directional diffusion of Cs and K atoms adsorbed on room temperature GaAs(110) and InSb(110). The geometric and electronic properties of both the naturally occurring and electric field-induced alkali metal structures observed on these semiconductor surfaces are discussed, including the possibility that the alkali metal overlayers are Mott insulators.

1. Introduction

The rapid growth in the number of operational scanning tunneling microscopes (STM) has been accompanied by many attempts to use the STM for nanoscale modification of surfaces [1]. There are four fundamental processes inherent to surface modification: removal of surface atoms; deposition of new material; lateral translation of adsorbed atoms; and induction of local chemistry (e.g. dissociation of an adsorbate or reaction between adjacent adsorbates). In order to achieve these modifications the scanning tunneling microscopist has three basic parameters to vary: the tip properties, including the material composition of the tip and the tip shape; the tunneling current, which can be varied via the tunneling gap or the tunneling bias; and the electric field in the tunnel junction, which is also dependent on the gap and bias.

A few examples from the recent literature elucidate how the above three parameters have been employed, usually in combination, to perform nanoscale surface modification with an STM. For instance, Lyo and Avouris have recently demonstrated the removal of surface atoms (small Si clusters) from Si(111) [2]. The process appears to depend on a combination of the electric field (field evaporpation) and the tip properties (via direct tip-surface interaction). The recent work of Mamin, Guethner, and Ruger provides a nice example of controlled deposition of material (Au) from tip to surface [3], with the deposition appearing to depend primarily on the electric field,

P. Avouris (ed.), Atomic and Nanometer-Scale Modification of Materials: Fundamentals and Applications, 25–35.
© 1993 *Kluwer Academic Publishers.*

although local heating due to the high current densities in the junction may also play a role [4]. Lateral translation of adsorbed atoms via the STM has been precisely demonstrated in the experiments of Eigler and Schweizer with Xe at 4 K, with the process dependent on the tip properties [5]. The high electric field and current density in the tunnel junction is a useful combination for the induction of local chemistry, such as the dissociation of individual decaborane molecules achieved on Si(111) [6].

In this paper we review our progress in using the electric field in the STM tunnel junction to translate adsorbed atoms (Cs and K) across the (110) surfaces of GaAs and InSb at room temperature [7]. Field ion microscopists demonstrated more than twenty years ago that polarizable atoms adsorbed on a stepped surface can be induced to diffuse by an electric field due to the field gradients that occur in the vicinity of the step edges [8-10]: we have exploited the large field gradients in the vicinity of the STM tip to induce a similar directional diffusion of these alkali metal adatoms. In addition, we have used the STM to explore the geometric and electronic properties of both the naturally occurring and induced structures [11-13]. Here we will first discuss the naturally occurring structures, then electric field-induced diffusion, and lastly the electronic properties of the observed structures. All experiments were performed in ultra-high vacuum at room temperature, as previously described [11-13]. Note that all STM images shown are of the filled electronic states (tunneling from surface to tip).

2. Results and Discussion

2.1 NATURALLY OCCURRING STRUCTURES

Cs and K adsorbed on room temperature GaAs(110) and InSb(110) form a variety of interesting structures, evolving from one- to two- to three-dimensional (3-D) with increasing coverage [11-13]. Following the adsorption of low coverages long 1-D chains oriented along the $[1\bar{1}0]$ direction are observed, as illustrated by the STM images of Cs on InSb(110) shown in Fig. 1. For Cs on both substrates and K on GaAs(110) these chains have a characteristic zig-zag structure, with the adatoms appearing to be adsorbed in the four-fold hollow-like sites between the substrate anions (As, Sb). The formation of these 1-D chains is consistent with the highly anisotropic barrier to diffusion: the calculated barrier along the $[1\bar{1}0]$ direction (0.2 eV) is approximately one fifth that expected along [001] (1.0 eV) [14]. However, this does not readily account for the unusual zig-zag structure within the chains. One factor in determining the internal structure of the chains may be the lattice mismatches between Cs (bulk nearest-neighbor distance a=0.52 nm) and GaAs(110) and InSb(110) (unit cell size along $[1\bar{1}0]$ a=0.40 nm and a=0.46 nm, respectively). The Cs atoms will not readily fit in adjacent hollows along $[1\bar{1}0]$ on these substrates. This is also true of K (a=0.46 nm) on GaAs(110).

In contrast to Cs and K on GaAs(110) and Cs on InSb(110), K is latticed-matched with InSb(110) and should therefore fit in every hollow along a 1-D chain. However, as observed in Fig. 2, although K also forms 1-D chains on InSb(110) the chains are a single atom wide with K atoms adsorbed in *every-other* hollow. Furthermore, the atomic-resolution image of Fig. 2(b) reveals substantial in-plane relaxation of the substrate Sb atoms surrounding each K atom, with the adjacent Sb atoms appearing to relax away from each K adatom. This has the effect of making the hollow sites adjacent to each adatom smaller than normal, which could result in the alternate spacing of the adatoms within each chain. Careful analysis of the apparent substrate anion atomic positions surrounding the Cs and K zig-zag chains reveals similar, although smaller, in-plane relaxations associated with these chains also. These results demonstrate that in addition to any kinetic effects associated with the initial formation of these 1-D chains, the alkali metal-substrate interactions also play an important role in determining the ultimate chain structures [14,15].

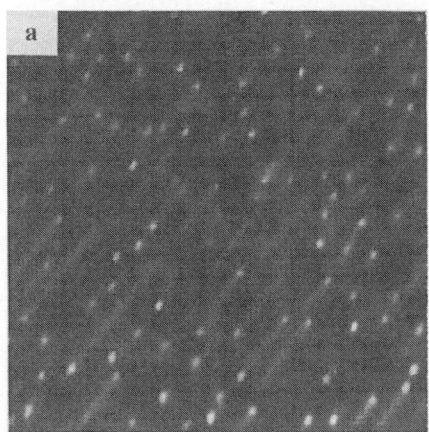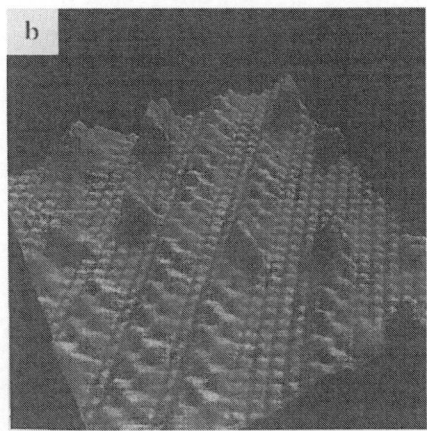

FIG. 1. STM topographic images of naturally occurring Cs zig-zag chains on InSb(110), with a Cs coverage ≈0.08 per substrate unit cell, recorded with negative sample bias. The chains run along the [1$\overline{1}$0] direction. (a) 60×60 nm^2 gray-scale view. (b) 12.5×12.5 nm^2 solid-rendered view with atomic resolution. Note that the substrate corrugation is due to the Sb atoms only at this bias voltage. The gray scale spans a range ≈0.15 nm.

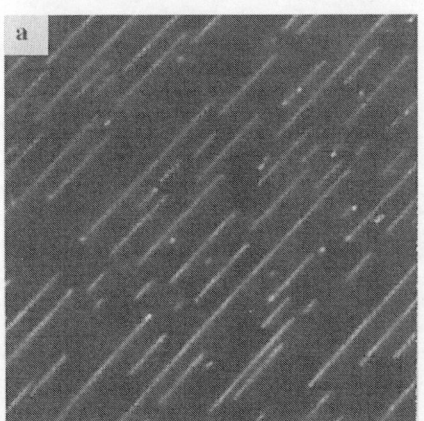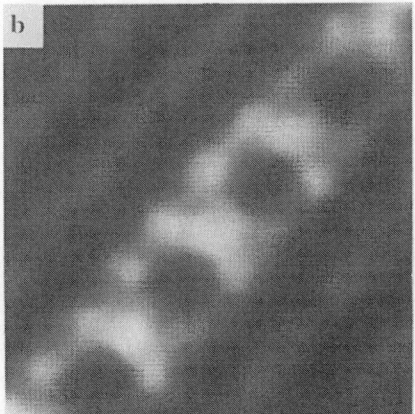

FIG. 2. Gray-scale topographic images of 1-D K chains on InSb(110). (a) 120×120 nm^2. (b) 3.5×3.5 nm^2 atomic-resolution view of a chain section. The brightest atomic-scale features appearing in every other hollow-like site along [1$\overline{1}$0] are believed to correspond to the K adatoms. Note the in-plane relaxation of the adjacent Sb atoms, observed as a slight dimerization along [1$\overline{1}$0], particularly along the chain and in the lower right corner of the image.

Following further adsorption of Cs or K on these substrates, the chains do not tile the surface, but rather become unstable, breaking up into disordered clusters. With increasing coverage the overlayers evolve into ordered 2-D structures composed of arrays of alkali metal (110)-like planar clusters. As previously reported [12,13], Cs forms a c(4×4) overlayer of five-atom clusters on GaAs(110) and a c(2×6) overlayer of four-atom clusters on InSb(110). We find that K forms a similar c(2×6) overlayer on both substrates, as shown for K on GaAs(110) in Fig. 3(a). Although the structure within the planar K clusters has not yet been resolved, the clusters appear to have the same four-atom structure as that observed for Cs on InSb(110), an atomic-resolution image of which is displayed in Fig. 3(b). Both the four- and five-atom planar clusters have an internal structure similar to that of an alkali metal bcc (110) surface [12,13]. In addition, these structures are close to those predicted to be stable for gas phase clusters of the same size [16]. These two observations indicate that the 2-D structures result from a balance between adsorbate-adsorbate interactions (which drive the formation of bulk-like clusters) and adsorbate-substrate interactions (which induce the commensurate superstructure).

When the 2-D overlayers are exposed to additional alkali metal, further adsorption occurs on top of the intial overlayers in a random fashion [12,13]. Since the first layer is not epitaxial, and the second layer has little interaction with the substrate, there is no mechanism for the formation of long-range order. The saturation structure observed for Cs or K adsorbed on either room temperature GaAs(110) or InSb(110) is a disordered bilayer.

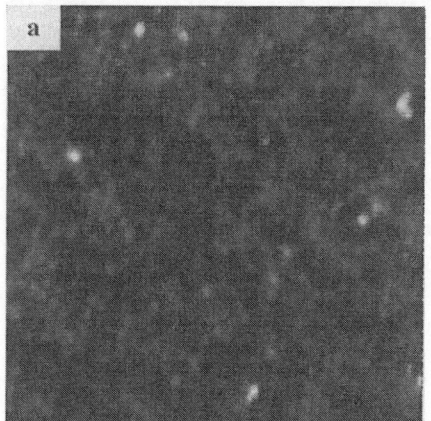 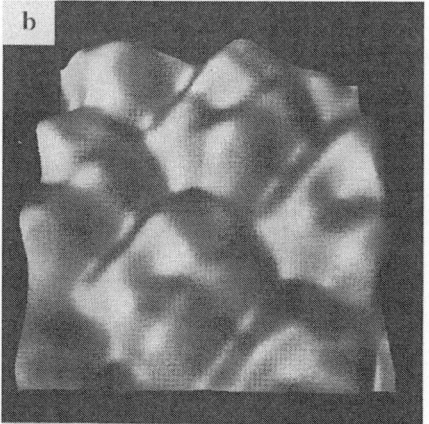

FIG. 3. 2-D alkali metal overlayers on InSb(110). (a) 40×40 nm^2 gray-scale image of a K overlayer. Each maximum in the image is a planar K cluster. The closely packed regions have a c(2×6) symmetry. (With slightly higher coverage the whole surface has this structure.) (b) 3×3 nm^2 solid-rendered view of similar Cs clusters with atomic resolution. The clusters appear to have four Cs atoms each.

2.2 FIELD-INDUCED DIRECTIONAL DIFFUSION

An atom or molecule with a static dipole moment μ and polarizability α placed in an electric field $\mathbf{E(r)}$ will have an induced dipole moment $\alpha\mathbf{E}$, and will experience a potential energy due to the field

$$U_E(\mathbf{r}) \cong -\mu \cdot \mathbf{E(r)} - \tfrac{1}{2}\alpha\mathbf{E(r)} \cdot \mathbf{E(r)} . \tag{1}$$

If the electric field varies as a function of $\mathbf{r}$, the atom or molecule will experience a potential energy gradient (that is, a force). Field ion microscopists have exploited this phenomenon to induce the directional diffusion of metal atoms adsorbed on the surfaces of field ion microscope (FIM) tips [8-10]. In an FIM a sharp tip, usually of refractory metal, terminated by a series of stacked low-index crystal planes is held a few centimeters away from an imaging plane at an electrostatic potential difference of many kilovolts. The electric fields at the surfaces on the end of the tip are typically 10^7 - 10^8 V cm^{-1}, with the intensity varying across each crystal terrace and enhanced near the terrace edges (due to the high curvature associated with a surface step). As a result of the field gradients near the step edges, adatoms with a positive static dipole in a positive electric field undergo thermally activated directional diffusion towards the region of greatest field (the terrace edge) [9,10]. For the transition metal adatoms studied in the FIM experiments, the induced dipole moments are typically an order of magnitude smaller than the static dipole moments ($\mu \sim 10^{-27}$ C cm), so the potential energy is dominated by the first term in Eq. (1) [9,10]. Hence the adatoms can be induced to diffuse toward or away from the step edges depending on whether the electric field is positive or negative, respectively.

In an STM the same ingredients necessary for electric field-induced directional diffusion are present, except that the electric field now results from having a sharp tip with an electrostatic potential difference of a few volts located less than a nanometer above the surface under study. In the STM the electric field intensity at the sample surface decreases with increasing radial distance from the end of the tip. For example, on a metal surface the field will fall to about half its maximum value at a distance approximately equal to the radius of the end of the tip [7]. Therefore, an adatom with a positive dipole moment on a surface at positive potential with respect to the tip will experience a potential energy well under the tip, and directionally diffuse towards the tip while the field is applied. (Recall that this is a thermally activated process.)

We have observed such diffusion for both Cs and K on room temperature GaAs and InSb (110) surfaces. As will be discussed in more detail below, the surfaces can be imaged with negative bias voltage without any apparent effect on the substrate or adsorbate overlayer. If the tip is positioned over the center of a previously imaged area and the sample bias is temporarily changed from negative to positive, subsequent images (with negative bias) reveal an increase in the alkali metal concentration in the region beneath the tip, as expected. One such sequence is presented in Fig. 4, showing a region of a GaAs (110) surface covered with Cs chains before and after a 0.35 s, +1 V pulse. As previously reported [7], during the voltage pulse alkali metal adatoms diffuse into the region beneath the tip, resulting in an increase in the number of alkali metal chains in the region, a shift in the chain length distribution towards longer chains, and an overall increase in local coverage. In general, the longer the positive bias is applied the greater the increase in local alkali metal coverage, as expected for a diffusion process. The effects of the induced diffusion, such as the appearance of new and longer alkali metal chains, are consistent with our understanding of the kinetics and thermodynamics that lead to the formation of 1-D chains following the initial adsorption (in the absence of an electric field).

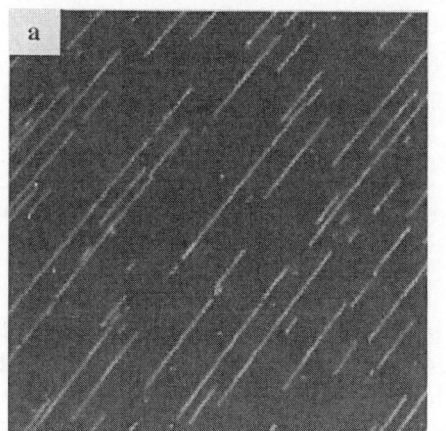 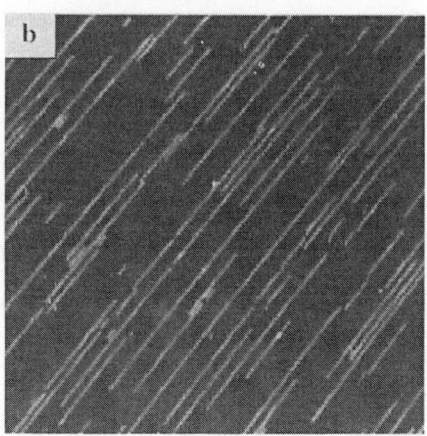

FIG. 4. Images of Cs chains in a 140×140 nm^2 area of p-type GaAs(110) recorded with negative sample bias (a) before, and (b) after positioning the tip over the center of the area and changing the bias to +1 V for 0.35 s.

The local effects of electric field-induced diffusion with the STM can be very dramatic, as illustrated by the images in Fig. 5. Prior to applying a 0.35 s, +3 V pulse, the area of GaAs(110) shown in Fig. 5(a) was covered with a uniformly sparse distribution of short (<20 nm) Cs chains. After inducing diffusion into the center of the imaged area, a region a few tens of nanometers across with very high local coverage is observed. Surprisingly, closer examination of this region reveals that a new 2-D phase has been created, an atomic-resolution image of which is displayed in Fig. 5(b). Unlike the naturally occurring 2-D phase for this system, composed of a c(4×4) array of five-atom planar clusters, this newly created phase of adsorbed Cs is primarily a c(2×2) arrangement of individual adatoms. Other, less well ordered structures that also do not naturally occur following room temperature adsorption have been created in this way on InSb(110) [7]. This demonstrates the potential of this technique to create and study novel configurations of adsorbed atoms and molecules at room temperature.

Although we have only briefly reviewed our results here, there are a number of puzzling observations worthy of discussion. The most striking observation is probably the polarity dependence of the directional diffusion; that is, diffusion occurs with a positive sample bias, but no apparent field effects are observed with the negative voltages typically used to obtain images of the filled states (−2 to −3 V). This is in contrast to the reversible directional diffusion observed in FIM, as discussed above. The explanation for the polarity dependence we observe lies in the relatively large polarizability of the alkali metal adatoms, and the resulting potential energy as expressed in Eq. (1). The polarizability of a Cs atom adsorbed on GaAs(110) is expected to be of similar magnitude to that of the free atom, ∼1×10^{-34} C cm^2 V^{-1} (100 Å^3) [17]. With such a polarizability, a static dipole moment $\mu \approx 1.6 \times 10^{-27}$ C m (as determined from the initial decrease in work function upon Cs adsorption [18]), and a sample bias of +3 V with a tunnel gap of 1 nm, a Cs adatom would experience a potential energy well ≈0.6 eV deep, with the static and induced dipoles both contributing equally beneath the tip. This is sufficiently deep compared with the expected diffusion barrier along [1$\bar{1}$0] to account for the directional diffusion we observe. In contrast, when the bias is reversed to –3 V, although the static dipole now leads to a *repulsive* potential energy barrier, the potential due to the induced dipole moment remains an *attractive* energy well. The net field-induced potential energy is approximately zero beneath the tip, accounting for the absence of induced diffusion under these conditions. Note that at larger negative voltages the induced dipole should dominate, making the potential energy again attractive. We have seen some evidence for diffusion towards the tip at –5 to –6 V, consistent with this expectation, although the results are inconclusive.

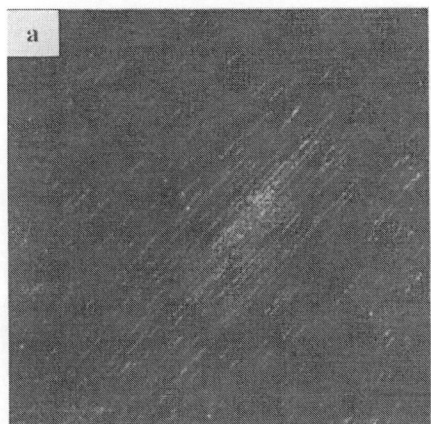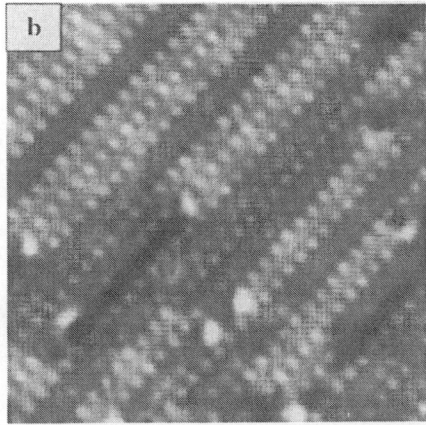

FIG. 5. The effect of a 0.35 s, +3 V pulse on Cs chains on *p*-type GaAs(110). (a) The 350×350 nm² region surrounding the pulse location. Prior to the pulse the region was sparsely covered with short Cs chains. (See Ref. [7].) (b) A 20×20 nm² image of the high-density area at the pulse location. This structure, which has a local c(2×2) symmetry, is not normally observed on GaAs(110).

In addition to the dependence on field polarity, on GaAs(110) we also observe a dependence of the threshold voltage (for observable diffusion) on sample dopant type (*n* versus *p*). While on *p*-type GaAs a sample bias of approximately +0.5 V is required to induce observable diffusion, a +2.0 V bias is necessary with *n*-type. This observation can be accounted for on the basis of the dielectric properties of the semiconductor, which allow the electric field to penetrate into the surface (band bending). Band bending has the effect of reducing the potential difference between the surface and the tip, thereby reducing the electric field at the surface and any resulting directional diffusion. With a *p*-type GaAs substrate at positive bias, upward band bending is limited by the creation of an inversion layer. In contrast, with an *n*-type substrate the bands can be bent upward through the bulk band gap, 1.45 eV, thereby decreasing the effective surface voltage by this amount [19].

The final observation which we will discuss is the large collection area beneath the tip observed on GaAs(110). Based on the electric field profile on a metal surface, the potential energy well beneath the tip is expected to be approximately a tip diameter wide. (Scanning

32

electron micrographs show our tips to be typically 20-40 nm in diameter.) However, the alkali metal adatom concentration increases within a region substantially wider than this on GaAs(110). We believe this also arises from the large field penetration that occurs on this wide band gap semiconductor, since such penetration will modulate both the magnitude and radial dependence of the electric fields perpendicular to the surface. Furthermore, this penetration will give rise to a component of the electric field *parallel* to the surface. Since the parallel polarizability of the Cs adatoms within 1-D chains is predicted to be even larger than that normal to the surface [17], an in-plane component of the field could contribute significantly to the potential energy gradient on this semiconductor surface.

2.3 ELECTRONIC PROPERTIES

In addition to characterizing the geometric properties of the naturally occurring and electric field-induced alkali metal structures, we have also investigated their electronic properties by measuring the voltage dependence of the tunneling current (*I-V* spectra) as a function of structure [12,13]. A comparison of typical *I-V* spectra recorded over 1- 2- and 3-D Cs structures on GaAs(110) and InSb(110) is displayed in Fig. 6. The apparent band gap associated with a 1-D zig-zag chain on GaAs(110) is ≈1.1 eV, indicating that these chains are not 1-D "wires." (We define the apparent band gap as the region of a spectrum with zero conductivity.) Note that the *I-V* spectrum recorded over a similar 1-D chain on InSb(110) has a much smaller gap width, ≈0.15 eV, approximately equal to the InSb band gap. We believe this is not necessarily a measure of the "chain gap," however, but rather an indication that tunneling is occurring into and out of the substrate states due to the open structure of the chain [13]. Similar results are obtained when spectra are recorded over 1-D K chains.

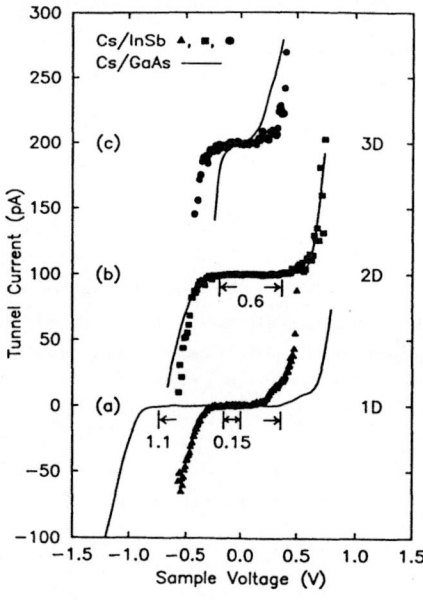

FIG. 6. A comparison of current vs voltage spectra recorded on various Cs structures on GaAs(110) (lines) and InSb(110) (symbols): (a) 1-D zig-zag chain; (b) 2-D overlayer; (c) 3-D saturation bilayer (see Refs. [12] and [13]). The band gaps indicated correspond to the regions of zero conductivity, determined on a more sensitive scale. Note that the GaAs spectra have been slightly smoothed, and that the spectra in (b) and (c) are offset from zero current.

The comparison between the electronic properties of the similar 2-D Cs overlayers on GaAs(110) and InSb(110), shown in spectra (b), reveals one of the most surprising results of our study: alkali metal monolayers on III-V (110) semiconductor surfaces are nonmetallic and have nearly identical *I-V* characteristics, with a characteristic band gap ~0.6 eV. This is true not only for the 2-D Cs overlayers, but for the 2-D K overlayers *and* the novel 2-D structures created via field-induced diffusion. This result is particularly striking for the case of InSb(110), where the overlayers open up an apparent gap *larger* than the substrate band gap. It is not until the formation of 3-D structures near saturation coverage (the disordered bilayers) that the measured gap narrows and these overlayers become metallic or near-metallic [spectra (c)].

An intriguing explanation for the unusual coverage dependence of the electronic properties of the alkali metal overlayers is that the 1-D and 2-D structures are Mott insulators [12,13,20]. If this is the case, the nonmetallic behavior arises from electron correlation effects caused by the low alkali metal density and coordination within these structures: the reduced electron wave function overlap between atoms as compared to that within bulk 3-D structures causes the electronic structure to remain atomic-like. The resulting coulomb repulsion between electrons (known as a correlation effect) creates a barrier to electron transport within the structures (the correlation energy), which we observe as a "band gap" in the *I-V* spectra. As the density and coordination increase with increasing dimensionality and the wave function overlap increases, the correlation energy decreases, lowering the observed gap. When 3-D structures form, with bulk-like density and coordination, a metallic band structure develops and the gaps in the spectra vanish. Note that a similar insulator-to-metal Mott transition occurs as a function of density in liquid alkali metals [21], supporting this explanation. In addition, recent band structure calculations for model alkali metal overlayers on GaAs(110) also offer evidence that electron correlation effects play a role in the electronic structure [14,22].

Although there is some support for the description of these systems as Mott insulators, an alternate explanation for their unexpected electronic properties may be offered in terms of more conventional band structure effects [13]. The similar chemical properties of the different alkali metals combined with the similar properties of the different III-V (110) semiconductor surfaces may result in a characteristic alkali metal-III-V(110) interface band structure for each similar interface structure. If this is the case, adsorbate-substrate interactions must be responsible for changing the electronic structure of the alkali metal overlayers from an expected metallic state to the observed insulating state. Note that there is at least one system exhibiting such behavior, the (1×1) Sb overlayer on GaAs(110): the isolated Sb overlayer is predicted to be metallic but the measured (and calculated) interface band structure is not [23-25]. Further experimental and theoretical work is required to resolve which mechanisms are responsible for the interesting electronic properties of these alkali metal overlayers [13].

3. Summary and Conclusions

The alkali metals (Cs and K) adsorbed on room temperature III-V (110) semiconductor surfaces (GaAs and InSb) form a rich variety of structures with surprising electronic properties. With low coverages 1-D chains oriented along [1$\bar{1}$0] are observed. These structures can be accounted for on the basis of anisotropic diffusion kinetics combined with the effects of adsorbate-substrate interaction. The 1-D chains are not metallic, with a gap in the *I-V* spectra observed for all chain types. At higher coverages the 1-D chains give way to ordered 2-D structures composed of arrays of alkali metal (110)-like planar clusters, showing the effects of both interadsorbate and adsorbate-substrate interactions. The 2-D overlayers all have nearly identical *I-V* characteristics, with an apparent band gap of ≈0.6 eV. Remarkably, on InSb(110) the 2-D alkali metal overlayers open up a gap *larger* than that of the substrate. At the highest attainable coverages at room temperature the absence of an epitaxial first layer results in the growth of disordered 3-D bilayers, with metallic or near-metallic characteristics. The evolution of the electronic structure with increasing dimensionality is consistent with the alkali metal overlayers behaving as Mott insulators. However, the electronic properties may alternately be attributed to simpler interface band structure effects. A more definitive explanation requires further investigation.

We have used the nonuniform electric field in the STM tunnel junction to induce the directional diffusion of the alkali metal adatoms toward the tip when a positive sample bias is applied. This thermally activated directional diffusion arises from the potential energy gradient created by the interaction of the static and induced dipole moments of the adatoms with the spatially varying electric field surrounding the tip. The absence of field-induced effects with negative bias (with which the images were recorded) can be attributed to the relatively large, free-atom-like polarizability of the alkali metal atoms when adsorbed on these semiconductor surfaces. The response of these systems to an applied electric field appears to be strongly affected by the dielectric properties of the semiconductor, in particular by the occurrence of band bending.

In the FIM, electric field-induced directional diffusion has been demonstrated for a variety of metal adatoms on FIM tip surfaces [8-10]. Given that electric fields of similar magnitudes are attainable in the STM tunnel junction, directional diffusion of a variety of polarizable atoms and molecules should be possible on semiconductor (and metal) surfaces other than those studied here. Of course an important caveat to this prediction is that one must always be on the lookout for unintended electric field-induced effects when studying surfaces with the STM, particularly in the presence of adsorbates with large static dipole moments or polarizabilities.

It is clear from this and other studies that the STM, along with the other scanned probe-type instruments, has great potential for controlled nanoscale modification of materials. If nanoelectronic devices fabricated via STM are ever to be employed in practical applications, however, at least two major hurdles must be overcome: the problem of interconnectivity - how can nanoscale devices be connected to more conventional circuit elements (without altering their properties)? and the problem of throughput - can such devices be fabricated reliably in parallel? While it is not certain whether STM-nanofabrication will soon lead to practical devices, it will clearly have a major impact probing the limits of miniaturization. The unprecedented ability to manipulate adsorbates into novel configurations, and subsequently study their geometric and electronic properties with the STM (or their mechanical and chemical properties with the atomic force microscope) should lead to a new understanding of how the fundamental properties of materials evolve atom-by-atom.

4. Acknowledgments

L. J. W. is grateful for the NRC-NIST Postdoctoral Fellowship which supported his work while at NIST. This work was also supported in part by the Office of Naval Research.

5. References

1. For recent reviews, see G. M. Shedd and P. E. Russell, Nanotechnology **1**, 67 (1990); C. F. Quate, in *Highlights in Condensed Matter Physics and Future Prospects,* edited by Leo Esaki (Plenum, New York, 1991), p. 573; and J. A. Stroscio and D. M. Eigler, Science **254**, 1319 (1991).
2. I.-W. Lyo and Ph. Avouris, Science **253**, 173 (1991).
3. H. J. Mamin, P. H. Guethner, and D. Rugar, Phys. Rev. Lett. **65**, 2418 (1990).
4. T. T. Tsong, Phys. Rev. B **44**, 13703 (1991).
5. D. M. Eigler and E. K. Schweizer, Nature **344**, 524 (1990).
6. G. Dujardin, R. E. Walkup, and Ph. Avouris, Science **255**, 1232 (1992).
7. L. J. Whitman, J. A. Stroscio, R. A. Dragoset, and R. J. Celotta, Science **251**, 1206 (1991).
8. E. V. Klimenko and A. G. Naumovets, Sov. Phys. Solid State **13**, 25 (1971) [Fiz. Tverd. Tela **13**, 33 (1971)]; Sov. Phys. Solid State **15**, 2181 (1974) [Fiz. Tverd. Tela **15**, 3273 (1973)].
9. T. T. Tsong and G. Kellogg, Phys. Rev. B **12**, 1343 (1975).
10. S. C. Wang and T. T. Tsong, Phys. Rev. B **26**, 6470 (1982).
11. L. J. Whitman, J. A. Stroscio, R. A. Dragoset, and R. J. Celotta, J. Vac. Sci. Technol B **9**, 770 (1991).
12. L. J. Whitman, J. A. Stroscio, R. A. Dragoset, and R. J. Celotta, Phys. Rev. Lett. **66**, 1338 (1991).
13. L. J. Whitman, J. A. Stroscio, R. A. Dragoset, and R. J. Celotta, Phys. Rev. B **44**, 5951 (1991).
14. J. Hebenstreit, M. Heinemann, and M. Scheffler, Phys. Rev. Lett. **67**, 1031 (1991).
15. K-induced relaxation of GaAs(110) is also discussed by C. A. Ventrice, Jr. and N. J. DiNardo, Phys. Rev. B **43**, 14313 (1991).
16. V. Bonacic-Koutecky, P. Fantucci, I. Boustani, and J. Koutecky, in *Studies in Physical and Theoretical Chemistry* (Elsevier, Amsterdam, 1989), Vol. 62, p. 429.
17. M. Krauss and W. J. Stevens, J. Chem, Phys. **93**, 8915 (1990).
18. Based on the results reported by D. Heskett, T. Maeda Wong, A. J. Smith, W. R. Graham, N. J. DiNardo, and E. W. Plummer, J. Vac. Sci. Technol. B **7**, 915 (1989).
19. R. M. Feenstra, J. A. Stroscio, and A. P. Fein, J. Vac. Sci. Technol. B **5**, 923 (1987).
20. N. J. DiNardo, T. Maeda Wong, and E. W. Plummer, Phys. Rev. Lett. **65**, 2177 (1990).
21. D. E. Logan and P. P. Edwards, in *The Metallic and Nonmetallic States of Matter,* edited by P. P. Edwards and C. N. R. Rao (Taylor & Francis, London, 1985), p. 78; W. Freyland and F. Hensel, *ibid,* p. 93, and references therein.
22. J. E. Klepeis, O. Pankratov, M. Scheffler, M. Methfessel, and M. Van Schilfgaarde, Bull. Am. Phys. Soc. **37**, 86 (1992).
23. R. M. Feenstra and P. Martensson, Phys. Rev. Lett. **61**, 447 (1988); P. Martensson and R. M. Feenstra, Phys. Rev. B **39**, 7744 (1989).
24. C. K. Shih, R. M. Feenstra, and P. Martensson, J. Vac. Sci. Technol A **8**, 3379 (1990).
25. C. Mailhiot, C. B. Duke, and D. J. Chadi, Phys. Rev. Lett. **53**, 2114 (1984); Phys. Rev. B **31**, 2213 (1985).

FIELD ION EVAPORATION FROM TIP AND SAMPLE IN THE STM FOR ATOMIC-SCALE SURFACE MODIFICATION

A. KOBAYASHI[1], F. GREY[1], H. UCHIDA[1], D.-H. HUANG[1], and M. AONO[1,2]

[1]Aono Atomcraft Project, Research Development Corporation of Japan (JRDC), Kaga 1-7-13, Itabashi, Tokyo 173, Japan
[2]The Institute of Physical and Chemical Research (RIKEN), Hirosawa 2-1, Wako, Saitama 351, Japan

ABSTRACT.   In a scanning tunneling microscope (STM) operated in ultrahigh vacuum, if we place a well-prepared Ag, W, Pt or Au tip above the Si(111)-7x7 surface at a separation of ~1 nm and apply an appropriate voltage pulse to it, we can extract a single Si atom from a predetermined position routinely. The extracted Si atoms are redeposited to the surface with a certain probability. The redeposited Si atom can be displaced intentionally on the surface. In case of the Si(001)-2x1 surface, usually two Si atoms forming a dimer are extracted at the same time. For both surfaces, Si atoms at crysallographically different sites including those at step edges are extracted with different probabilities. The microscopic mechanisms of these processes are discussed.

1. Introduction

The scanning tunneling microscope (STM) which was invented by Binnig and Rohrer [1] for imaging the topographic and electronic structures of solid surfaces on the atomic scale is also a promising method for processing materials at the atomic level. In fact, several preliminary demonstrations [2-8] suggest the power of this approach, although challenges remain in clearly understanding the physical mechanisms involved.

The present paper is concerned with the manipulation of single Si atoms on the Si(111)-7x7 and Si(001)-2x1 surfaces using Ag, W, Pt and Au STM tips in ultrahigh vacuum. A similar study of interest has been made by Lyo and Avouris [6] for the combination of the Si (111)-7x7 surface and a W tip. They have demonstrated that a single Si atom can be extracted from the surface and the extracted Si atom can be redeposited onto the surface, by applying a voltage pulse between sample and tip in an appropriate polarity. The tip-sample

37

P. Avouris (ed.), Atomic and Nanometer-Scale Modification of Materials: Fundamentals and Applications, 37–47.
© 1993 *Kluwer Academic Publishers.*

separation in these experiments was as small as ~0.3 nm, which is almost in the point-contact regime, so that not only a field effect due to the applied voltage but a direct chemical interaction between tip and sample is of importance (because of this, they refer to this method as the "chemically-assisted field evaporation" method). In contrast to this, in our experiments discussed in the present paper, the tip-sample separation is as large as ~1 nm at which no chemical interaction is significant. The reason why we have selected this condition is that we want to reveal the effect of field separately as one of steps in clearly understanding the physical mechanisms involved in the transfer of atoms between sample and tip in the STM.

As demonstrated in the present paper, we have revealed the values of experimental parameters at which we can routinely extract a single Si atom from a predetermined position of the Si(111)-7x7 surface in our experimental condition. We have also found that the extracted Si atom is redeposited onto the surface with a certain probability and that the redeposited Si atom can be displaced intentionally. In case of the Si(001)-2x1 surface, however, usually two Si atoms forming a dimer are extracted at the same time. In any case, the probability of extraction of Si atoms is closely related to the chemical stability of the Si atoms. The purpose of the present paper is to give a brief review of these results.

2. Experimental

Si(111) and Si(001) samples were cut from n-type wafers and each of them was mounted on a sample holder in an STM (VG-STM 2000 or JEOL-JSTM 4000XV) operated in UHV. The sample was cleaned by flash heatings at 1250 °C in UHV ($\leq 5 \times 10^{-8}$ Pa) to obtain the 7x7 or 2x1 reconstructed structure for the Si(111) and Si(001) samples, respectively.

The STM tip was a Ag, W, Pt or Au wire with a diameter of 0.2-0.3 mm which was sharpened by electrolytic etching using an appropriate solution. The tip was mounted on a tip holder in the STM and cleaned by flash heatings at 600-1500 °C in UHV ($\leq 1 \times 10^{-7}$ Pa). Although this step of cleaning was indispensable for our purpose, the tip was still contaminated at the atomic level and therefore not good for well-controlled atomic-scale processing. By applying an appropriate voltage pulse many times between tip and sample, the residual contamination was removed though field evaporation [9], and at the same time the most stable mini-tip was remained [9], as suggested by actual atom manipulation experiments.

3. Extraction of Single Si Atoms

If we place an STM tip above the Si(111)-7x7 surface at separation s of ~1 nm and apply an appropriate positive or negative voltage, V_t, to it ($|V_t|$ depends on the polarity of V_t and s), Si atoms are ex-

tracted from the surface one after another [10]. When the duration
of applied voltage, t, is appropriately short (~10 ms, although it
depends on V_t and s), a single Si atom is extracted [11]. Examples
of such single Si atom extraction and larger fabrication (creation
of grooves) are shown in Figs.1 and 2, respectively.
   With respect to the mechanisms of the extraction of Si atoms,
all of our extensive experimental results are consistent with the

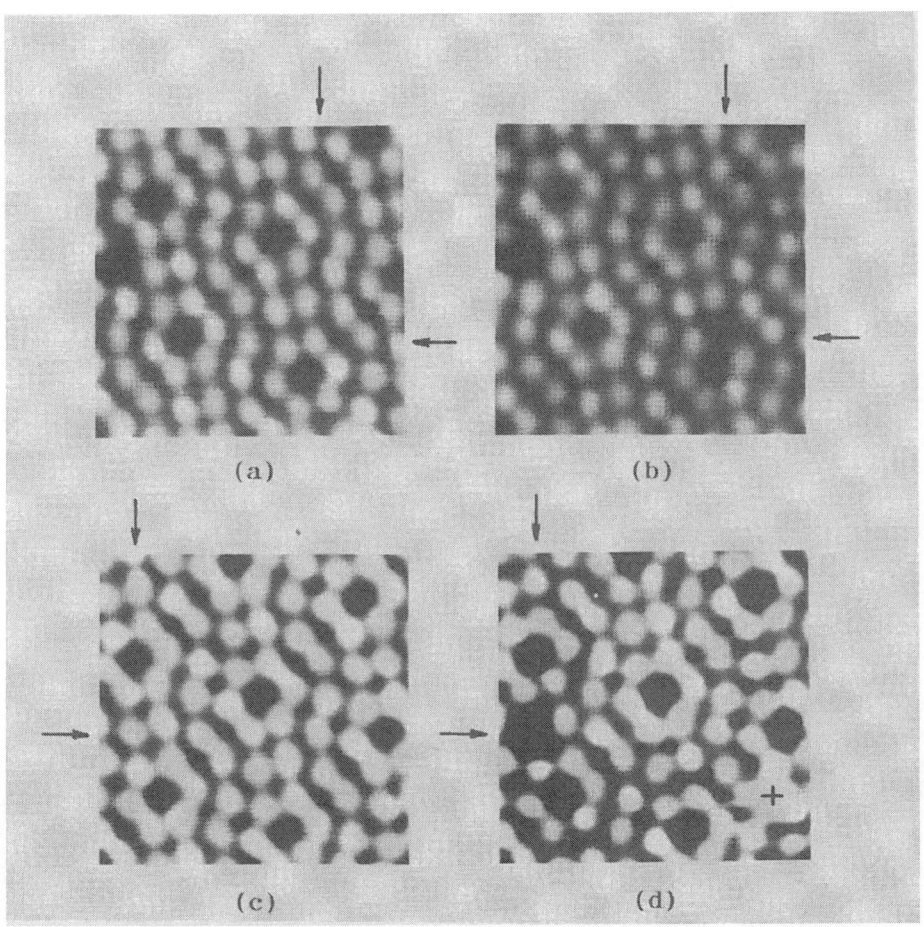

Fig.1 STM images showing the extraction of single Si
atoms from the Si(111)-7x7 surface using a W tip in an
STM. Si atoms indicated by arrows in (a) and (c) were
extracted as shown in (b) and (d), respectively; a voltage of +6 V was applied to the tip for 10 ms to extract
the Si atoms. In (d), an extra atom indicated by a
cross has appeared, which is identified to be a redeposited Si atom.

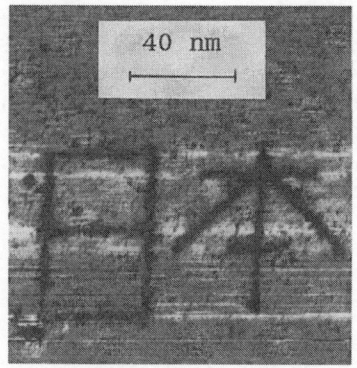

Fig.2 Characters written on the Si(111)-7x7 surface by extracting Si atoms using a W tip in an STM, which stand for "Japan" in Japanese.

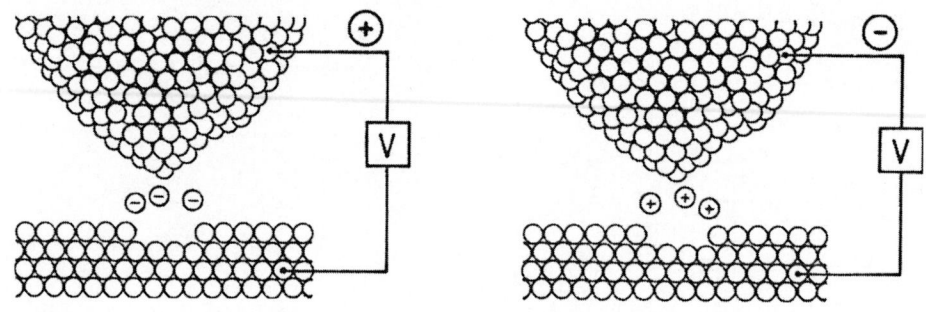

Fig.3 Schematic figures showing the mechanisms of the extraction of Si atoms from the Si(111)-7x7 surface using an STM tip (Ag, W, Pt or Au) at a tip-sample separations as large as ~1 nm. If an appropriate positive (negative) voltage is applied to the tip, surface Si atoms evaporate as negative (positive) ions, as illustrated in (a) and (b).

picture [10] schematically illustrated in Fig.3. Namely, when we apply an appropriate positive (negative) V_t, surface Si atoms are extracted as negative (positive) ions through field evaporation. The "symmetry" with respect to the polarity of V_t seen in Fig.3 is observed in various experimental results. An example is shown in Fig.4. This shows the relationship between the threshold of V_t necessary for extracting Si atoms and the logarithm of the tunneling current I_t (log I_t is roughly proportional to s), which was measured using Ag, W, Pt and Au tips. The symmetry mentioned above is clearly observed; a quantitative deviation from the symmetry is due to a difference between the critical fields for positive and negative Si

ion evaporation. According to recent theoretical calculations by
Lang [12], a Si atom placed on one of closely separated two jellium
surfaces can evaporate as a positive or negative ion depending on
the polarity of a voltage applied between the two surfaces, although
the magnitude of ionic charge is only a fraction of the unit charge
|e| in both cases. As we see in Fig.4, as the work function of tip
increases in order of Ag, W, Pt and Au, the magnitude of the thresh-
old voltage, $|V_t|$, decreases for negative V_t but increases for posi-
tive V_t. This is also consistent with Fig.3. Namely, the field
created at the sample surface is determined not only by V_t but by
the work function of tip, and the latter enhances the field for
negative V_t but cancels it for positive V_t.

    There are four kinds of Si atoms in the top layer of the Si
(111)-7x7 surface, i.e., the "corner" and "center" Si adatoms in
both the "faulted" and "unfaulted" halves of the 7x7 unit cell. In
order to see if these crystallographically different Si adatoms are
extracted with different probabilities, we extracted more than 230
Si adatoms randomly and observed how the created vacancies are distributed among these different sites. The results are shown in
Table 1 [13], where (a) and (b) were obtained for positive and
negative V_t, respectively.

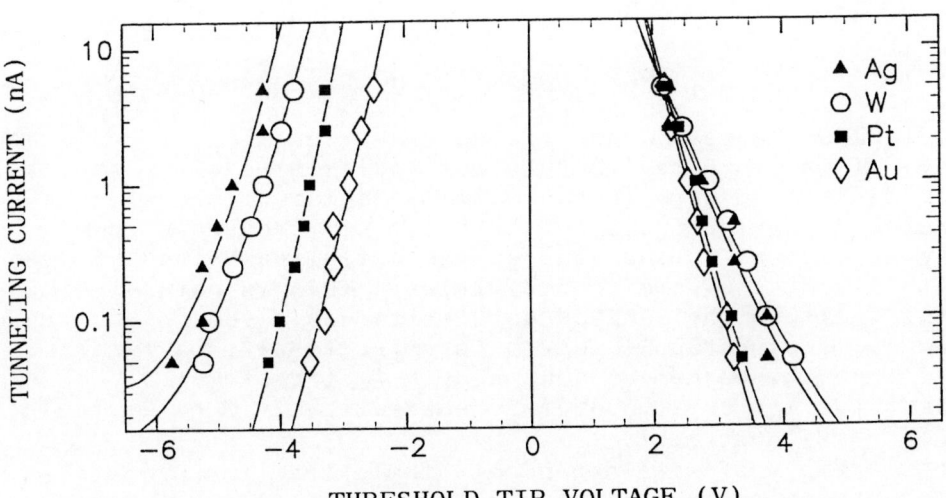

Fig.4 Relationship between the threshold tip voltage
necessary for extracting Si atoms from the Si(111)-7x7
surface and the tunneling current (the logarithm of the
tunneling current is roughly proportional to the tip-
sample separation). The measurements were made using
Ag, W, Pt and Au tips.

Table 1   Frequencies of Si atom extraction from four different crystallographic sites of the Si(111)-7x7 surface
using a W tip in an STM.   (a) and (b) show the results for
positive and negative tip voltages (in both cases, the magnitude of tip voltage was 6 V).

(a)

Corner Si Adatom		Center Si Adatom	
Faulted Half	Unfaulted Half	Faulted Half	Unfaulted Half
51 (21%)	48 (19%)	73 (30%)	74 (30%)
99 (40%)		147 (60%)	

(b)

Corner Si Adatom		Center Si Adatom	
Faulted Half	Unfaulted Half	Faulted Half	Unfaulted Half
41 (18%)	46 (20%)	79 (34%)	64 (28%)
87 (38%)		143 (62%)	

If we look at Table 1, there is no essential difference between
(a) and (b).   This indicates that in both polarities of V_t the extraction of Si atoms occurs in the same mechanism, i.e., field evaporation, which is again consistent with the picture shown in Fig.3.
Second, the Si adatoms in the faulted and unfaulted halves are extracted with almost the same probabilities.   This is natural because
the stacking fault in the faulted half exists in a subsurface region.
Third, however, the center Si adatom is extracted with a larger
probability than the corner Si adatom by a factor of ~1.5.   This
suggests that the center Si adatom is energetically more unstable
than the corner Si adatom.
Another example of site-dependent atom extraction probabilities was observed for the Si(001)-2x1 surface [14].   Fig.5(a)
shows an STM image of the Si(001)-2x1 surface, in which an "A-type"
atomic step exists running from the upper left to the lower right.
We scanned a W tip perpendicular to the step at V_t of -3 V.   The
resulting surface is shown in Fig.5(b).   An interesting aspect of
Fig.5(b) is that no Si atoms at the step edge were extracted at all
whereas Si atoms on the terraces on the both sides were extracted
(Si atoms at the so-called "C-defect" were preferentially extract-

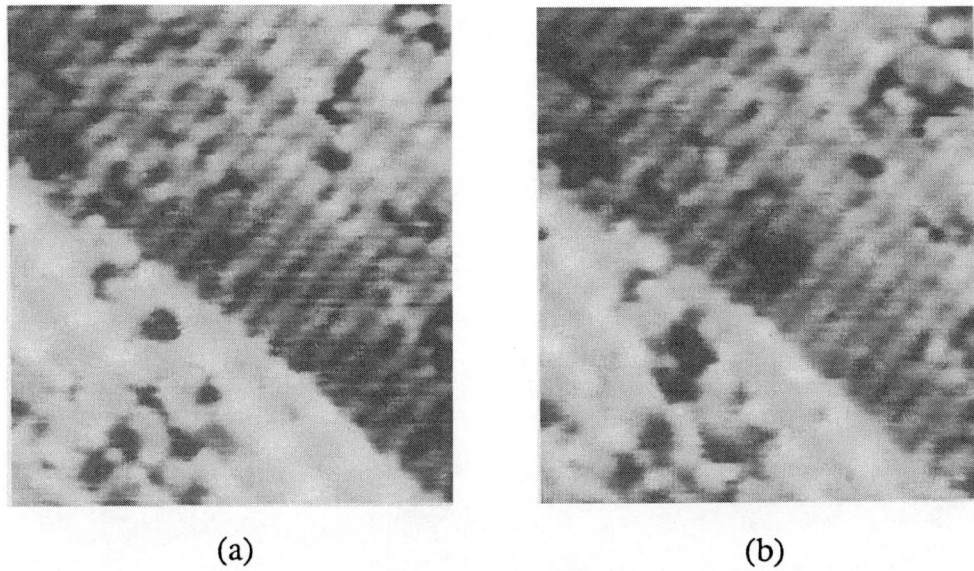

(a)   (b)

Fig.5   STM images of the Si(001)-2x1 surface (a) before
and (b) after extracting Si atoms using a W tip in an STM.
In order to extract Si atoms, a voltage of +3 V was applied
to the tip, and the tip was scanned from the lower left to
the upper right across the edge of an "A-type" atomic step
running from the upper left to the lower right.   Although
some Si atoms on the terrace were extracted, no Si atoms
at the step edge were extracted at all.

ed).   However, this is true only for the A-type step.   In fact, Si
atoms at the edge of the "B-type" step are more easily extracted.
This is consistent with theoretical calculations by Chadi [15],
which indicate that the B-type step is energetically more unstable
than the A-type step.
        It is of interest to point out that although we can extract
single Si atoms from the Si(111)-7x7 surface as already discussed,
usually adjacent two Si atoms forming a dimer are extracted at the
same time in case of the Si(001)-2x1 surface as found in experimental results such as Fig.5.

4. Deposition of Single Si Atoms

When we extract a single Si atom from the Si(111)-7x7 surface, an
extra atom sometimes appears near the created vacancy.   Such an
example is shown in Figs.1(c) and (d): When a Si atom indicated by
arrows in Fig.1(c) was extracted as shown in Fig.1(d), an extra atom

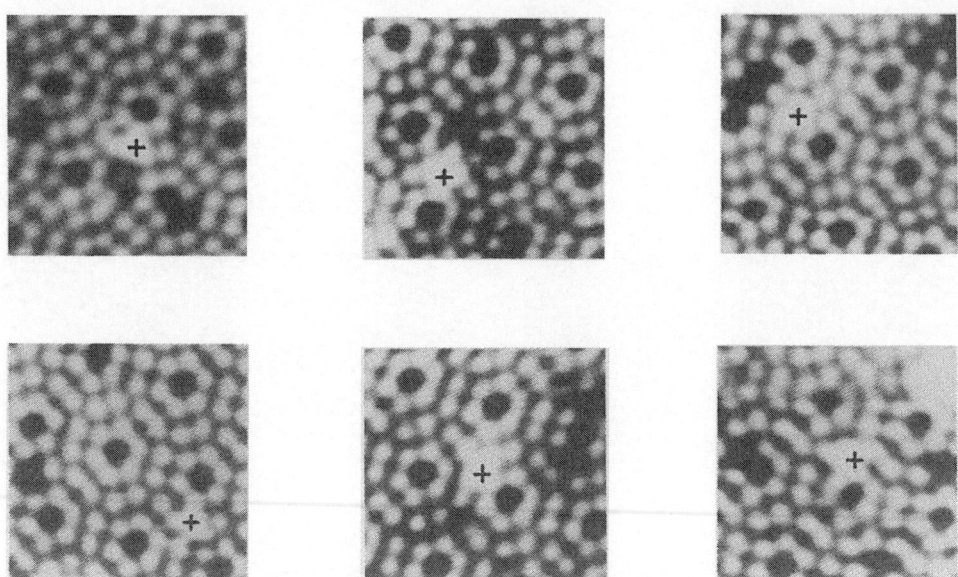

Fig.6 STM images showing that an extra atom that sometimes appears when a single Si atom is extracted from the
Si(111)-7x7 surface always occupies the center of one
corner adatom and adjacent two center Si adatoms as
indicated by a cross (the extra atom is identified to be
a Si atom).

appeared as indicated by a cross in Fig.1(d). Interestingly, the
extra atoms always appear in the center of one corner Si adatom and
adjacent two center Si adatoms (see Fig.6). The extra atom is
identified to be a Si atom because any other possibilities can be
reasonably excluded [9,11]. This identification is in fact consistent with an experimental result by Koehler et al.[16]: Si atoms
deposited onto the Si(111)-7x7 surface from an evaporation source
occupy the site mentioned above. In this way, we can conclude that
the extra atoms are created by the redeposition of extracted Si
atoms.

However, according to experiments in which more than 280 Si
atoms were extracted using a W tip, the extra Si atoms appeared only
with a small probability of 0.19. This suggests that only a small
fraction of extracted Si atoms stuck to the tip at most, where the
"tip" means an atomic-scale mini-tip which was responsible for the
extraction of Si atoms of concern.

5. Displacement of Single Si Atoms

The extra Si atom on the Si(111)-7x7 surface discussed in the previous section can be displaced intentionally. An example is shown
in Fig.7. Fig.7(a) shows an STM image of an area of the Si(111)-7x7
surface containing such an extra Si atom. We placed a W tip at a
upper left position of Fig.7(a), being distant from the extra Si
atom by ~5 nm, and applied +6 V to it for 10 ms. By this procedure, the extra Si atom was displaced to another crystallographically equivalent position as shown in Fig.7(b).
     The mechanism of this displacement is not clear yet. A plau-
sible interpretation is field-induced migration, but the extra atom
does not necessarily move toward the tip, at least in experiments
which we have done so far. Further investigations are required to
clarify the mechanism of the displacement.

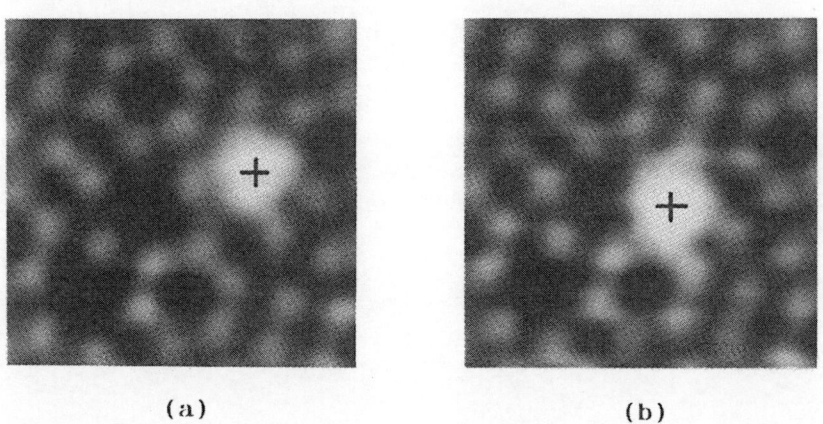

(a)   (b)

Fig.7 STM images showing that the extra Si atom created
on the Si(111)-7x7 surface can be displaced intentionally
using an STM tip. A W tip was placed in an upper left
position of (a) being distant from the extra Si atom (indicated by a cross) by ~5 nm, and a voltage of +6 V was
applied to it for 10 ms. The extra Si atom was displaced
to another crystallographically equivalent site.

6. Summary

We have studied the interaction between the Si(111)-7x7 or Si(001)-
2x1 surface and a Ag, W, Pt or Au tip in an STM operated in ultra-
high vacuum, under the condition that the tip-sample separation is
as large as ~1 nm as in usual STM imaging but a considerably large
positive or negative voltage are applied between them. The follow-

ing conclusions were obtained:

(1) All of our experimental results are consistent with a picture that when an appropriate positive (negative) voltage is applied to the tip, Si atoms of the surfaces are extracted as negative (positive) ions through field evaporation.

(2) By applying the voltage mentioned in (1) for an appropriately short time, we can extract a single Si atom from the Si(111)-7x7 surface at a predetermined position. Larger fabrication such as the creation of grooves is of course possible by scanning the tip with an appropriate voltage beeing applied to it. In case of the Si(001) -2x1 surface, usually adjacent two Si atoms forming a dimer are extracted at the same time.

(3) The easiness of Si atom extraction from the surfaces depends on the crystallographic sites of Si atoms. For example, in case of the Si(111)-7x7 surface, the center Si adatom is more easily extracted than the corner Si adatom. For the Si(001)-2x1 surface, Si atoms at the A-type step edge are more easily extracted than those at the B-type step edge.

(4) When we extract Si atoms from the Si(111)-7x7 surface using a W tip, the extracted Si atoms are redeposited onto the surface with a small probability of ~0.19. This indicates that only a small fraction of extracted Si atoms sticks to the tip at most, where the "tip" means an atomic-scale mini-tip responsible for the extraction of Si atoms of concern.

(5) The redeposited Si atom mentioned in (4) can be displaced intentionally by placing a tip near the Si atom and applying an appropriate voltage pulse to it.

References

[1] Binnig, G. and Rohrer H. (1982), Helvetica Phys. Acta 55, 762.

[2] Becker, R. S., Golovchenko, J. A. and Swartzentruber, B. S. (1987), Nature 325, 419.

[3] Mamin, H. J., Gurthener, P. H. and Gugar, D. (1990), Phys. Rev. Lett. 65, 2418.

[4] Eigler, D. M. and Schweizer E. K. (1990), Nature 344, 524.

[5] Lyo, I.-W. and Avouris, P. (1991), Science 253, 173.

[6] Hosoki, S., Hosaka, S. and Hasegawa, T. (1992), Appl. Surf. Sci. (in press).

[7] Iwatsuki, M., Kitamura, S. and Mogami, A. (1990), Proc. XII-th Intern. Congress for Elec. Microscopy (San Francisco Press Inc., San Francisco), p.322.

[8] Aono, M., Kobayashi, A., Uchida, H., Hejoh, H. and Nomura, E. (1991), J. Crystallog. Soc. Jpn. 33, 158.

[9] Huang, D.-H., Uchida, H. and Aono, M. (1992), Jpn. J. Appl. Phys. (in press).

[10] Kobayashi, A., Grey. F. and Aono, M. (to be published).

[11] Uchida, H., Huang, H. and Aono, M. (to be published)
[12] Lang, N. D. (1992), Phys. Rev. B45, 13559.
[13] Uchida, H., Huang, D.-H., Grey, F. and Aono, M. (to be published).
[14] Kobayashi, A., Grey, F. and Aono, M. (to be published).
[15] Chadi, D. J. (1987), Phys. Rev. Lett. 59, 1691.
[16] Koehler, U., Demuth, J. E. and Mamers, R. J. (1989), J. Vac. Sci. Technol. A7, 2860.

WRITING OF LOCAL, ELECTRICALLY ACTIVE STRUCTURES IN AMORPHOUS SILICON FILMS BY SCANNING TUNNELING MICROSCOPY

E. Hartmann[a] and R. J. Behm[b]
Institut für Kristallographie und Mineralogie der Universität München
Theresienstr. 41
D-8000 München 2
Germany

F. Koch
Physik Department E16, Technische Universität München
James-Franck-Str.
D-8046 Garching
Germany

ABSTRACT. Phosphorus doped, hydrogenated amorphous Si films [a-Si:H(P)], deposited on heavily doped, p-type crystalline Si substrates can locally be modified in scanning tunneling microscopy (STM) by creating electrically active structures on a nanometer-scale. The intense electron beam provided by the tip of the STM leads to characteristic changes of the layer conductivity (electrically active structures), which is interpreted as to arise from local, electronically induced changes of the coordination of Si and dopant atoms within the a-Si:H(P) film and the resulting shift of the Fermi level position. Depending on the modifying conditions (mainly tunnel voltage and current), the lateral dimensions of the electrically active structures range between 5 nm and several hundred nm. Three different types of electrically active structures are distinguished according to different modifying conditions: (a) breaking of weak Si-Si bonds at the surface, (b) structural modifications in a near-surface region of the a-Si:H(P) layers, and (c) changes in the bonding configuration throughout the entire film down to the a-Si:H(P)/c-Si interface. In the latter case, the electrical properties of the p^+-n heterojunction at the interface are modified.

---

a) present address: *Physik Department E16, Technische Universität München, James-Franck-Str., D-8046 Garching, Germany*

b) new present address: *Abteilung für Oberflächenchemie und Katalyse, Universität Ulm, Postfach 4066, D-7900 Ulm, Germany*

P. Avouris (ed.), Atomic and Nanometer-Scale Modification of Materials: Fundamentals and Applications, 49–64.

1. Introduction

Scanning tunneling microscopes (STMs) have increasingly been used as an active tool to create nanometer- to atomic-scale structures on various materials. Mamin et al. deposited tip material onto Au-samples by applying voltage pulses of several 100 ns [1]. Removal of sample material on a nanometer-scale was demonstrated by van Loenen and co-workers upon touching Si surfaces with the probing tip [2]. Finally, Lyo and Avouris removed and re-deposited individual Si atoms by reducing the tip/sample distance during operation [3].

In contrast to these efforts of material transfer we used the STM to produce electrically active structures by modifying phosphorus doped, hydrogenated amorphous Si [a-Si:H(P)] films [4,5]. These layers were deposited on highly doped, p-type Si substrates forming a p^+-n heterojunction. Depending on the tunneling conditions (voltage, current, and the duration of the exposure to the highly localized, intense electron beam from the tip), the STM can be operated in a 'read' or 'write' mode. Under reading conditions, at voltages between + 3 V and + 4 V applied to the sample and 0.1 to 0.5 nA tunnel current, STM imaging does not alter the surface topography or the electrical properties of the amorphous Si layers, and the same area can be repeatedly scanned without any changes in the images. Under conditions of high current densities (V_t > + 5 V with I_t > 1 nA ; 'write cycle'), with the electron flow directed towards the sample, structural modifications within the amorphous layers can take place which result in changes of the electrical film properties. As the STM probes electronic characteristics, these induced modifications can be detected with the same instrument in the constant current imaging mode ('read cycle') as well as in measurements of the effective tunneling barrier height.

The correlation between electrical and structural modifications of these films, and the interpretation of our results relies strongly on previous studies of hydrogenated amorphous Si films. These studies already demonstrated that the electrical properties of the a-Si:H films can be modified after deposition. For instance, exposure to light leads to a degradation of amorphous Si due to an increase of defect states and, correlated, to a worsening of the opto-electronical properties (Staebler-Wronski effect, SWE) [6,7]. As a second example, changes of the electrical characteristics of a-Si:H(P) layers are induced also in so-called bias-annealing experiments using a field-effect transistor (FET) configuration [8,9]. Upon varying the gate-voltage at elevated temperatures ($T \approx 200$ °C), the density of electrons within the a-Si:H film could be changed. The resulting changes in the layer conductivity were demonstrated by characteristic shifts of the current-voltage behavior. In both cases - SWE and bias-annealing experiments - the coordination defects can be healed up again by heating the sample to around 200 °C without an electric field applied. This treatment leads to a recovery of the former (as-grown) layer properties. The above modifications are explained in terms of the 'generalized autocompensation model' [10].

In the first part of this article we will review our previous results, then present and discuss our model for explaining these modifications and finally, in the last part, we confirm our concept with further experiments.

2. Experimental

The a-Si:H(P) layers were grown on heavily doped, p-type Si (111) wafers (0.01 Ω cm) by rf glow-discharge deposition techniques. For comparative measurements, also n-type Si substrates were included. Prior to the deposition the Si substrates were immersed into hydrofluoric (HF) acid in order to remove the native oxide and subsequently rinsed thoroughly in high quality de-ionized and filtered water (resistivity 10^7 Ω cm). The amorphous layers were doped by adding 1 at. % phosphine (PH$_3$) to the silane (SiH$_4$) gas flow and grown at a deposition temperature of 250 °C. This results in a bulk hydrogen concentration of nearly 8 at. %. The film thicknesses ranged between 30 nm and 500 nm; the layer resistivity ran up to roughly 100 Ω cm - low enough to perform STM experiments. A hydrogen content of approximately 12 at. % in the near surface region inhibits rapid growth of a native oxide layer.

Freshly prepared samples were transported in a desiccator filled with helium gas to prevent appreciable contamination of the sample surface, e.g. by exposure to hydrocarbons and atmospheric humidity. Auger electron spectra (AES) recorded on samples which were stored for two days in the laboratory atmosphere revealed that the surface contained only small amounts of carbon and oxygen. The relative peak intensities for elemental Si and C were comparable to those recorded on amorphous Si films which were sputter-cleaned under ultra-high vacuum conditions (UHV) [11]. A slightly enhanced oxygen signal compared to the sputter-cleaned samples was attributed to adsorbed water.

STM experiments were performed at room temperature in a high-vacuum chamber which was vented with N$_2$ for sample loading. After pumping down the system with a turbomolecular pump, this was switched off and a vibration-free ion getter pump maintained high vacuum conditions ($p < 10^{-5}$ Pa). The probing tips were electrochemically etched from polycrystalline tungsten wires.

3. Results and Discussion

The 'as-deposited' films exhibit a characteristic granular structure with a peak to peak corrugation height of about 2 nm superimposed on a surface that is perfectly flat on a larger scale. This structure was reproducibly observed in images obtained at low tunnel voltages and currents (imaging conditions: $I_t = 60$ pA and $V_t = +3$ V). If we increase the tunnel voltage and current above a threshold value with the tip standing on a spot, prominent features like that displayed in Fig. 1 were created which could subsequently be detected in the imaging mode. The actual values of the threshold parameters depend on the layer thickness of the a-Si:H films which in this case was 30 nm. The structure is 260 nm in diameter and the maximum height amounts to 25 nm. In the surrounding region, the granular structure of the unperturbed surface morphology is resolved.

In order to verify the electronic nature of this structure we traversed it with the tip at constant height. For these investigations, the vertical position of the tip was 'frozen in' by interrupting the feedback loop. The distance was adjusted such that it is small

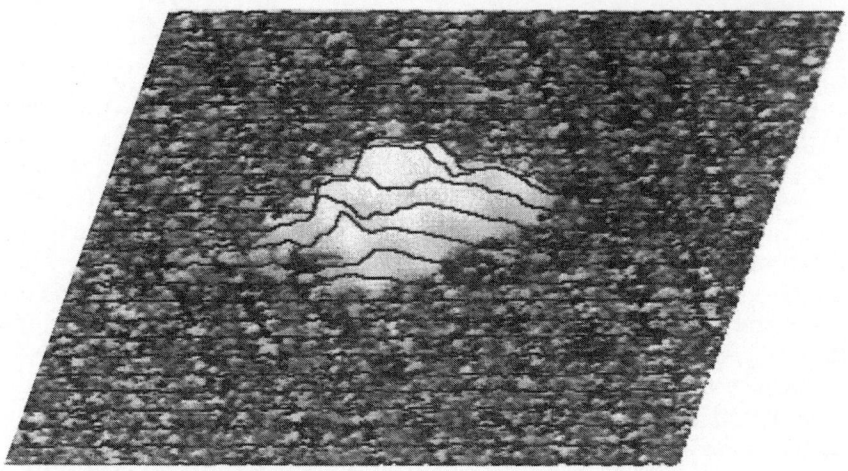

Figure 1. Prominent structure (260 nm in diameter and 25 nm in height) written in a 30-nm-thick *a*-Si:H(P) (*n*-type) film deposited on *p*⁺-type Si substrate.

compared to the total height of the elevation. Following such 'traversing' experiments no visible changes of the 'structures' were ever observed. The surrounding surface topography was resolved as before demonstrating that the tip did not undergo geometrical modifications due to a crash damage. Hence, these structures in the STM images must reflect changes in the electric surface properties rather than actual topographic protrusions.

This implies that above the modified areas the tunnel current rose to a degree that, for maintaining constant current, the tip had to be retracted by about 25 nm. At first sight, this large retraction appears to be implausible, but the apparent discrepancy can be resolved by analyzing the actual electron transfer mechanism. The different possible mechanisms 'tunneling', 'field emission' and 'Schottky emission' exhibit different current-voltage relations, which are given in the upper part of Fig. 2. *I-V* curves on the modified areas were determined from the respective maximum currents measured during traversing experiments at different tunnel voltages. These values are plotted in the diagrams depicted in the lower part of Fig. 2. The scales of the axes are chosen so that if one of these mechanisms describes the electron transfer correctly, the data will fall on a straight line in the respective diagram. Evidently, electron transfer above modified regions follows most closely the dependence predicted by Schottky emission. The two other possibilities for electron transfer, 'tunneling' and 'field emission', do not show a straight-lined dependence.

Schottky emission results from extremely low values of the effective barrier height between the two electrodes [12]. In STM measurements, the effective barrier height can be determined by modulation techniques, i.e. by modulating the tip/sample distance [13-15]. Above modified areas, values of around 0.4 eV were measured which are very low compared to several eV obtained on atomically flat surfaces under UHV

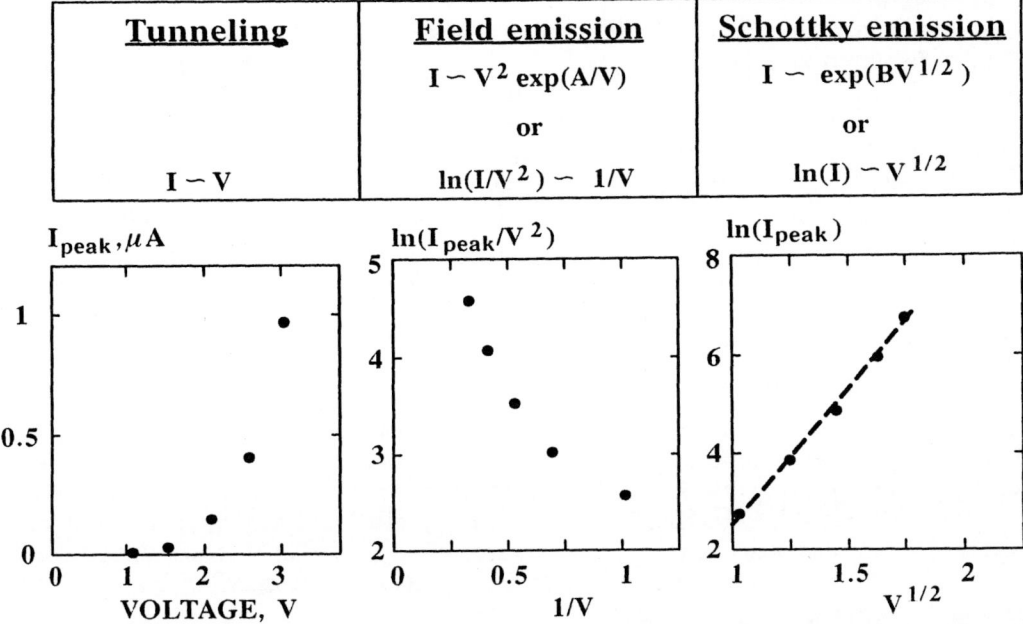

Tunneling	Field emission	Schottky emission
	$I \sim V^2 \exp(A/V)$	$I \sim \exp(BV^{1/2})$
	or	or
$I \sim V$	$\ln(I/V^2) \sim 1/V$	$\ln(I) \sim V^{1/2}$

Figure 2. Determination of the electron transfer mechanism above modified regions. The maximum current (I_{peak}), recorded during pass through experiments is plotted vs. voltage so that a straight-lined dependence reflects the correct mechanism.

conditions. Low effective barrier heights were characteristic even on unmodified areas of the amorphous films. For this behavior, three effects are responsible in STM measurements. First of all, the high density of occupied states at the surface of the a-Si:H(P) layers ($6*10^{12}$ cm^{-2}) varies with applied voltage, resulting in a change of the width of the depletion region in the sample area beneath the tunneling tip [16]. When applying a more positive bias at the sample, this effect leads to a lowering of the effective barrier height and hence to an additional increase in the tunnel current. Secondly, the surface corrugation amplitude of 2 - 4 nm leads to a non-zero angle between the gradient of the electron density outside the sample surface and the direction of the tip motion during the modulation procedure, which reduces the measured value of $(dI)/(dz)$ [14,17]. Finally, tip and sample are likely to be covered by a contamination layer which is also known to reduce the effective tunnel barrier [18]. Consequently, under imaging conditions we believe that on both modified and unmodified surface regions the electron transfer mechanism is predominantly determined by Schottky emission. The distance between tip and sample surface was estimated at 2 - 3 nm (see below) which limits the lateral resolution of the surface topography shown in the STM images.

In order to explain the electrical modifications of the amorphous layers, it is necessary to consider the density of states within the mobility gap of as-grown a-Si:H(P). This is displayed as a function of energy in Fig. 3a. In crystalline material, localized bonding states (valence band) are separated from de-localized antibonding states (conduction band) by a well defined band gap (E_C - E_V). This is characteristic for covalently bonded, semiconducting material like Si, and results from the long-range order of the network. In amorphous Si, which equally consists of a network of tetrahedrally bonded Si atoms, preserving the short-range order, and therefore exhibits similar nearest neighbor ordering as the crystalline material, the Si-Si bond lengths vary by about 10 % of the nearest neighbor distance of 0.235 nm. Also the tetrahedral bond angle of 109°28' is distorted by up to $\pm$ 10°. Both effects together result in a long-range disorder. The modifications in the local binding environment of Si atoms lead to localized states ('band-tail' states) within the gap region, where the conduction band-tail states lie just below the conduction band mobility-edge (E_C), while the valence band-tail states lie above the valence band mobility-edge (E_V). In addition, the diagram shows the energetic position of both electronically active, four-fold coordinated, positively charged P donors (P_4^+) and doping-induced, three-fold coordinated, negatively charged defect states (Si_3^-). The energy distribution of these states is supposed to be Gaussian-shaped. In the case of a gas-phase concentration ratio [PH_3]/[SiH_4] of 10^{-2} during the layer deposition process, corresponding to a density of nearly 10^{21} P atoms cm^{-3} in the Si network, the peak value of both distributions amounts to around 10^{18} cm^{-3} [19]. Consequently, most of the P atoms are incorporated in the electronically inactive, three-fold coordinated, neutral state (P_3^0).

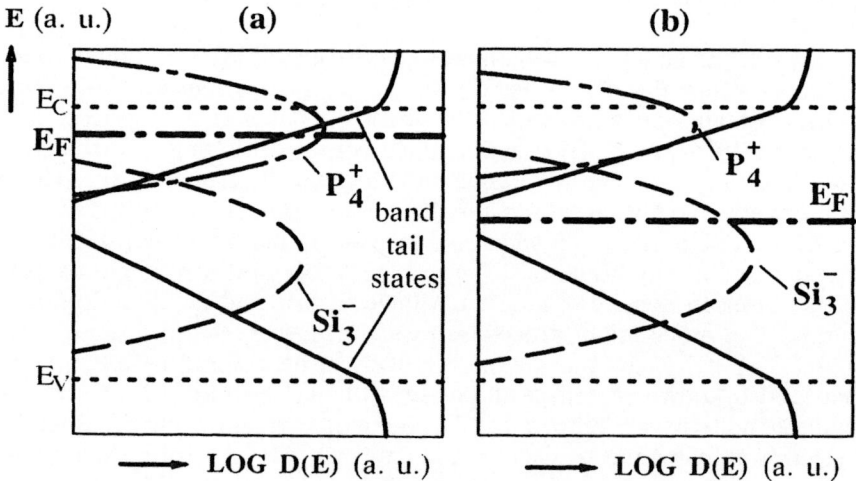

Figure 3. Density of states vs. energy of a-Si:H(P) (a) before and (b) after electronically induced structural modifications within the amorphous material.

For explaining the electronic changes, we proposed a simple model which is de-scribed in the following [4,5]. The electronic behavior of these films is closely related to the structural coordination of Si **and** dopant atoms within the film. Structural modifications of the atoms can be induced at slightly elevated temperatures to around 150 °C by changing the electron density within the localized band-tail states. Upon in-creasing the electron density, the concentration of electronically active P-dopant sites will be reduced (P_4^+ --> P_3^0) and the density of dangling bond defect sites will be increased (Si_4^0 --> Si_3^-) due to an enhanced mobility of atomic hydrogen within the network [9,10]. Consequently, the conductivity of the amorphous coating is reduced equivalent to a shift of the Fermi level towards mid-gap position, as shown in Fig. 3b. Similar thermal effects will occur also during local film modifications by the STM. Under conditions of high electron injection by the tunneling tip, the temperature will rise locally caused by an energy transfer from the tunneling electrons gained across the vacuum gap and by the high current densities within the sample material. Taking these effects into account, the local temperature increase was estimated at approximately 200 °C [15] using the thermal diffusion equation [20]. These electronically induced changes of bonding configurations and the resulting decrease of the layer conductivity were described in terms of the 'generalized autocompensation model' [10]. An example for such behavior will be presented later.

This situation will change drastically if the modifications reach down to the interface of the p^+-n heterojunction formed by a-Si:H(P)/c-Si. At that point, the local reduction of the layer conductivity will be overcompensated by a second, counteracting effect.

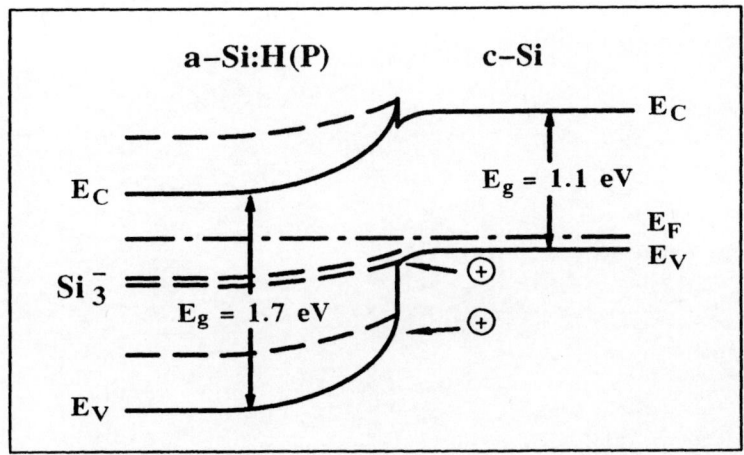

Figure 4. Energy-band diagram for an a-Si:H(P) film deposited on p^+-type c-Si in thermal equilibrium. Two different situations are indicated: (1) before modification (full lines) and (2) after structural modifications which extend down to the p^+-n heterojunction (broken lines).

This can be explained more precisely using the energy-band diagram displayed in Fig. 4 which demonstrates the situation in thermal equilibrium. Before the modification (solid lines), a potential barrier in the valence band (E_V) hinders the penetration of holes from the highly doped substrate material into the amorphous film and most of the voltage between tip and sample is dropped across the layer. After modification (broken lines), this potential barrier is reduced, allowing an exponentially increased supply of holes. An additional pathway for holes is opened by the enormous increase of negatively charged defect sites (Si_3^-) (see also Fig. 3b). These holes can easily recombine with electrons injected by the tunneling tip. As a consequence of this mechanism, the electric field within the amorphous layer is diminished, and the main part of the applied voltage drops across the vacuum gap. This results in a substantial increase in the effective tunnel voltage between tip and sample surface and, hence, also in the electron flow from the tip to the sample. Under constant current imaging conditions, this is compensated by a retraction of the tip as experimentally observed.

A further example for modifying the electrical properties of the buried p^+-n heterojunction is given in Fig. 5. A line pattern was written into a 60-nm-thick a-Si:H(P) film with a writing speed of 20 nm/s under feedback control. The writing current was 300 nA and a voltage of + 10 V was applied to the substrate. The average line width amounts to 60 nm and the apparent height is 45 nm. (The total apparent height is dependent on the thickness of the amorphous layer.) Here we demonstrate the possibility of controlled writing also in the immediate vicinity of previously modified areas.

The effect of changing the electrical properties of the buried p^+-n heterojunction could also be observed in current-voltage (I-V) measurements. In order to follow high current intensities, a special I-V converter with variable amplification factors was used. For these measurements, the tip was positioned locally above the sample surface

Figure 5. Topographic image (1.2 μm x 1.0 μm) after writing of a line pattern in a 60-nm-thick a-Si:H(P) film by modifying the electrical characteristics of the buried p^+-n junction.

under feedback control. After stabilizing the tip/sample distance by applying appropriate values for tunnel voltage (V_{stab}) and tunnel current (I_{stab}), the feedback loop was interrupted to 'freeze' the tip/sample distance. Fig. 6 shows the sudden transition ('switching') from the high-impedance (curve 1) to the low-impedance state (curve 2) of the layer. For this experiment, the voltage was ramped from 0 V to + 13 V (rise time $t_r = 0.25$ s) above an unmodified surface area of a 70-nm-thick a-Si:H film. At the switching-point of 12.7 V, the tip/sample system changes from the highly resistive into the highly conducting state and the current rapidly increases up to the saturation value of the I-V converter (broken line). At this point, the high conductance state, caused by the modification of the p^+-n junction, is reflected by the second I-V curve. The following I-V measurements show the characteristics of the low-impedance behavior; i. e. this state is preserved even if no voltage is applied between tip and sample. The switching appears to be irreversible and the sample arrangement exhibits a memory effect.

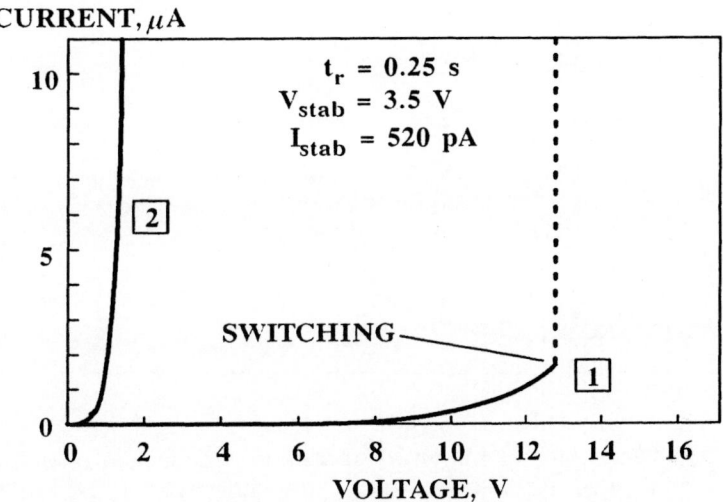

Figure 6. Current-voltage curves showing the abrupt transition (broken line) from the high-impedance (curve 1) into the low-impedance state (curve 2).

In the following, more experiments supporting our model are presented. In a first set of experiments we varied the thicknesses of the amorphous films. Modifications in the electrical characteristics of the p^+-n heterojunction could be observed up to a thickness of 130 nm. In contrast, such modification experiments performed on 500-nm-thick films were not successful. This we explain by a reduced current density in deeper regions of the layer material due to scattering processes of carriers within the amorphous film, i.e. a spreading of the electron beam. Hence, the heating effect in the junction region is too low for structural changes described above to take place.

Secondly, we performed similar modification experiments using P-doped amorphous Si films deposited on n^+-type Si substrate materials. These samples have no interior p-n junction but represent a resistive n^+-n structure. As expected, these experiments, carried out on a 100-nm-thick film, were not successful.

In a third series of experiments we reduced the electron exposure to a degree that structural modifications could be induced only in the near-surface region. This leads to a situation described by the energy-band diagram depicted in Fig. 7. This type of modification leaves the potential barrier in the valence band unchanged and the characteristics of the interior p-n junction are maintained. Therefore, we anticipate a local reduction of the layer conductivity resulting in a local depression in STM images.

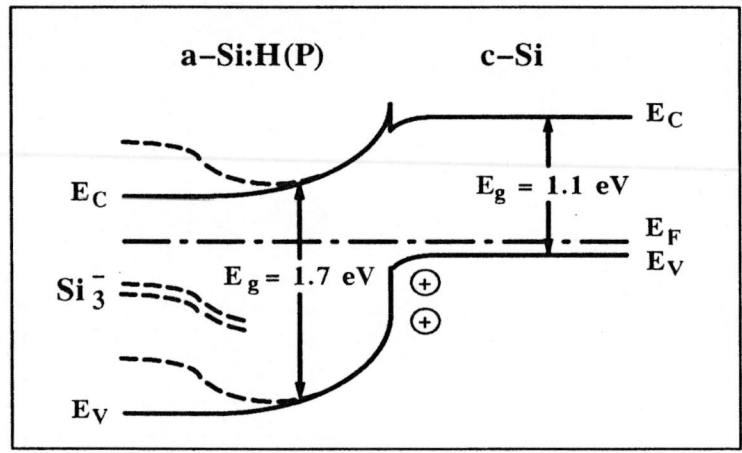

Figure 7. Energy-band diagram for an a-Si:H(P) film deposited on p^+-type c-Si in thermal equilibrium. Two different situations are indicated: (1) before modification (full lines) and (2) structural modifications only in a near-surface region (broken lines).

Such behavior is indeed observed, as demonstrated in Fig. 8. The topography image (Fig. 8a) shows an (electronic) trench which was created in a 70-nm-thick a-Si:H(P) film with a voltage of + 10 V and a current of 250 nA written under feedback control (line 1). The writing speed was chosen as 26 nm/s and the length of the line extends to 640 nm. To elucidate the electronic nature of this structure, we simultaneously measured the effective tunneling barrier height, which is displayed in Fig. 8b. At the same location, a reduction of the barrier height is observed. A decrease of the effective barrier height can be attributed to a shift of the Fermi energy towards the valence band mobility-edge (E_V) [21] (see also Fig. 7) and/or with a collapse of the tunneling barrier for very close distances between tip and sample [3,22,23]. The trench shown in

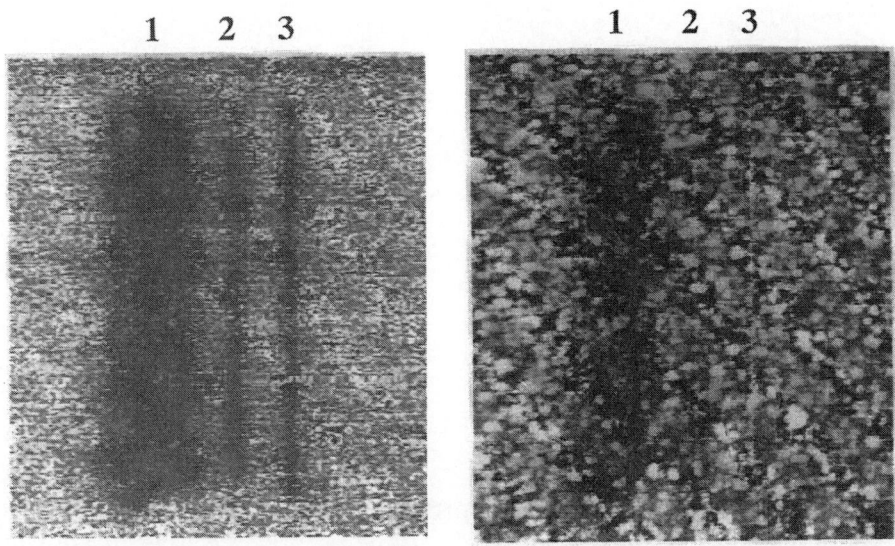

Figure 8. (a) Topography and (b) effective barrier height image (680 nm x 800 nm) after the creation of three different electronic structures: line 1 was written with $V_t = + 10$ V, $I_t = 250$ nA, line 2 with $V_t = + 10$ V, $I_t = 41$ nA, and line 3 with $V_t = + 9$ V, $I_t = 17$ nA.

Fig. 8a has an apparent depth of around 2 - 3 nm. In many studies on such modified a-Si:H(P) films, the tip quality changed very often during imaging likely due to a crash damage with the sample surface. Based on these experimental observations, the tip motion of 2 - 3 nm towards the sample surface above electrical depths can be considered as the separation between tip and sample surface under reading conditions. This type of modification is independent of the conduction type of the substrate material. The gradual reduction in the current owing to a decrease of the layer conductivity was also observed in I-V curves [24].

A further decrease of the writing parameters led to another type of modification (lines 2 and 3 in Fig. 8) [25]. The second line was written with a voltage of + 10 V and a current of 41 nA, and line 3 was created with $V_t = + 9$ V and $I_t = 17$ nA. In both cases, the writing speed was 26 nm/s as previously used for the creation of line 1. Line 2 is only visible in the barrier image (Fig. 8b), line 3 is detectable also as a very small elevation in the topographic image with an average width of around 5 nm.

These observed modifications result from a reduction of the writing parameters. Hence, only structural changes located directly at the layer surface are expected to be responsible. We attribute these comparatively small elevations to modifications in the surface electronic structure, caused by the formation of additional occupied surface states. This is illustrated in Fig. 9 by the respective energy-band diagrams (a) before and (b) after the modification process of clean amorphous Si coatings. In the case of

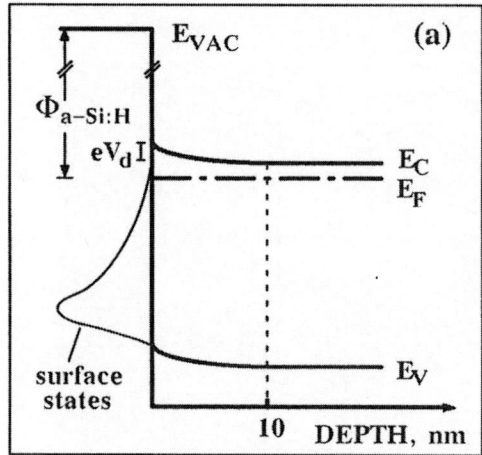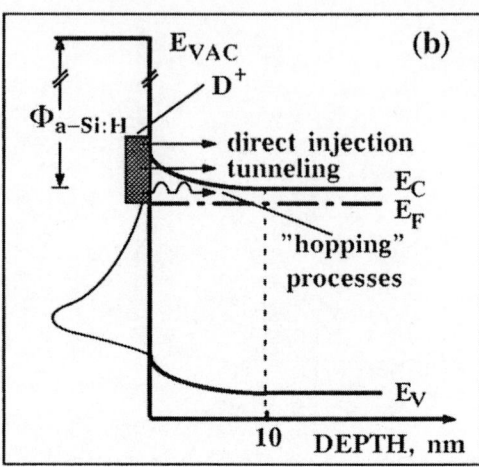

Figure 9. Energy-band diagram of *a*-Si:H(P) showing the surface region (a) before and (b) after disruption of weak Si-Si bonds. Three possible mechanisms are indicated leading to a small increase of the conductivity.

the highly P-doped films, the Fermi level (E_F) lies about 150 meV below the conduction band mobility-edge (E_C). For as-grown material (Fig. 9a), the depletion region extends less than 10 nm from the surface into the amorphous layer and a surface potential eV_d of 180 meV was estimated [15]. The relatively large concentration of occupied surface states stems from an enhanced distortion of the amorphous network at the surface compared to the 'bulk' film material, leading to a large number of streched bonds ('weak Si-Si bonds') as well as from Si_3^- states. These weak Si-Si bonds have a rather low binding energy and can therefore be easily broken by light illumination (SWE) or by electron bombardment, as in the present case. Consequently, additional dangling bond defect sites are formed around the (surface) Fermi energy (Fig. 9b). These electronically produced surface states have a donor-like character (D^+). Therefore, defect states located below E_F cause a downwards shift of the surface Fermi level which in turn results in a steeper band bending characteristic. This is the reason for the local reduction in the effective tunneling barrier height, observed in Fig. 8b. Also a disruption of Si-H bonds in the near surface region would be conceivable, but the probability for this process is negligible due to the relatively strong binding energy of this configuration.

Electrons which are injected by the probing tip into these surface states can easily be transfered from these states to the conduction band mobility-edge which explains the locally increased conductivity in line 3 (Fig. 8). For the actual transfer, which is supported also by the electric field in the depletion region (upward band bending), three different mechanisms can contribute and lead to a (small) increase of the conductivity. They are indicated in Fig. 9 and include direct injection, tunneling through the potential barrier, and carrier hopping within localized band-tail states. Of course, this effect is always in competition with the local conductivity reduction due to structural changes in the near-surface region (see above). While in the case of line 3 in Fig. 8 the former effect dominates, they almost compensate each other in the case of line 2.

This type of modification depends on the polarity of the applied voltage. Experiments, with the electron flow directed from the sample towards the tip showed no modification of the layer surface. Applying a negative voltage to the sample, the resulting electric field within the amorphous film is too low to gain sufficient kinetic energy for creating additional surface defects. As a consequence, only electrons emitted from the tip and accelerated across the vacuum gap have sufficient energy to disrupt weak Si-Si bonds. These observations made with inverse voltage polarity also exclude the possibility that these structures result from polymerization of hydrocarbons adsorbed on the sample surface. As demonstrated by several groups, modifications of contamination layers can be induced at both polarities [26].

In our last example we want to show that the extent of these surface modifications can also vary widely, depending on the respective writing conditions. The image in Fig. 10 shows five lines which were written in a 70-nm-thick amorphous Si film deposited on p^+-type Si substrate material. A writing speed of 64 nm/s was used and the respective values for writing voltage and current were: line 1: $V_t = + 6.0$ V, $I_t = 1$ nA; line 2: $V_t = + 7.0$ V, $I_t = 6$ nA; line 3: $V_t = + 8.0$ V, $I_t = 14.7$ nA; line 4: $V_t = + 9.0$ V; $I_t = 14.7$ nA; and line 5: $V_t = + 9.0$ V, $I_t = 90$ nA, respectively. With increasing writing parameters, a gradual extension of the lateral and vertical

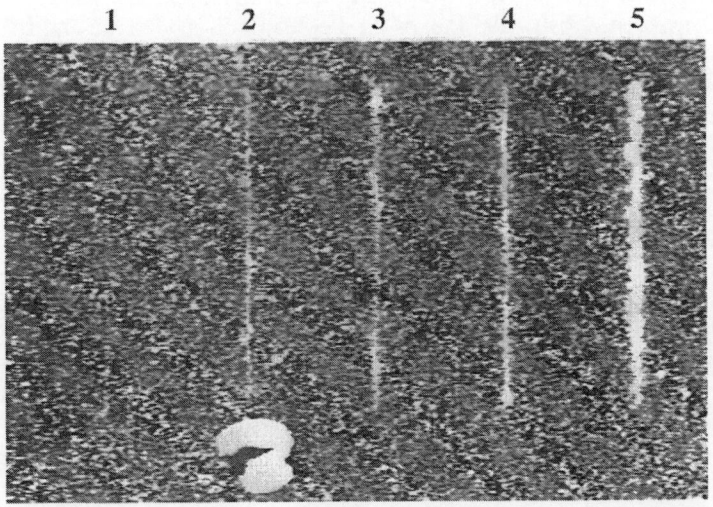

Figure 10. Topography image (1.4 μm x 0.85 μm) after the creation of five differently pronounced lines (line 1 not visible) in a 70-nm-thick a-Si:H(P) film, by the disruption of weak Si-Si bonds. The lines differ by writing voltage and current (see text for details).

(electronic) dimensions takes place. In the case of line 1, the tip-induced changes of the electronic properties are too small to be visible in the topographic illustration. In the barrier image (not shown here), however, this line is indicated by a local lowering in the tunnel barrier height. For the creation of this set of lines, the writing speed was higher than the modification experiments displayed in Fig. 8. As a result, the exposure of a surface unit area to the electron beam from the tip and, therefore, the local temperature increase within the a-Si:H(P) film were too low in order to induce a large number of structural modifications in deeper regions of the coating. Hence, in this experiment no electronic depressions were created. These results demonstrate the great sensitivity of producing electronically active structures as a function of writing conditions. (The elevation in the lower part of the image results from surface contaminations).

4. Summary

We have demonstrated the feasibility of locally modifying a-Si:H(P) films on nanometer-scale regions (5 nm to several 100 nm) by STM. The resulting electrically active, local structures can be explained in terms of the 'generalized autocompensation model'. With increasing writing parameters we find three different types of modifications: (a) bond breaking of weak Si-Si bonds directly at the layer surface which leads to a higher surface conductivity, (b) near-surface modifications resulting in a reduced layer conductivity, and (c) changes of bonding configurations throughout the entire amorphous film down to the interface a-Si:H(P)/c-Si. In the latter case, on heavily doped, p-type substrates, the p^+-n heterojunction can be modified, allowing an exponential increase of hole-injection into the amorphous layer. The associated, marked increase in conductivity is directly observed in current-voltage curves showing a abrupt transition (switching) from a high-impedance (unmodified p^+-n heterojunction) into a highly conducting state (electrically modified p^+-n heterojunction).

Acknowledgements

We gratefully acknowledge G. Krötz and G. Müller (DASA, Munich, Germany) for fabricating the samples used in this work and for fruitful discussions concerning the autocompensation model. We are indebted to M. Enachescu (Physics Department E16, Techn. Univ. of Munich, Germany) for the assistance during the STM measurements.

References

[1]   Mamin, H. J., Guethner, P. H., and Rugar, D. (1990) 'Atomic emission from a gold scanning-tunneling-microscope tip', *Phys Rev. Lett.* **65**, 2418-2421.

[2]   van Loenen, E. J., Dijkamp, D., Hoeven, A. J., Lenssinck, J. M., and Dieleman, J. (1989) 'Direct writing in Si with a scanning tunneling microscope', *Appl. Phys. Lett.* **55**, 1312-1314.

[3]   Lyo, I.-W. and Avouris, Ph. (1991) 'Field-induced nanometer- to atomic-scale manipulation of silicon with the STM', *Science* **253**, 173-176.

[4]   Hartmann, E., Behm, R. J., Krötz, G., Müller, G., and Koch, F. (1991) 'Writing electronically active nanometer-scale structures with a scanning tunneling microscope', *Appl. Phys. Lett.* **59**, 2136-2138.

[5]   Hartmann, E., Enachescu, M., Koch, F., Krötz, G., Müller, G., and Behm, R. J. (1991) 'Electronically induced modifications of a-Si:H(P) films by scanning tunneling microscopy', *J. Non-Cryst. Solids* **137&138**, 1067-1070.

[6]   Staebler, D. L. and Wronski, C. R. (1977) 'Reversible conductivity changes in discharge-produced amorphous Si', *Appl. Phys. Lett.* **31**, 292-294; Staebler, D. L. and Wronski, C. R. (1980) 'Optically induced conductivity changes in discharge-produced hydrogenated amorphous silicon', *J. Appl. Phys.* **51**, 3262-3268.

[7]   Brandt, M. S. and Stutzmann, M. (1991) 'Kinetics of light-induced defect creation in amorphous silicon: The constant degradation method', *J. Non-Cryst. Solids* **137&138**, 211-214.

[8]   Street, R. A., Kakalios, J., and Tsai, C. C. (1987) 'Thermal-equilibrium processes in amorphous silicon', *Phys. Rev. B* **35**, 1316-1333; Street, R. A., Hack, M., and Jackson, W. B. (1988) 'Mechanisms of thermal equilibration in doped amorphous silicon', *Phys. Rev. B* **37**, 4209-4224.

[9]   Krötz, G., Wind, J., Stitzl, H., Müller, G., Kalbitzer, S., and Mannsperger, H. (1991) 'Experimental tests of the autocompensation model of doping', *Phil. Mag. B* **63**, 101-121; Krötz, G. and Müller, G. (1991) 'Structural equilibration in intrinsic, single-doped, and compensated TFTs - experiments and calculations', *J. Non-Cryst. Solids* **137&138**, 163-166.

[10]   Müller, G., Kalbitzer, S., and Mannsperger, H. (1986) 'A chemical-bond approach to doping, compensation, and photo-induced degradation in amorphous silicon', *Appl. Phys. A* **39**, 243-250.

[11]   Meier, J. E. (1992), Ph.D. thesis, University of Konstanz, Germany.

[12]   Simmons, J. G. (1964) 'Potential barriers and emission-limited current flow between closely spaced parallel metal electrodes', *J. Appl. Phys.* **35**, 2472-2481.

[13]   Binnig, G. and Rohrer, H. (1983) 'Scanning tunneling microscopy', Surf. Sci. **126**, 236-244; Binnig, G. and Rohrer, H. (1986) 'Scanning tunnel microscopy', *IBM J. Res. Dev.* **30**, 355-369.

[14]   Wiesendanger, R., Eng, L., Hidber, H.-R., Oelhafen, P., Rosenthaler, L., Staufer, U., and Güntherodt, H.-J. (1987) 'Local tunneling barrier height images obtained with the scanning tunneling microscope', *Surf. Sci* **189/190**, 24-28.

[15]   Hartmann, E. (1992), Ph.D. thesis, Technical University of Munich, Germany.

[16]   Maeda, K., Umezu, I., Ikoma, H., and Yoshimura, T. (1990) 'Nonideal J-V characteristics and interface states of an a-Si:H Schottky barrier', *J. Appl. Phys.* **68**, 2858-2867.

[17]   Nakagiri, N. and Kaizuka, H. (1990) 'Simulations of STM images and work function for rough surfaces', *Jap. J. Appl. Phys.* **29**, 744-749.

[18]   Jahanmir, J., West, P. E., and Rhodin, T. N. (1988) 'Evidence of Schottky emission in scanning tunneling microscopes operated in ambient air', *Appl. Phys. Lett.* **52**, 2086-2088.

[19]   Street, R. A. (1982) 'Doping and Fermi energy in amorphous silicon', *Phys. Rev. Lett.* **49**, 1187-1190.

[20]   Staufer, U., Scandelle, L., and Wiesendanger, R. (1989) 'Direct writing of nanometer-scale structures on glassy metals by the scanning tunneling microscope', *Z. Phys. B - Condensed Matter* **77**, 281-286.

[21]   Weimer, M., Kramer, J., and Baldeschwieler, J. D. (1989) 'Band bending and the apparent barrier height in scanning tunneling microscopy', *Phys. Rev. B* **39**, 5572-5575; Hosaka, S., Sagara, K., Hasegawa, T., Takata, K., and Hosoki, S. (1990) 'Tunneling barrier height imaging and polycrystalline Si surface observations', *J. Vac. Sci. Technol. A* **8**, 270-274.

[22]   Gimzewski, J. K. and Möller, R. (1987) 'Transition from the tunneling regime to point contact studied using scanning tunneling microscopy', *Phys. Rev. B* **36**, 1284-1287.

[23]   Lang, N. D. (1988) 'Apparent barrier height in scanning tunneling microscopy', *Phys. Rev. B* **37**, 10395-10398; Ciraci, S. and Tekman, E. (1989) 'Theory of transition from the tunneling regime to point contact in scanning tunneling microscopy', *Phys. Rev. B* **40**, 11969-11972.

[24]   Hartmann, E., Enachescu, M., Behm, R. J., and Koch, F., in preparation.

[25]   Hartmann, E., Enachescu, M., Behm, R. J., and Koch, F., submitted to *Appl. Phys. Lett.*

[26]   McCord, M. A. and Pease, R. F. W. (1986) 'Lithography with the scanning tunneling microscope', *J. Vac. Sci. Technol. B* **4**, 86-88; Ringger, M., Hidber, H.-R., Schlögl, R., Oelhafen, P., and Güntherodt, H.-J. (1985) 'Nanometer lithography with the scanning tunneling microscope', *Appl. Phys. Lett.* **46**, 832-834.

ATOMIC–SCALE IMAGING AND MODIFICATION OF SPINS USING A MAGNETIC–SENSITIVE SCANNING TUNNELING MICROSCOPE

R. WIESENDANGER
University of Basel
Dept. of Physics
Klingelbergstrasse 82
CH–4056 Basel
Switzerland

ABSTRACT. The main objective of the present investigation has been the atomic–resolution STM investigation of a well–characterized surface of a magnetic sample with a clean and sharp ferromagnetic probe tip under ultra–high vacuum conditions. By using in–situ prepared iron tips and a magnetite (001) substrate, a clear contrast between $Fe(2+)$ and $Fe(3+)$ on the octahedrally coordinated B–sites of magnetite has been obtained, most likely due to the difference in the spin configuration of the two different magnetic ions. In contrast, by using non–magnetic tungsten tips, which probe the spin–averaged density–of–states, any corrugation due to the difference in the electronic structure of $Fe(2+)$ and $Fe(3+)$ sites has remained below the noise level of the experiment. Based on the ability to get a clear magnetic contrast at the atomic level with the ferromagnetic iron tips, we have observed modifications of the surface spin configuration which are localized to a scale of one nanometer.

1. INTRODUCTION

The development of scanning tunneling microscopy (STM) has triggered the invention of various STM-related scanning probe methods, some having magnetic sensitivity. In particular, magnetic force microscopy (MFM) [1,2] and tunneling–stabilized magnetic force microscopy (TSMFM) [3] have become important experimental techniques to non–destructively probe the stray field distribution above a magnetic sample under ambient conditions with a spatial resolution of typically 20–100 nm [4]. On the other hand, it has been found that the magnetic stray field of a magnetized tip can cause local changes of the substrate surface magnetization at sufficiently small tip–substrate separations as a consequence of magnetic dipole interactions. While such modifications are undesirable for the characterization of the as–prepared magnetic sample, they

P. Avouris (ed.), Atomic and Nanometer-Scale Modification of Materials: Fundamentals and Applications, 65–73.
© 1993 *Kluwer Academic Publishers.*

can advantageously be exploited to write magnetic bits. This
has first been demonstrated by means of TSMFM with a
magnetized iron thin film tip, which also served to read the
magnetically recorded information [3]. The size of the
written bits were on the order of several hundred nanometers.

To increase the spatial resolution in magnetic imaging
and writing considerably, the long-range magnetic dipole force
interaction between tip and substrate being probed has to be
replaced by a short-range tip-sample interaction.
Spin-polarized electron tunneling or magnetic exchange force
detection appear to be the most promising experimental methods
to achieve the ultimate resolution in magnetic imaging and
writing. Therefore, we have recently started with a
systematic experimental work towards this goal.

2. METHOD OF SPIN-POLARIZED SCANNING TUNNELING MICROSCOPY (SPSTM)

Spin-polarized STM (SPSTM) at room temperature can be based on
the 'magnetic valve effect', leading to a spin-dependence of
the tunneling current flowing between two magnetic electrodes,
as theoretically predicted [5] and experimentally verified for
the STM by the direct observation of vacuum tunneling of
spin-polarized electrons between a ferromagnetic chromium
dioxide tip and a Cr(001) test surface [6-8]. This initial
SPSTM work allowed to deduce a local polarization of the
tunnel junction on a nanometer length scale. However,
in-plane atomic resolution has not been obtained with the
chromium dioxide thin film tips used in those experiments
because they were not as sharp as required.

More recently, improvements in the in-situ preparation of
atomically sharp and clean magnetic sensor tips under
ultra-high vacuum (UHV) conditions [9,10] have allowed for
routine atomic resolution studies of non-magnetic
(e.g. Si(111)7x7, Si(100)2x1) as well as magnetic
(e.g. Fe_3O_4(001)) sample surfaces. As a consequence, it has
become possible to extend the SPSTM studies down to the atomic
level [11-13].

3. MAGNETITE: CRYSTAL, ELECTRONIC AND SPIN STRUCTURE

Magnetite, the best-known natural magnetic material, has
played a key role in the development of magnetism. It has
cubic inverse spinel structure and is ferrimagnetic with a
Curie temperature of 860 K (Fig. 1). There exist
tetrahedrally coordinated Fe 'A-sites' as well as octahedrally
coordinated Fe 'B-sites'. The A-sites are occupied by Fe(3+)
whereas the B-sites are half occupied by Fe(3+) and half by
Fe(2+) with spin configurations 5x(3d↑) and 5x(3d↑)/1x(3d↓),
respectively. At room temperature, fluctuations of the sixth
electron (3d↓) among B-sites are rapid in the bulk, explaining

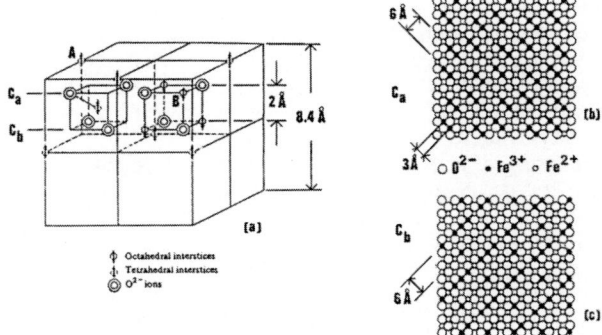

Fig. 1: a) Crystal structure of magnetite with tetrahedrally coordinated A-sites occupied by Fe(3+) and octahedrally coordinated B-sites occupied by Fe(3+) as well as Fe(2+). The Ca and Cb planes are shown in (b) and (c) respectively with an ordering of Fe(3+) and Fe(2+) according to a model [19] for the low-temperature phase of magnetite.

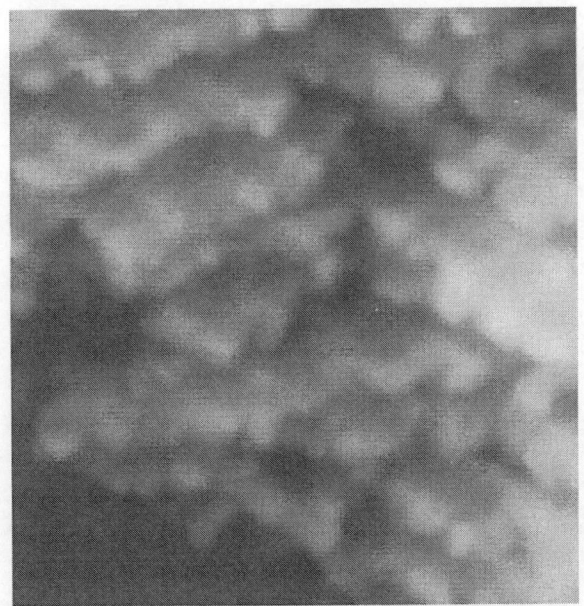

Fig. 2: Topographic STM image (155 Å x 165 Å) of the Fe-O (001) planes of magnetite obtained with a non-magnetic tungsten tip. The terraces are separated by 2-Å high steps. On top of these terraces atomic rows of Fe B-sites can be seen with a spacing of 5.9 Å, changing their orientation by 90° from one terrace to the next. (Tunneling current: I = 1 nA, sample bias voltage: U = +3.0 V).

the relatively high electrical conductivity of about
$100/(\Omega \, cm)$ compared with other iron oxides. Below the
so-called Verwey transition [14] at about 120 K, these charge
fluctuations cease and a static periodic arrangement of $Fe(3+)$
and $Fe(2+)$ ions sets in, leading to a decrease in conductivity
by about two orders of magnitude. The driving force is
interatomic Coulomb interaction, so the Verwey transition in
magnetite is considered to be an example of a Wigner
crystallization [15,16] in three dimensions, albeit modified
by electron-phonon coupling.

4. EXPERIMENTAL STM AND SPSTM RESULTS ON THE MAGNETITE
   (001) SURFACE: ATOMIC-SCALE IMAGING AND LOCAL MODIFICATION

For the STM and SPSTM studies, magnetite (001) surfaces were
prepared by mechanical polishing followed by in-situ annealing
of the single crystals up to about 1000 K in UHV.
Well-ordered and clean surfaces were obtained as verified by
low-energy electron diffraction (LEED) and Auger electron
spectroscopy (AES) [12,17]. The correct surface stoichiometry
was checked by x-ray photoelectron spectroscopy (XPS). The
STM experiments were performed in the same multichamber UHV
system (Nanolab-I) containing the surface preparation and
analysis facilities. The pressure during the STM experiments
remained at 1x10E(-11) mbar or below. Several different probe
tips were used to exclude tip-specific imaging artifacts.
     At first, the topographic structure of the magnetite
(001) surface was studied by STM with non-magnetic
electrochemically etched tungsten tips. A step-and-terrace
structure has been revealed with the terraces separated by
either 4.2-Å or 2-Å high steps, depending on the sample
surface location. On top of the terraces being separated by
2-Å high steps, parallel atomic rows with a spacing of 5.9 Å
were observed which change their orientation by 90° from one
terrace to the next (Fig. 2). These atomic rows have been
identified with the rows of octahedrally coordinated Fe
B-sites in the Fe-O planes of magnetite (Fig. 1) which are
indeed separated by 2 Å. The oxygen sites remain invisible in
the STM images because the corresponding O-1s and p-states lie
well outside the energy window accessible with the STM [13].
No atomic-scale structure has been resolved along the rows of
Fe B-sites by using non-magnetic tungsten tips indicating a
smooth spin-averaged density-of-states corrugation along these
rows.
     As a second step, SPSTM experiments have been performed
by using ferromagnetic iron probe tips prepared in-situ
according to a procedure described previously [9,10]. In
contrast to the experimental results obtained with the
tungsten tips, a strong corrugation (0.3 Å - 0.8 Å) along the
rows of octahedrally coordinated Fe B-sites was measured with
the ferromagnetic iron tips (Fig. 3). Several different
periods can be found along these rows which are, however,

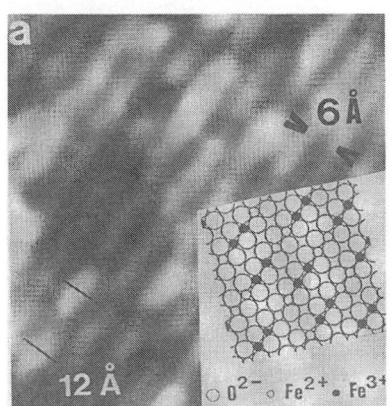

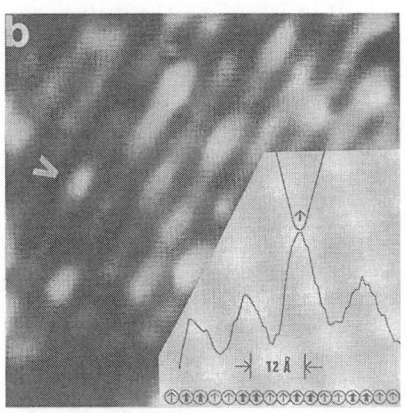

Fig. 3: a) Magnetic STM image obtained with an iron tip showing the rows of Fe B-sites with the 5.9-Å spacing and a clear modulation along these rows. (I = 1 nA, U = +3.0 V). Inset: schematic structure of the Fe-O (001) plane of magnetite. b) Magnetic STM image obtained after the one presented in a). A new feature (marked by an arrow) appears which fits into the predominantly observed 12-Å periodicity between Fe(3+) and Fe(2+) sites. Inset: single measured line section showing the dominant 12-Å periodicity due to the different spin configuration of Fe(3+) and Fe(2+) sites which are indicated by arrows of different size.

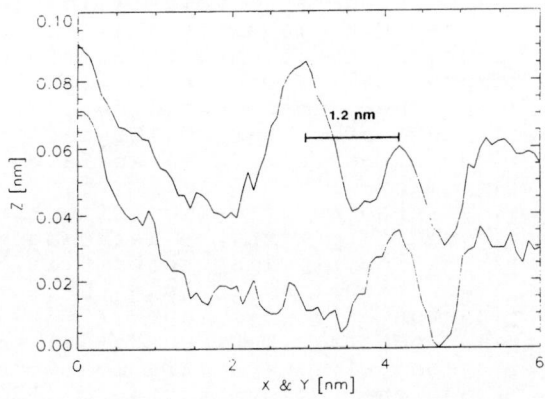

Fig. 4: Comparison of the line sections taken from Figs. 3a and 3b along the particular row of Fe B-sites which becomes modified in the STM image of Fig. 3b. The new maximum fits exactly into the 12-Å periodicity characteristic for the surface spin structure, in contrast to the 3-Å periodicity of the atomic structure.

always multiples of 3 Å, the nearest-neighbour distance of Fe-sites along the rows. A statistics of the observed periods in about twenty different STM images [18] has revealed a clear dominance of a 12-Å period (Fig. 3b inset), which is equal to the $Fe(3+)-Fe(2+)$ periodicity observed in the bulk for the low-temperature phase of magnetite below the Verwey transition as shown in Figs. 1b and 1c [19].

The surface structure of magnetite (001), which apparently shows only short-range order in the spatial distribution of $Fe(3+)$ and $Fe(2+)$, can be imaged over extended time periods up to an hour without any observable changes. However, occasionally local modifications between two successive SPSTM images were observed. For instance, an additional feature (marked by an arrow) appears in Fig. 3b, which is not present in Fig.3a. Remarkably, the new feature fits exactly into the dominant 12-Å periodicity along the rows of Fe B-sites, as verified by a comparison of individual line sections along the particular modified row, taken from the SPSTM images of Figs. 3a and 3b. The corresponding two line sections are presented in Fig. 4.

5. DISCUSSION

The static pattern with a dominant 12-Å period, as observed with the iron tips on a long time scale, is remarkable because it implies that for the Fe-O (001) surface plane of magnetite, a Wigner crystallization into an ionic state is likely to occur above room temperature. An independent proof for the non-metallic state of the Fe-O (001) surface plane comes from local tunneling spectroscopy data from which an energy gap of about 1.5 eV has been extracted. This is in clear contrast to the metallic state of bulk magnetite at room temperature. However, an increase of the Verwey transition temperature at the surface can theoretically be expected as a result of a band narrowing due to the reduced coordination [16]. The lack of long-range order indicates that the Wigner crystallization of the 3d↓ electrons on the Fe-O (001) surface plane takes the form of a 2D Wigner glass.

The fact that a strong corrugation with the dominant 12-Å periodicity is observed only with the ferromagnetic iron tips, but not with tungsten tips, implies that it is the difference in the spin configuration, rather than the difference in the electronic configuration of $Fe(3+)$ and $Fe(2+)$, which gives rise to the clear contrast between $Fe(3+)$ and $Fe(2+)$ sites obtained with the iron tips. Local modifications of the surface, involving the characteristic 12-Å period can therefore primarily be attributed to local changes of the surface spin configuration which, of course, are coupled with changes in the local surface electronic structure. For instance, charge transfer of one or more electrons to convert $Fe(3+)$ into $Fe(2+)$, or vice versa, at an adjacent Fe B-site could easily explain the experimental observation of local modifications, involving the 12-Å period. Alternatively, a local modification of the spin configuration of an atom or group of atoms may be associated with non-collinear spin structures of the surface.

Trivial explanations for the observed local modifications have definitely been ruled out. For instance, material deposition from the tip would not reproducibly occur exactly at those surface sites fitting into the 12-Å periodicity along the particular rows of B-site iron. Surface contamination could be another trivial explanation which could probably lead to the characteristic 12-Å period assuming a different adsorption behaviour of Fe(3+) and Fe(2+), as would be the case for a 'magnetocatalyst'. However, by repeating the experiment under poor vacuum conditions, it has been found that adsorption occurs at the oxygen sites, rather than the Fe B-sites of the magnetite (001) substrate. Furthermore, surface contamination is seen with tungsten tips as well as with iron tips. Therefore, any local modification as the one shown in Fig. 3 would have been observable with tungsten tips as well. However, this was not found experimentally.

6. CONCLUSIONS

In conclusion, it has been possible to image the magnetic structure of the magnetite (001) surface on an atomic scale by using ferromagnetic iron tips. This constitutes the first experimental proof that in-plane atomic resolution with magnetic tips on magnetic samples can be achieved. Based on the ability to get magnetic contrast between Fe(3+) and Fe(2+) at the atomic level, local modifications in the surface spin configuration have been observed which are localized to a scale of one nanometer. It is most likely that such localized changes are induced by the exchange interaction between tip and sample at close tip-surface distances. Future experiments will focus on the influence of the tip-to-substrate spacing on the observed local magnetic surface modifications. Potential applications to ultra-high resolution magnetic recording and ultra-high density magnetic data storage are evident. In particular, the magnetite (001) surface may allow for a new type of recording ('Wigner glass recording') being based on the many nearly-degenerate configurations of B-site Fe(3+) and Fe(2+) ions that can exist on this surface.

7. ACKNOWLEDGEMENTS

I would like to thank my collaborators in the SPSTM work on magnetite: I.V. Shvets, D. Bürgler, G. Tarrach, T. Schaub, H.-J. Güntherodt, and J.M.D. Coey. Financial support from the Swiss National Science Foundation and the Kommission zur Förderung der wissenschaftlichen Forschung is gratefully acknowledged.

8. REFERENCES

[1] Martin, Y. and Wickramasinghe, H.K. (1987). Appl. Phys. Lett. 50, 1455.

[2] Saenz, J.J., Garcia, N., Grütter, P., Meyer, E., Heinzelmann, H., Wiesendanger, R., Rosenthaler, L., Hidber, H.R., and Güntherodt, H.-J. (1987). J. Appl. Phys. 62, 4293.

[3] Moreland, J. and Rice, P. (1990). Appl. Phys. Lett. 57, 310.

[4] Grütter, P., Mamin, H.J., and Rugar, D. (1992). In: Scanning Tunneling Microscopy II, eds. Wiesendanger, R. and Güntherodt, H.-J., Springer Series in Surface Sciences Vol. 28, p. 151. Springer, Berlin/Heidelberg.

[5] Slonczewski, J.C. (1989). Phys. Rev. B 39, 6995.

[6] Wiesendanger, R., Güntherodt, H.-J., Güntherodt, G., Gambino, R.J., and Ruf, R. (1990). Phys. Rev. Lett. 65, 247.

[7] Wiesendanger, R., Bürgler, D., Tarrach, G., Wadas, A., Brodbeck, D., Güntherodt, H.-J., Güntherodt, G., Gambino, R.J., and Ruf, R. (1991). J. Vac. Sci. Technol. B 9, 519.

[8] Wiesendanger, R., Bürgler, D., Tarrach, G., Güntherodt, H.-J., and Güntherodt, G. (1992). In: Scanned Probe Microscopy, ed. Wickramasinghe, H.K., AIP Conf. Proc. Vol. 241, p. 504. AIP, New York.

[9] Wiesendanger, R., Bürgler, D., Tarrach, G. Schaub, T., Hartmann, U., Güntherodt, H.-J., Shvets, I.V., and Coey, J.M.D. (1991). Appl. Phys. A 53, 349.

[10] Wiesendanger, R., Bürgler, D., Tarrach, G., Shvets, I.V., and Güntherodt, H.-J. (1992). Mat. Res. Soc. Symp. Proc. Vol. 231, 37.

[11] Wiesendanger, R., Shvets, I.V., Bürgler, D., Tarrach, G., Güntherodt, H.-J., and Coey, J.M.D. (1992). Z. Phys. B 86, 1.

[12] Wiesendanger, R., Shvets, I.V., Bürgler, D., Tarrach, G., Güntherodt, H.-J., Coey, J.M.D., and Gräser, S. (1992). Science 255, 583.

[13] Wiesendanger, R., Shvets, I.V., Bürgler, D., Tarrach, G., Güntherodt, H.-J., and Coey, J.M.D. (1992). Europhys. Lett. 19, 141.

[14] Verwey, E.J.W. (1939). Nature 144, 327.

[15] Wigner, E. (1938). Trans. Far. Soc. 34, 678.

[16] Mott, N.F. (1974). Metal-Insulator Transitions, Taylor & Francis Ltd., London.

[17] Shvets, I.V., Wiesendanger, R., Bürgler, D., Tarrach, G., Güntherodt, H.-J., and Coey, J.M.D. (1992). J. Appl. Phys. 71, 5489.

[18] Wiesendanger, R. (1992). In: New Concepts for Low-Dimensional Electronic Systems, ed. Bauer, G.,

Springer Series in Solid State Sciences,
Springer, Berlin/Heidelberg/New York.

[19] Iida, S., Mizushima, K., Mizoguchi, M., Kose, K.,
Kato, K., Yanai, K., Goto, N., and Yumoto, S.
(1982). J. Appl. Phys. 53, 2164.

PHYSICS AND CHEMISTRY IN HIGH ELECTRIC FIELDS

H.J. Kreuzer
Department of Physics, Dalhousie University
Halifax, N.S. B3H 3J5 Canada

ABSTRACT. Progress is reviewed in our understanding of the effects of high electrostatic fields (of the order of volts per angstrom) on the adsorption and reaction of atoms and molecules on metal and semiconductor surfaces.

1. Introduction

Electrostatic fields of the order of volts per angstrom, as they occur over macroscopic distances at field emission tips, in zeolite cavities and at the electrode-electrolyte interface, are of the same order as the fields inside atoms and molecules. They are thus strong enough to induce *re*-arrangement of electronic orbitals of atoms and molecules, in particular at metal surfaces, leading to new phenomena that can be summarily described as field-induced chemisorption and field-induced chemistry. Several review articles have been written in recent years about these topics [1-4].

Electric field effects on matter can be classified, rather arbitrarily, into two categories: (*i*) in low fields, i.e. below roughly $10^{-1}V/\text{Å}$, atoms, molecules and condensed matter only get polarized; we will call such effects physical. (*ii*) In fields larger than typically $10^{-1}V/\text{Å}$ chemical effects come into play in addition in that the electronic orbitals get distorted to such a degree as to effect the chemical characteristics of an atom or molecule e.g. by establishing new bonding orbitals. In this way, molecules, unstable in field free situations, may be stabilized by a strong electric field. Also, new pathways in chemical reactions, e.g. in heterogeneous catalysis and in chemical vapor deposition, may be established.

To discuss field effects qualitatively, we look, in Fig. 1, at a molecule AB adsorbed on a metal. Far from the surface and in the absence of a field, the atomic orbitals of A and B hybridize into molecular orbitals which we take to be a lower-lying bonding orbital and an empty antibonding orbital. As the molecule approaches a metal surface, additional hybridization with the conduction electrons occurs leading to shifts and broadening of these orbitals. As illustrated in Fig. 1, the antibonding orbital gets partially occupied resulting in (*i*) bonding to the surface and (*ii*) weakening of

75

P. Avouris (ed.), Atomic and Nanometer-Scale Modification of Materials: Fundamentals and Applications, 75–86.
© 1993 *Kluwer Academic Publishers.*

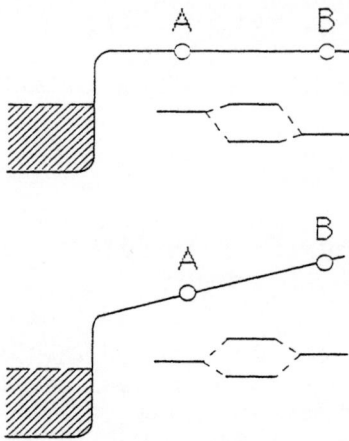

Fig.1: Schematic view of the electronic level structure of an AB molecule adsorbed on a metal without and with an external electric field.

the $A–B$ bond and stretch vibration. Molecules such as CO and N_2 are examples. Applying an electric field, F_0, pointing away the surface, adds the potential energy, eF_0z, for the electrons outside the metal (assuming, for this simplified discussion, total expulsion of the field from the metal). This raises the atomic levels of the atoms A and B by amounts of the order eF_0z_a and eF_0z_b, respectively, resulting in a substantial *re*-arrangement of the molecular orbitals. For the situation depicted in Fig. 1, the anti-bonding orbital empties out again, leading to restabilization of the molecule and probably a weakening of the surface bond. If we increase the field strength to the point where the bonding orbital is lifted above the Fermi energy of the metal, it will drain as well, leading to field-induced dissociation. Note that in the absence of the field the bonding orbital of the AB molecule is more B-like whereas the anti-bonding orbital has more A character. As the electric field is increased, these characteristics are changed in a continuous manner into a situation where the bonding orbital is more A-like and the anti-bonding orbital has B character. This possibility of changing the relative position of orbitals of the constituent atoms in a molecule with respect to each other, leads to new, field-induced chemistry. As an example, local electric fields, generated by a sharp metal tip, can be used to preferentially dissociate certain species that one wants to deposit locally under the tip, e.g. in field-assisted chemical vapor deposition.

In this paper we will first review our present understanding of electrostatic fields at metal surfaces. Then we will look at field evaporation and field-induced chemisorption. Finally we will discuss in some detail field-induced chemical reactions at surfaces as the topic of most relevance to atomic and nanoscale manipulation of materials.

2. Electrostatic Fields at Metal Surfaces

Classical electromagnetic theory assumes that the surface of a metal is a mathematical plane with excess charges and a dipole layer at which the normal component of the electric field drops discontinuously to zero, at least for a perfect conductor. On real surfaces, however, the electron distribution and also electric fields vary smoothly over distances of a few angstroms. A simple model [5-7] that bears out these features is the jellium model of a metal in which we assume that the ionic lattice can be smoothed into a uniform positive charge density n_+ that drops to zero abruptly half a lattice constant above the topmost lattice plane. Within the framework of density functional theory, the electron density and the local field distribution can be determined from the selfconsistent solution of a Schrödinger-like equation and of Pois-

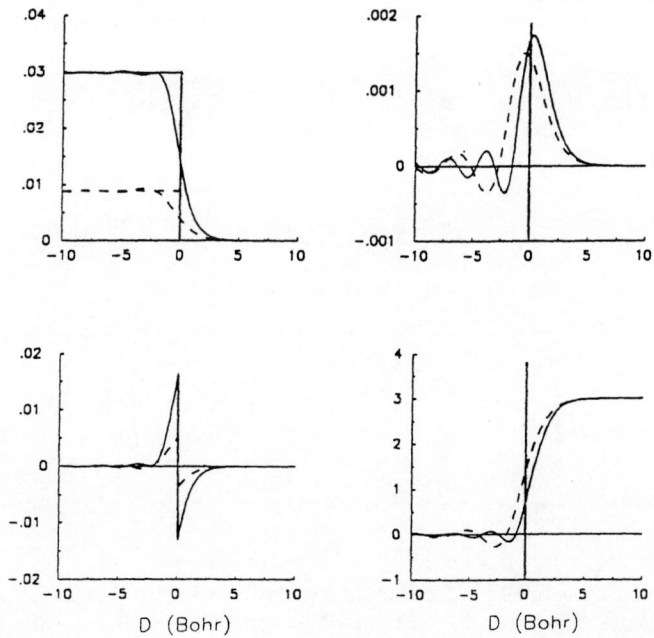

Fig.2: A schematic view of (a) the charge distribution at a metal surface without a field ($F_0=0$), the positive jellium background is indicated, solid lines: $r_s=2.0$, dashed lines: $r_s=3.0$; (b) the surface dipole layer $\rho^s=n_+^s-n_-^s$ for $F_0=0$, constructed from (a); (c) the field-induced surface charge $\delta\rho$; and (d) the applied electric field F at a metal surface.

son's equation. In Fig.2 we present the results of a density functional calculation for a jellium surface. In panel (a) we show the selfconsistent electron distribution in the absence of an external field with the local deviation from charge neutrality, i.e. the dipole layer, given in panel (b). In panel (c) we have added some excess charge, $\delta\rho$, that gives rise to the external field in panel (d). We note that the field decays smoothly into the metal with appreciable strength left at the position of the top most ion layer. This can be viewed as partial penetration of the field into the metal, or as incomplete expulsion of the field from the metal. To compare these quantum mechanical calculations with classical results from Maxwell's theory, we note that the plane at which boundary conditions are imposed on the classical fields, i.e. the discontinuous drop of the normal component of the electric field to zero, is given by the center of gravity of the excess charge $\delta\rho$, i.e. roughly the point where the field has dropped to half its value at infinity. For future reference we note here that this plane does not remain constant but moves towards the ion cores as the asymptotic field strength increases, due to the fact that the electrons are pushed into the metal increasing the field penetration and the Friedel oscillations.

Density functional calculations have recently been performed to deal with the situation with a lonely metal atom on a flat, close-packed metal surface.[8] The latter was modeled by a flat jellium surface with the metal atom treated ab initio. The calculations are based on the chemisorption programme of Lang and Williams [9], extended to account for the external field selfconsistently as

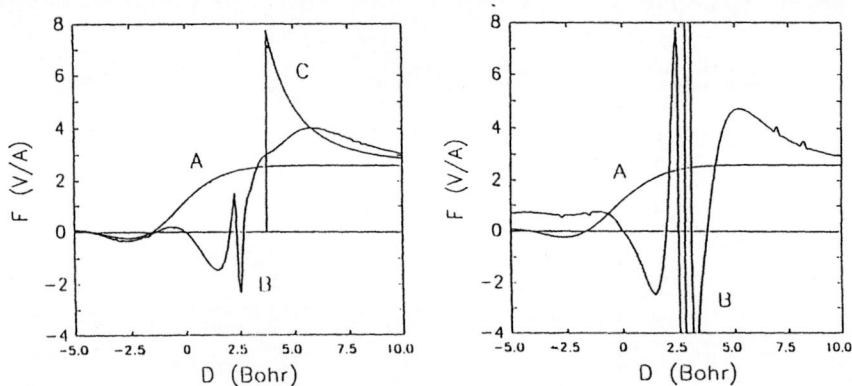

Fig. 3: Electrostatic field strength along line through the center of the adatom for *Ti* (left) and *Rh* (right) on a metal. Curves *A* and *B*: without and with the adatom; curve *C*: classical result.

well. In Fig.3 we plot, curve *A*, the external electric field laterally far from the lonely atom. Curve *B* is the field along a line through the center of the atom, and curve *C* is the classical result approximating the atom by a hemispherical boss. Compared with the field in the absence of the adatom, we note the expulsion of the field from the adatom region, which results in an enhancement of the field just outside the adatom, however, not as much as classical theory predicts. Rather, the partial field penetration into the adatom region reswults in a smearing out of the field as a reflection of the adjustability of the electronic distribution at the surface. Evaporation field strengths, i.e. the fields at which metal atoms desorb at zero temperature, and dipole moments of the adadatom have also been calculated in this model yielding good agreement with experimental values. In closing we note that the local field enhancements predicted in Fig.3, have recently been confirmed in precision measurements of field ion appearance energies.[10]

3. Field evaporation

Field evaporation is the removal of lattice atoms as singly or multiply charged positive ions from a metal in a strong electric field F of the order of several $V/\mathring{A}$, as it occurs at field ion tips [18,19]. The term field desorption is usually reserved for the process of removing field-adsorbed atoms or molecules from a field ion tip [11-12]. Field evaporation and field desorption are thermally activated processes; as such, their rate constants can be parametrized according to Frenkel-Arrhenius as

$$r_d = \alpha(T,F)\nu(T,F)\,e^{-Q(F)/k_B T} \tag{1}$$

Here $Q(F)$ is the field dependent height of the activation barrier to be overcome by the desorbing particle. The field and temperature dependent prefactor $\alpha\nu$, contains a factor α accounting for the efficiency of the charge and energy transfer mechanism, and an effective "attempt" frequency ν to overcome the activation barrier. The minimum field strength beyond which at low temperatures the metal tip evaporates is termed the evaporation field strength; it varies from $2.5V/\mathring{A}$ for Ti to $6.1V/\mathring{A}$ for W with a typical experimental error margin of 10-15% [12]. Ernst has measured $Q(F)$ and $\nu(F)$ for Rh [13], and Kellogg [14] has presented data for W in the field range 4.7 to $5.9V/\mathring{A}$.

Two phenomenological models have been used in the past to calculate the activation energy $Q(F)$, the "image-force" model [15] and the "charge-exchange" model [16]. A recent assessment has been given by Kellogg [14]. Field evaporation is a dramatic demonstration of the limitation of classical concepts in solid state physics. Maxwell's theory says that the electric field drops to zero at the image plane, i.e. just outsice the metal. Thus the electric field has, classically, no effect on the ion cores of the metal, and thus, classically field evaporation is not possible. On the other hand, we have seen in section 2 that field expulsion from the metal is not complete and that the field strength at the topmost ion cores in a metal can be substantial, and in particular strong enough to cause field evaporation.

Experiment suggests that field evaporation of metal atoms occurs most likely at steps, kinks and edges or for small clusters of atoms on larger planes. Theory should calculate the electric field, the electron density, and the geometry of the ion cores for such configurations selfconsistently. In an early attempt Kreuzer and Nath [17] have used the ASED-MO cluster method taking the electric field again from jellium calculations, i.e. foregoing the selfconsistency requirement. Still, their calculation of evaporation field strengths reproduced trends in experimental data remarkably well. Selfconsistency was recently achieved within the context of density functional calculations for an adatom on a jel-

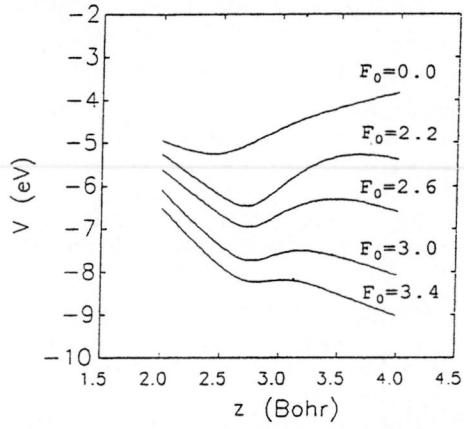

Fig.4: Potential energy for Nb on jellium with r_s=3.0 bohr as a function of nuclear distance, z, from the jellium edge. F_0 in $V/\overset{\circ}{A}$.

lium surface.[8] In Fig.4, we show the adiabatic energy curves for niobium on jellium for several field strengths. In zero field the depth of the surface potential is 5.2eV, which is, not surprisingly, somewhat less than the cohesive energy of *Nb*, 7.47eV, because the jellium lacks *p*- and *d*-orbitals. This deficiency is also the cause of the rather weak repulsion at short distances. With a field applied, the ground state energy curves must assume an asymptotic form $-eF_0z$ for large z, appropriate for a singly charged positive ion. Ionization of the adatom can occur when its highest occupied level is lifted by the field energy term, eF_0z, above the Fermi level of the jellium. If the ionization level were not shifted and broadened by the interaction with the metal, the changeover to the asymptotic form would happen abruptly at the apex of the potential energy curve, i.e. at the point where the diabatic energy curves for the neutral and ionic species cross. For the fields chosen for Fig. 3, the apex is so close to the metal surface that considerable interaction between the adatom and the metal is still in effect. This results in a considerable broadening of the ionization level of the adatom so that only partial charge draining occurs in the apex region.

We will call the energy difference between the minimum of the surface potential and the local maximum at its apex the activation energy, $Q(F_0)$, for ionization. In zero field, the activation energy is equal to the binding energy of the adatom. We note that the activation energy becomes zero for the evaporation field strength which we estimate for Nb on jellium to be $3.6V/\AA$, which compares very favorably with the experimental value for Nb, $F_{ev}=3.5V/\AA$. Similar results are obtained for other metals.

It has been suggested [17] that the field dependence of the activation energy for different metals obeys a universal scaling law if one plots $Q(F_0)/Q(F_0=0)$ as a function of $f=F_0/F_{ev}$. In a simple model, this scaling law is given by

$$Q(F_0)/Q(F_0=0) = \sqrt{1-f} + \tfrac{1}{2}f \, \ell n \, [\,(1-\sqrt{1-f})/(1+\sqrt{1-f})\,] \qquad (2)$$

Experimental data on tungsten [14] and theoretical results obtained in the jellium model [8] and by the ASED-MO method [17] have confirmed this conjecture.

4. Field-induced Chemisorption

It has been known since the invention of the field ion microscope that the imaging gas, usually helium or neon, adsorbs on the tip in electric fields of the order of the best image field strength, i.e. several volts per angstroms, at standard operating temperatures around 80K [18-20]. In particular, field-adsorbed helium on a tungsten tip is bound by about 250meV in fields of the order of 4-5V/$\AA$. Two mechanisms have been proposed to explain field adsorption: (i) polarization (dipole-dipole) forces [21-22] and (ii) field-induced chemisorption [23-24]. A consensus has emerged over the past few years that polarization forces are not sufficient to explain field adsorption of rare gases [25-26]. We will therefore skip this subject and procede with a review of field-induced chemisorption.

As discussed earlier, electric fields of the order of volts per angstroms are comparable to those experienced by valence electrons in atoms and molecules. One should therefore expect that in external fields of that magnitude a redistribution of the valence electrons in the coupled adsorbate-solid system takes place which effects both the orbitals of the surface bond as well as internal bonds in an adsorbed molecule. Whether this redistribution leads to enhanced or reduced binding depends on whether bonding or anti-bonding orbitals are more strongly affected. We will refer to this phenomenon as field-induced chemisorption. Very surprisingly it is important even for the most inert atom, namely helium, i.e. in fields of the order of 5V/$\AA$ polarization induces the occupation of excited states at the level of a few percent. Thus even helium cannot, in such fields, be regarded as a closed shell atom with the consequence that it forms weak covalent bonds as it approaches a metal surface.

In our early work [23-24,27] on field-induced chemisorption of rare gases we have used the atom superposition and electron delocalization molecular orbital method (ASED-MO) [28]. The resulting field dependence of the activation energy, the position of the minimum in the adsorption potential and the vibrational frequency of

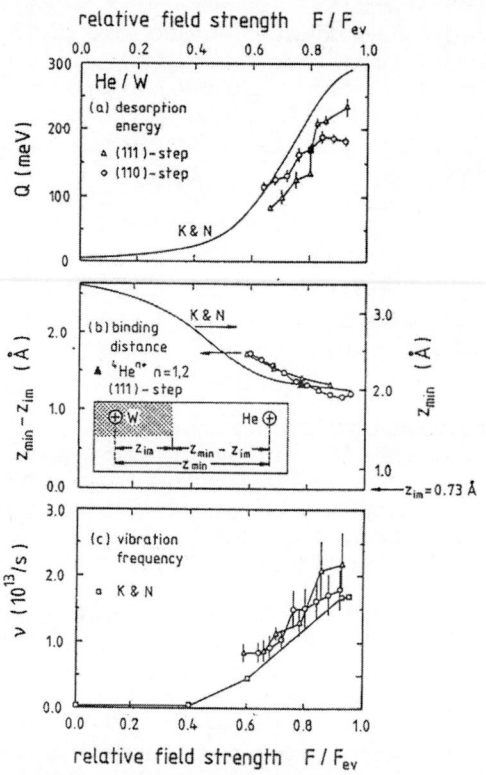

Fig.5: Activation energy Q, binding position z_{min}, and vibration frequency of helium field-adsorbed on tungsten.

the He atom in the surface potential is depicted in Fig.5 together with experimental results [27].

Exciting as these results may be one must be aware of a number of caveats with the ASED-MO method as discussed in Ref.29. We have therefore performed a selfconsistent calculation for the field-adsorption of helium and neon on a jellium surface using density functional theory in the local density approximation. Not only are the earlier results based on the semi-empirical ASED-MO theory reproduced, but new insight into the transition from physisorption to covalent chemisorption is obtained; for details see Ref.29. it should be obvious that field effects are also important in understanding the manipulation of xenon atoms with a scanning probe.

5. Field-induced Chemical Reactions

To estimate the field strength needed to affect the chemistry of a reaction, we note that the equilibrium constant, K, of a reactive system depends on the field strength, F, via a van't Hoff equation [30]

$$\frac{\partial \ell n K}{\partial F} = \frac{\Delta M}{RT} \qquad (3)$$

where ΔM is a partial molar energy related to the change in electric moment in the reaction, i.e. $\Delta M = \Delta p F + \frac{1}{2}\Delta \alpha F^2 + \dots$ Here Δp is the difference in the permanent dipole moments of the products and the reactants, and $\Delta \alpha$ is the change in their polarizability. In order to achieve values for ΔM comparable with typical reaction enthalpies or volumes, one needs fields in excess of $0.1V/\overset{\circ}{A}$.

Block and coworkers, following earlier work by Inghram and Gomer [31] have developed a field pulse technique in the field ion microscope that allows the investigation of the field effect on chemical reactions; a detailed account has been given by Block.[32] Systems that have been studied by this technique are the formation of metal subcarbonyls, the polymerization of acetone, the reaction of sulphur on metal surfaces, the decomposition of methanol on metal surfaces, hydride formation on semiconductors, NO reactions on metals and many more.

As an example we will look, in this section, at the adsorption, dissociation and reaction of NO on a Pt(111) surface in high electric fields. Although NO adsorbed on various planes of platinum does not dissociate at room temperature, applying electric fields in excess of $0.4V/\overset{\circ}{A}$ causes rapid decomposition. Employing pulsed field desorption mass spectrometry, Kruse et al [33] observed N_2O^+, N_2^+ and, to a lesser extend, O^+ ions from the stepped Pt(111) regions of a field emitter tip as the field is increased, with decreasing amounts of NO^+ being recorded. Beyond $1.2V/\overset{\circ}{A}$ no NO^+ could be desorbed.

In zero field NO adsorbs on Pt(111) [34] (and on Ni(111) [35] and on Ru(001) [36]) in bridge (on top) sites at low (high) coverage with an outward negative (positive) dipole moment. This flexibility is due to the lone electron in the 2π antibonding level, consisting mainly of the $2p_x$ levels of O and N. This picture has been confirmed by ASED-MO calculations which are also used to study the field effects.[37]

As an electric field is applied, the levels on the O atom, being further away from the surface, are raised up relatively higher than those on the N atom. This in particular effects the $2p_x$ levels on O and N resulting in a shift of the electronic charge of the 2π level to the O atom. As a result the overlap between the 2π level of NO and the levels of the metal decreases with a subsequent decrease in the electron transfer from the metal to the 2π level stabilizing the adsorbed NO molecule. Also note that the dipole moment of adsorbed NO, i.e. O^-N^+, is opposite to the field direction, thus the total energy increases as the field is increased.

Using the ASED-MO method [37] one can study the effect of an electric field on the bending mode of adsorbed NO. One finds that for fields larger than about 0.4 $V/\overset{o}{A}$ the activation energy for dis-

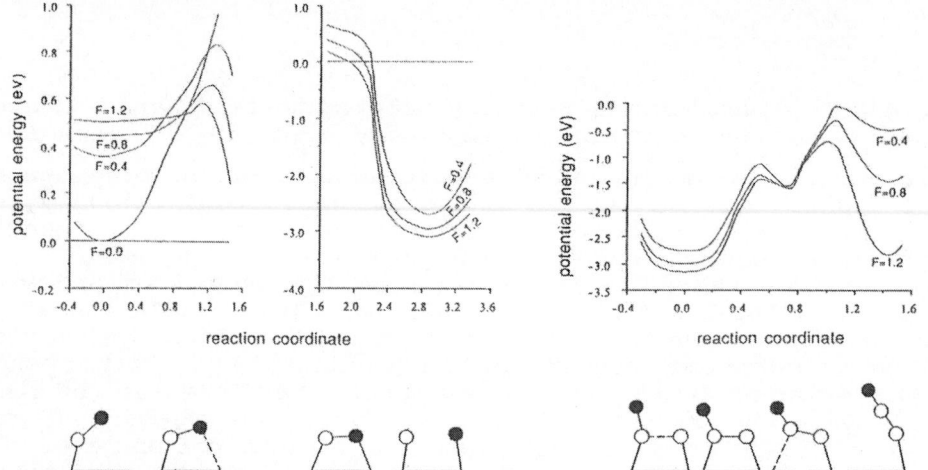

Fig.6: Potential energy curves and adsorption geometries of NO on Pt(111) as a function of reaction coordinate.

sociation becomes negligible, see Fig.6. In fields larger than 1.2 $V/\overset{o}{A}$ we finally observe the reaction $2NO \rightarrow N_2O + O$. A detailed discussion of the electronic structure of the species involved and the reaction pathways can be found in Ref.[37].

Outlook

Electric fields of the order of volts per angstrom affect the valence electrons on atoms shifting their energies by electron volts with respect to each other. In recent years a microscopic theory has emerged that looks at the new physics and chemistry from a microscopic point of view. We now understand field adsorption as field-induced chemisorption, a mechanism that affects even the lightest rare gases. We also now have a microscopic theory to describe kinetic effects in high electric fields at surfaces such as thermal field desorption and field evaporation [38-39].

The electronic structure calculations have been based so far on two vastly different approaches, namely the semi-empirical tigh binding (ASED-MO) method and density functional theory. Where comparisons are possible, an astonishing degree of agreement in the predictions is found. The reason for this lies in the fact that fields of the order of volts per angstrom shift electronic levels in atoms and molecules by several eV relative to each other. Such large shifts can obviously be handled quite reliably by semi-empirical methods.

Chemistry in high electric fields is an even more exciting field. Because fields of the order of $V/Å$ affect the valence electrons of atoms, new molecular species are stabilized in high fields thus opening up new reaction pathways in heterogeneous catalysis. Most work so far has been concentrating on static electric fields; however, many new phenomena are to be expected in alternating fields as well, as the work on photon-induced field desorption suggests.

What is important to learn from the example of field-induced chemistry in the last section is the fact that high electric fields can significantly alter reaction pathways, and indeed promote or hinder entire reactions. The fields involved are typically less than 1 $V/Å$, i.e. of the order of those generated between a tunneling tip and a substrate metal or semiconductor under conditions of atomic manipulation.

Acknowledgment

This work was supported in part by a grant from the Office of Naval Research. Funding was also provided by the Network of Centres of Excellence in Molecular and Interfacial Dynamics, one of the fifteen Networks of Centres of Excellence supported by the Government of Canada.

References

[1]   H.J. Kreuzer, in *Physics and Chemistry at Solid Surfaces VIII*; Vanselow, R., Ed.; Springer-Verlag: Berlin, Germany, 1990; *pp* 133-158.

[2]   H.J. Kreuzer, Surface Science 246, 336, 1991.

[3]   H.J. Kreuzer, in *Surface Science of Catalysis: In situ Probes and Reaction Kinetics*; D.J. Dwyer and F.M. Hoffmann, Eds., ACS Symposium Series 482 (American Chemical Society, Washington, 1992).

[4]   J.H. Block, in *Surface Science of Catalysis: In situ Probes and Reaction Kinetics*; D.J. Dwyer and F.M. Hoffmann, Eds., ACS Symposium Series 482 (American Chemical Society, Washington, 1992).

[5]   N.D. Lang and W. Kohn, Phys. Rev.B *1*, 4555 (1970); *3*, 1215 (1971); *7*, 3541 (1973).

[6]   P. Gies and R.R. Gerhardts, Phys. Rev. B *33*, 982 (1986).

[7]   F. Schreier and F. Rebentrost, J. Phys. C: Solid State Phys. *20*, 2609 (1987).

[8]   H.J. Kreuzer, L.C. Wang, and N.D. Lang, Phys. Rev. 45, 12050 (1992).

[9]   N.D. Lang, A.R. Williams, Phys. Rev. B 18, 616 (1978.

[10]   W.A. Schmidt, N. Ernst, and Yu. Suchorski, Surface Science (in press).
[11]   E.W. M{ller and T.T. Tsong, Progr. Surf. Sci. *4*, 1 (1973).
[12]   T.T. Tsong, Surf. Sci. *70*, 211 (1978).
[13]   N. Ernst, Surf. Sci. *87*, 469 (1979).
[14]   G.L. Kellogg, Phys. Rev. *B 29*, 4304 (1984).
[15]   E.W. M{ller, Phys. Rev. *102*, 618 (1956).
[16]   R. Gomer, *Field Emission and Field Ionization* (Harvard University Press, Cambridge, Mass, 1961).
[17]   H.J. Kreuzer and K. Nath, Surf. Sci. *183* (1987) 591.
[18]   E.W. M{ller, Quart. Rev., Chem. Soc. *23*, 177 (1967).
[19]   E.W. M{ller, S.B. MacLane, and J.A. Panitz, Surface Sci. *17*,439 (1967).
[20]   E.W. M{ller, Naturwissenschaften *57*, 222 (1968).
[21]   T.T. Tsong and E.W. M{ller, Phys. Rev. Lett. *25*, 911 (1966).
[22]   T.T. Tsong and E.W. M{ller, J. Chem. Phys. *55*, 2884 (1967).
[23]   D. Tomanek, H.J. Kreuzer, and J.H. Block, Surface Sci. *157*, L315 (1985).
[24]   K. Nath, H.J. Kreuzer, and A.B. Anderson, Surface Sci. *176*, 261 (1986).
[25]   K. Watanabe, S.H. Payne, and H.J. Kreuzer, Surface Sci. *202*, 521 (1988).
[26]   R.G. Forbes, Surface Science **246**, 386 (1991).
[27]   N. Ernst, W. Drachsel, Y. Li, J.H. Block, and H.J. Kreuzer, Phys. Rev. Lett. *57*, 2686 (1986).
[28]   A.B. Anderson and R.G. Parr, J. Chem. Phys. *53*, 3375 (1970); A.B. Anderson, J. Chem. Phys. *60*, 2477 (1974), *62*, 1187 (1975), *63*, 4430 (1975).
[29]   R.L.C. Wang, H.J. Kreuzer and R.G. Forbes, Surface Science (in press).
[30]   K. Bergmann, M. Eigen, L. de Maeyer, Ber. Bunsenges. Phys. Chem.67, 819 (1963).
[31]   M.G. Inghram, R. Gomer, Z. Naturforsch. Teil *A* 10, 864 (1955).
[32]   J.H. Block, in *Physics and Chemistry at Solid Surfaces IV*; Vanselow, R. and Howe, R., Eds.; Springer-Verlag: Berlin, Germany, 1982.
[33]   N. Kruse, G. Abend, J.H. Block, J. Chem. Phys. **88**, 1307 (1988).
[34]   M. Kiskinova, G. Pirug, H.P. Bonzel, Surface Sci. **136**, 285 (1984).
[35]   M.J. Breitschafter, E. Umbach, and D. Menzel, Surface Sci. **109**, 493 (1981).
[36]   P. Feulner, S.Kulkarni, E. Umbach, D. Menzel, Surface Sci. **39**, 489 (1980).
[37]   Kreuzer, H.J.; Wang, L.C. J. Chem. Phys. **1990**, *93*, 6065.
[38]   H.J. Kreuzer, K. Watanabe and L.C. Wang, Surface Sci. **232**, 379 (1990).
[39]   L.C. Wang and H.J. Kreuzer, Surface Sci. **237**, 337 (1990).

FIELD–INDUCED TRANSFER OF AN ATOM
BETWEEN TWO CLOSELY SPACED ELECTRODES

N. D. Lang
IBM Research Division
Thomas J. Watson Research Center
Yorktown Heights, New York 10598

ABSTRACT. The transfer of an adsorbed atom from one electrode to another in close proximity, with a potential difference between the electrodes, is analyzed theoretically. We consider in particular the cases of silicon and sodium atoms.

There have recently been several experiments in which an atom is transferred between tip and sample in the scanning tunneling microscope (STM) under the influence of an applied bias of several volts. [1-3] This process has been analyzed previously [4] for the case of the transfer of a Si atom using a model consisting of two planar metallic electrodes, represented using the jellium model [5], with a single atom in the region between them. Here we review that work, and discuss also the case of a sodium atom, which is broadly representative of the behavior of alkalis and other electropositive atoms. [6]

The solution proceeds as follows: First, within the framework of the density-functional formalism, the single-particle wave functions, electron density distribution, and potentials are found for the pair of bare metallic electrodes, assuming them for simplicity to be identical ($r_s = 2$ jellium model), in the presence of the bias voltage. Next the method of Lang and Williams [7] is used to find these quantities for the total system consisting of the two electrodes plus the atom. From the density distribution, the electrostatic force on the nucleus of the atom can be determined immediately.

87

P. Avouris (ed.), Atomic and Nanometer-Scale Modification of Materials: Fundamentals and Applications, 87–96.

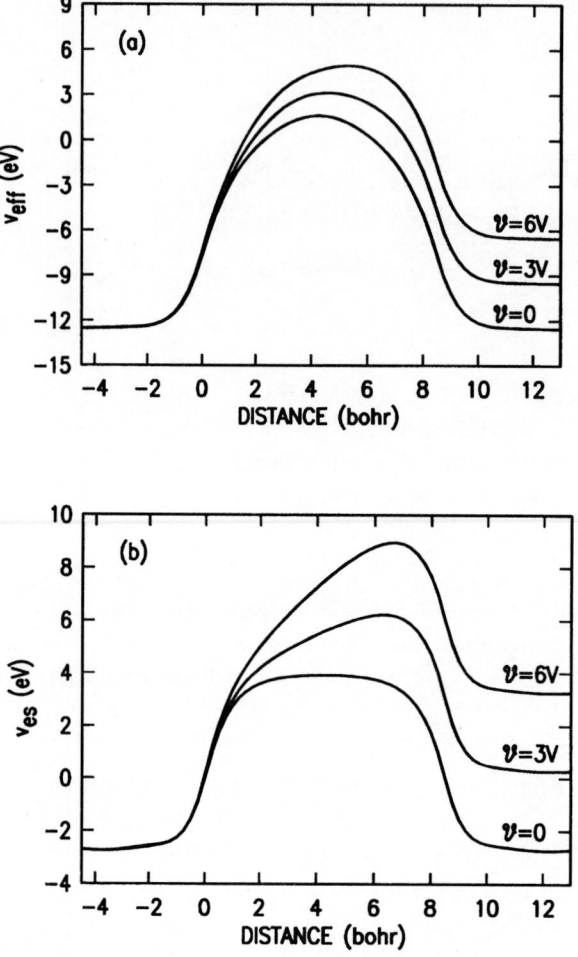

Figure 1. Potentials for two bare jellium-model electrodes ($r_s = 2$) with left electrode positive-background edge at distance 0 and right electrode background edge at distance 8.5 bohrs. (1 bohr = 0.529 Å.) The left electrode is at a bias $\mathcal{V}$ relative to the right electrode. Note that the potentials are those seen by an electron, which is why the potentials are higher at the right for positive $\mathcal{V}$. (a) Total effective potential v_{eff}, which is the sum of electrostatic (v_{es}) and exchange-correlation (v_{xc}) potentials. Left-electrode Fermi level is at 0 on energy scale, right-electrode Fermi level is at $|e|\mathcal{V}$ (which puts it slightly above the tunneling barrier for $\mathcal{V}=6$ V). (b) Electrostatic potential v_{es}.

Figure 1 gives an example of results for the potentials of the bimetallic junction. (The term "potential" here really denotes the potential energy of an electron, or in units where $|e| = 1$, the potential "seen" by an electron.) The total effective potential v_{eff}, which appears in the single-particle wave equations of the density-functional theory, is the sum of the electrostatic and exchange-correlation potentials:

$$v_{eff}(z) = v_{es}(z) + v_{xc}(z) , \tag{1}$$

where v_{xc} is taken in the local-density approximation. [8] Figure 1a gives v_{eff} for biases $\mathscr{V}$ of 0, 3 and 6 V, for an electrode spacing (distance between the positive background edges of the jellium models for the two electrodes) of 8.5 bohrs, and with the left electrode taken positive. Fig. 1b gives just the electrostatic part v_{es}.

Once the atom is introduced into the region between the electrodes, the polarization of all core orbitals on the atom is included in the self-consistent calculation for the charge distribution, in order to obtain an accurate value for the total electrostatic force F on the nucleus. It is also convenient to define an energy for discussion purposes as [9]

$$E(z) = - \int_{z_0}^{z} dz' F(z') , \tag{2}$$

where z is the coordinate along the surface normal and z_0 is arbitrary.

Consider first the case of a silicon atom. The electrode spacing used in the calculation [4] was 8.5 bohrs (as in Fig. 1), which is in the range relevant to experiment (see below). The electrostatic force on the nucleus as a function of atom position was computed, and then the total energy as a function of position was derived using Eq. (2). Figure 2 shows the energy curve for the $\mathscr{V}=0$ case (a symmetric double-well potential). It also shows the energy curve for the case of a single electrode and no applied field (i.e. the simple chemisorption case) for comparison. The energy barrier for an atom to leave the surface in the chemisorption case (i.e. the heat of adsorption of Si on a simple metal represented by the $r_s = 2$ jellium model) is 3 eV [7]; bringing up the second electrode to the distance shown lowers the energy barrier to transfer the atom away from the surface to ~0.8 eV. It is this activation barrier that will be of primary interest to us in this discussion.

Table 1 gives the actual values of the activation barrier Q extracted from the curves of energy *versus* atom position for bias values of 0, ±3, and ±6 V. Note the non-linear dependence of the barrier lowering on bias (for positive sample bias).

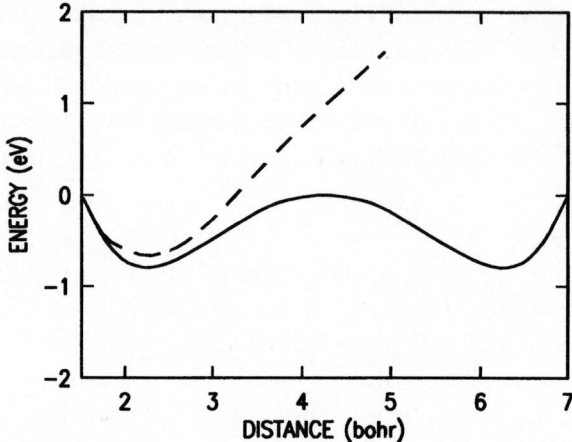

Figure 2. Solid curve: energy for Si atom between two electrodes specified in Fig. 1 with zero bias, as a function of distance of atom from left electrode. Electrode separation held fixed at 8.5 bohrs. Zero of energy is taken to be value at atom distance of 1.5 bohrs [i.e. z_0 in Eq. (2) defining energy is set at this distance]. Dashed curve: same as for solid curve except right electrode is absent. Energy value far to right is 3 eV above the minimum in the curve (i.e. the heat of adsorption is 3 eV).

TABLE 1. Activation barrier Q for transfer of Si atom between two jellium-model electrodes. The electrode from which the atom is transferred is at a bias $\mathscr{V}$ relative to the other, and the positive-background edges of the two electrodes are separated by 8.5 bohrs. The electrostatic field $\mathscr{E}$ is that midway between the electrodes in the absence of the atom.

$\mathscr{V}$ (V)	Q (eV)	$\mathscr{E}$ (V/Å)
6	0.43	1.8
3	0.78	0.9
0	0.84	0
-3	0.97	-0.9
-6	0.51	-1.8

Since a positive bias here leads to more of an activation barrier lowering than a negative bias of the same magnitude, our simple symmetric model predicts a net transfer of Si atoms from the positive to the negative electrode: the potential energy well on the positive-electrode side of the barrier will not be as deep as that on the negative-electrode side. This is in agreement with the results of Lyo and Avouris. [1] Note from Table 1 that putting a negative bias of moderate size ($\lesssim$ 3 eV in magnitude) on a given electrode leads in fact to an increase of the activation barrier for atom transfer from this electrode, and it is only for larger biases that the barrier starts to decrease.

The above results indicate that most of the activation barrier lowering that permits atom transfer between tip and sample results simply from the proximity of the two electrodes (which might be called a chemical effect), and that the effect of the bias for the distances discussed here is smaller. The proximity, in the case shown, decreases the activation barrier from 3 eV to 0.8 eV, as noted, and the positive sample bias decreases it some several tenths of an eV further, in addition to providing a directional driving force. Figure 3 shows the strong dependence of the zero-bias activation energy on electrode separation: a separation change of 2 bohrs ($\sim$1 Å) leads to a change in activation barrier of more than 1 eV. The experiments [1-3] are in accord with this, in the sense that they show a very sharp dependence of transfer rate on tip-sample separation.

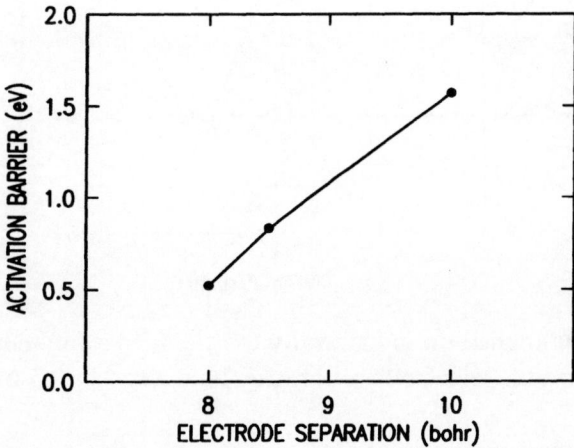

Figure 3. Activation barrier for transfer of Si atom from one electrode to another as a function of electrode separation, with V=0. Dots, which have been connected, correspond to the computed values.

To gain some insight into the force acting on the atom, it is useful to consider the related problem of field desorption. The usual explanation for desorption in that case is that due to field-induced charge transfer between atom and surface, the atom becomes an ion whose sign is the same as that of the electric field (taking the direction of the outward surface normal to be positive). This means that the field will exert a force on the ion that tends to move it away from the surface. This suggests the importance of the charge induced on the atom in the two-electrode problem by the applied bias.

It is difficult to define a net charge on the atom from charge-density maps, because the value depends on the volume over which the charge distribution is integrated. Another, approximate approach relates to the fact that charge transfer to or from an atom changes the electrostatic potential within the atom, which in turn shifts the core-level eigenvalues (see Ref. [4] for details).

The charge δq induced on the atom by the field as a function of atom position, calculated from core-level shifts, is shown in Fig. 4 for $V = 6$ V (left electrode positive). Note that δq is at most several tenths of a unit charge ($|e|$) and changes sign near the midpoint between the electrodes.

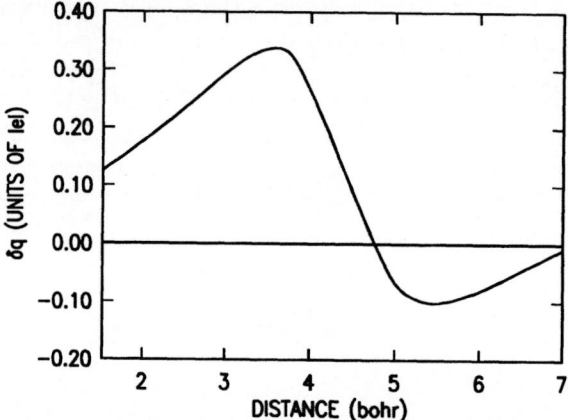

Figure 4. Charge δq induced on the atom by the field, as a function of distance of Si atom from left electrode. Electrode separation held fixed at 8.5 bohrs.

It should not be surprising that $\delta q / |e|$ does not attain values like $+2$ or $+3$ here, as it can at large distances in a field desorption experiment. [10] It is seen in the analysis of field desorption given in Ref. [11], that for fields comparable with those in the present case (see Table 1), the charge on the desorbing atom is also no greater than a few tenths $|e|$ within ~5 bohrs of the surface, but in that problem, the charge continues to rise as the atom moves to greater distances. In the present problem, of course, when the atom moves well away from the positive electrode, it encounters the negative electrode, which tends to drive electrons *onto* the atom, giving δq a negative value. The atom never gets very far, on the appropriate distance scale, from a charged surface; this is the reason δq is never very large.

The calculations of Ref. [4] were compared with the results of the scanning-tunneling-microscope experiment done by Lyo and Avouris. [1] In this experiment, done in vacuum at room temperature, Si was transferred from a Si surface to the tip only when the sample was positive, and could be transferred back by reversing the polarity of the bias. It was found that for a 10 ms, +3 V pulse, the probability of atom transfer just reached unity when the distance between tip and sample was decreased to a value s_t specified below in terms of the measured apparent tunneling barrier height. It should of course be emphasized that the calculation is appropriate for Si on a metal surface, but such data are not yet available, and so only a qualitative comparison was made with these measurements.

As Müller notes in his work on field desorption [12], the time required for an atom at the surface to overcome an energy barrier Q by thermal motions is

$$\tau = \tau_0 e^{Q/kT} , \tag{3}$$

with τ_0 a vibration time for the adsorbed atom, which he takes as 10^{-13} sec. For this same choice of τ_0, and with $\tau=10$ ms, the pulse length in the experiment, this equation implies for T = 300 °K that Q=0.66 eV.

Consider now the tip-sample separation s_t that corresponds to this Q value. The apparent tunneling barrier height ϕ_A measured in the microscope was found in Ref. [1] to decrease from its full value to near zero over a range of tip-sample separations of ~3 Å, as has been discussed theoretically. [13] It is convenient to specify separations by the corresponding value of ϕ_A (which provides an absolute scale). The measured ϕ_A was ~2 eV for the separation s_t.

Now ϕ_A can be calculated for the model of two electrodes with the Si atom at equilibrium on one electrode. The atom was kept fixed at the equilibrium distance for the chemisorption case, and ϕ_A was calculated in Ref. [4] for small bias as $\frac{1}{8}(d\ln I/ds)^2$ atomic units. [13] It was found in this way that $\phi_A \sim 2$ eV when the separation of the electrodes (measured between the two positive background edges) was ~8½ bohr. It is for this separation that the numbers in Table 1 were computed; the fact that the values of the activation barrier given in Table 1 are comparable to the value extracted from experiment suggests that the calculation of this quantity is reasonable.

It has been seen that for the Si case, a large part of the activation barrier lowering required for transfer of an atom between tip and sample in the STM arises simply from the close proximity of the two electrodes (a chemical-bonding effect). Field-induced charge transfer to or from the atom is significant to the force exerted on it, and hence to further changes in the activation barrier. The magnitude of the charge on the atom, however, depends strongly on atom position, and is only a few tenths of a unit charge. The calculated results show the same direction of atom transfer [14] relative to bias polarity as the experiment of Ref. [1], a similarly sharp dependence of activation barrier (or transfer rate) on distance, and a comparable magnitude for the activation barrier.

The case of field-induced transfer of a Na atom provides an interesting comparison with Si transfer. The calculations of Ref. [6] for Na were done for an electrode spacing of 12 bohrs, which was taken as a reasonable choice in the absence of any experimental data with which to compare.

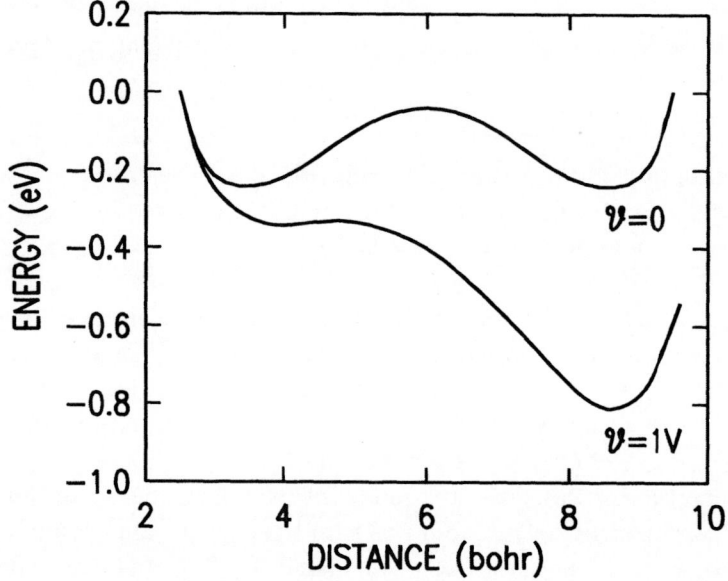

Figure 5. Energy for Na atom between two jellium-model electrodes, as a function of distance of atom from left electrode, for bias values $\mathcal{V}$ of 0 and 1 V. Electrode separation held fixed at 12 bohrs. Zero of energy is taken to be value at atom distance of 2.5 bohrs [i.e. z_0 in Eq. (2) is set at this distance].

Curves of energy as a function of atom position are given in Fig. 5 for bias values of 0 and 1 V. We see immediately from this figure that raising the bias V to 1 V decreases the barrier for transfer from the left electrode nearly to zero. The figure also shows that the barrier to transferring a Na atom in the opposite direction is ~0.5 eV at this bias. Thus the model predicts a net transfer of Na atoms from the positive to the negative electrode, as in the Si case.

The striking contrast between the Na and Si cases is that the bias required to give a substantial decrease of the activation barrier is so much smaller for Na than for Si. The important differences are that the Si atom is more strongly bound to a metal surface than the Na atom, and that the zero-bias net charge on the Na is appreciably greater [6,15] than it is on the Si, when each atom is midway between the electrodes (where the zero-bias energy barrier that must be overcome has its maximum).

References

1. I.-W. Lyo and Ph. Avouris, Science **253**, 173 (1991); Ph. Avouris and I.-W. Lyo, Appl. Surf. Sci. (to be published).

2. D. M. Eigler, C. P. Lutz, and W. E. Rudge, Nature **352**, 600 (1991).

3. H. J. Mamin, P. H. Guethner, and D. Rugar, Phys. Rev. Lett. **65**, 2418 (1990).

4. N. D. Lang, Phys. Rev. B **45**, 13599 (1992).

5. See e.g. N. D. Lang, "Density-Functional Approach to the Electronic Structure of Metal Surfaces and Metal-Adsorbate Systems," in *Theory of the Inhomogeneous Electron Gas*, ed. S. Lundqvist and N. H. March (Plenum Press, New York, 1983), pp. 309-389.

6. The work on Na transfer is described in more detail in N. D. Lang, "Field-Induced Transfer of an Electropositive Atom Between Two Closely Spaced Electrodes", in *Manipulations of Atoms under High Fields and Temperatures: Applications*, eds. B. Vu Thien, N. Garcia, and K. Dransfeld (Kluwer Academic Publishers, Dordrecht, to be published).

7. N. D. Lang and A. R. Williams, Phys. Rev. B **18**, 616 (1978).

8. W. Kohn and L. J. Sham, Phys. Rev. **140**, A1133 (1965).

9. The Hellmann-Feynman theorem allows us to define a total energy for the system in terms of the electrostatic force on the nucleus. This theorem is discussed

in, for example, J. I. Musher, Amer. J. Phys. **34**, 267 (1966); S. T. Epstein, *The Variation Method in Quantum Chemistry* (Academic Press, New York, 1974), p. 107.

10. See e.g. T. T. Tsong, *Atom-Probe Field Ion Microscopy* (Cambridge University Press, Cambridge, 1990).

11. H. J. Kreuzer, L. C. Wang, and N. D. Lang, Phys. Rev. B **45**, 12050 (1992); N. D. Lang, Solid State Commun. **84**, 155 (1992).

12. E. W. Müller, Phys. Rev. **102**, 618 (1956).

13. N. D. Lang, Phys. Rev. B **37**, 10395 (1988).

14. Note that the direction of atom transfer in the experiment of Eigler *et al.* [2] for Xe is opposite to that found in the experiment of Lyo and Avouris [1] for Si, i.e. Xe is transferred off of a surface of negative polarity. J. R. Cerdá, F. Flores, P. L. de Andres, and P. M. Echenique (to be published) and R. E. Walkup, D. M. Newns, and Ph. Avouris (to be published) have suggested that for the Xe case, where there is negligible charge transfer to or from the atom, the atom transfer has a rather different explanation from that discussed here.

15. See N. D. Lang, "Theory of Alkali Adsorption on Metal Surfaces," in *Physics and Chemistry of Alkali Metal Adsorption*, eds. H. P. Bonzel, A. M. Bradshaw, and G. Ertl (Elsevier, Amsterdam, 1989), pp. 11-24, for a discussion of the degree of ionization of alkali atoms adsorbed on metal surfaces. See also M. Scheffler, Ch. Droste, A. Fleszar, F. Máca, G. Wachutka, and G. Barzel, Physica B **172**, 143 (1991); G. Pacchioni and P. S. Bagus, Surf. Sci. **269/270**, 669 (1992).

VIBRATIONAL HEATING AND ATOM TRANSFER WITH THE STM

R. E. WALKUP, D. M. NEWNS, and PH. AVOURIS
IBM Research Division,
IBM Thomas J. Watson Research Center,
Yorktown Heights, NY 10598, USA.

ABSTRACT. The high current density obtainable with the STM can be used to promote multiple vibrational excitation of an adsorbate via inelastic electron tunneling. The resulting vibrational distribution is approximately Maxwell-Boltzmann, with a characteristic vibrational temperature. A simple expression is given for this vibrational temperature as a function of the tunneling current. Calculations indicate that vibrational heating by inelastic tunneling plays an important role in the transfer of Xe atoms between a sample and tip. A simple expression is given for the rate of atom transfer over a barrier. Directionality of the atom-transfer process and applications to other systems are discussed.

1. Introduction

The scanning tunneling microscope (STM) is widely used for examining the electronic structure of surfaces with atomic resolution. Early on it was realized that the STM might also provide the capability of local vibrational spectroscopy via measurement of changes in the conductance associated with the opening of channels for inelastic tunneling. Inelastic tunneling had previously been extensively studied in metal-insulator-metal tunnel junctions [1]. However, efforts to use the STM for local vibrational spectroscopy have not been very successful. Recently we pointed out that inelastic tunneling may play a very important role in efforts to manipulate atoms and molecules with the STM by the application of current (voltage) pulses [2]. In this case, the very high current density obtainable with the STM can be used to promote a sequence of inelastic vibrational excitations, thus localizing substantial energy in a bond. In this article we discuss a simple model of multiple vibrational excitation by inelastic tunneling; and we apply this model to the recent experiments of Eigler and co-workers [3] on the current-driven transfer of Xe atoms between a sample surface and the STM tip.

P. Avouris (ed.), Atomic and Nanometer-Scale Modification of Materials: Fundamentals and Applications, 97–109.
© 1993 Kluwer Academic Publishers.

First we review the two mechanisms for inelastic tunneling that have been discussed in the STM literature: dipole excitation [4], and resonance tunneling [5]. In the dipole mechanism, the tunneling electrons interact with the transition dipole moment of a vibrational mode. Upon inelastic tunneling, the vibrational level changes from $|i>$ to $|f>$, and the electron's kinetic energy changes by $\hbar\omega = E_f - E_i$. Excitation of the vibration can occur when the bias voltage exceeds the energy change, $\hbar\omega$. The inelastic tunneling fraction, i.e. the ratio of the inelastic to elastic tunneling conductance, is $\simeq (\mu_{if}/ea_0)^2$, where μ_{if} is the transition dipole moment and a_0 is the bohr radius [4]. One can make an estimate of the inelastic tunneling fraction for the $|0> \rightarrow |1>$ transition by noting that the dipole matrix element is $\mu_{01} \simeq x_{01}\partial\mu_{el}/\partial x$, where $x_{01} = <1|x|0> = [\hbar/2m\omega]^{1/2}$, m is the reduced mass for the vibration, and $\partial\mu_{el}/\partial x$ is the derivative of the electric dipole moment with respect to the vibrational coordinate, x. For a dipole derivative of $\simeq 1$ Debye/bohr, and $x_{01} \simeq 0.1$ bohr, the inelastic tunneling fraction would be $(\mu_{if}/ea_0)^2 \sim 10^{-3}$. This is a typical order-of-magnitude number. In the resonance mechanism, the electron can be regarded as temporarily hopping onto the adsorbate. The resulting force along the vibrational coordinate leads to a small probability of vibrational excitation. For a resonance that is far from the Fermi level compared to the resonance width, the inelastic tunneling fraction [5] is $\simeq |F x_{01}/(E_a - E_F)|^2$, where F is the force along the vibrational coordinate when the resonance is occupied, x_{01} is defined above, and $E_a - E_F$ is the displacement of the resonance from the Fermi level. For an order-of-magnitude estimate, one can take $F \simeq 1$ eV/bohr, $x_{01} \simeq 0.1$ bohr, and $E_a - E_F \simeq 3$ eV, which yields an inelastic tunneling fraction of $\sim 10^{-3}$. This estimate is comparable to that for the dipole mechanism. The resonance mechanism can be much stronger when the resonance overlaps the Fermi level [5].

2. Vibrational heating

Now consider the heating of a vibrational mode by inelastic electron tunneling. The simplest useful model is to regard the vibrational mode as a harmonic oscillator. Inelastic tunneling induces transitions among the oscillator levels, and one must also account for damping of the vibration. In both the dipole and resonance mechanisms, inelastic tunneling primarily induces transitions between adjacent levels on the vibrational ladder. In the dipole mechanism this results from dipole selection rules, and in the resonance mechanism this results from the fact that the average energy transfer per electron is usually much smaller than $\hbar\omega$ [5]. Also, the transition rate between $|n - 1>$ and $|n>$ is proportional to the upper quantum number, n. This scaling relation results from a linear expansion along the vibrational coordinate, and the fact that for a harmonic oscillator $|<n - 1|x|n>|^2 \propto n$. Damping of the vibration is also expected to occur via transitions between adjacent levels on the vibrational ladder, provided that the average energy loss per cycle is small compared to $\hbar\omega$. Similarly, the vibrational relaxation rate for $|n> \rightarrow |n-1>$ is expected to be proportional to the upper quantum number, n. Relaxation is discussed in more detail below. Given these simple relations for the transition rates, one can write a set of master equations, with neighboring oscillator levels coupled by net upward (u) and downward (d) rates, both proportional to the upper quantum number. It is known from the theory of stochastic processes [6] that the general steady-state solution to such master equations is $p_n \propto (u/d)^n$, where p_n is the probability of finding the oscillator in state $|n>$, and (u/d) is the ratio of upward to downward rates, which is independent of the quantum number, n. This steady-state distribution has the Boltzmann form, with a vibrational temperature of $kT_{vib} = \hbar\omega/ \log(d/u)$. To apply this to the case of heating by inelastic tunneling, we must specify the net upward (u) and downward (d) rates. In general both of these rates contain contributions from inelastic tunneling and from relaxation (coupling to the surrounding bath). To specify the inelastic tunneling contribution, we assume that the STM bias voltage exceeds $\hbar\omega$ by an amount that is large compared to the electron temperature. In this case, the inelastic tunneling rates for upward and downward transitions are essentially equal, and can be expressed as $(I/e)f_{in}$, where I is the tunneling current flowing through the adsorbate and f_{in} is the inelastic tunneling fraction for the $|0> \rightarrow |1>$ transition. We assume that the relaxation rates for coupling to the bath satisfy detailed balance, so that thermodynamic equilibrium is reached in the absence of a tunneling current. Writing the downward rate for $|1> \rightarrow |0>$ as γ, the relaxation process must have an upward rate $\gamma e^{-\hbar\omega/kT_s}$, where T_s is the sample temperature. Using these results, the steady-state vibrational temperature in the presence of the tunneling current is:

$$kT_{vib} = \frac{\hbar\omega}{\log\left[\dfrac{(I/e)f_{in} + \gamma}{(I/e)f_{in} + \gamma e^{-\hbar\omega/kT_s}}\right]} \qquad (1)$$

With this expression one can estimate the vibrational temperature as a function of the tunneling current for given values of the inelastic tunneling fraction, f_{in}, and the vibrational damping rate, γ.

As an example, consider vibrational heating of a Xe atom adsorbed on a metal surface. The vibrational degree-of-freedom of interest is the displacement of Xe along the surface normal. The flow of current through the Xe atom should result in vibrational heating via both the dipole and resonance mechanisms. The adsorption of Xe on a metal surface lowers the work function, implying a dipole moment per atom of $\simeq 0.2 - 0.3$ Debye [7]. From the calculations of Lang [8] one can estimate the change in electric dipole moment with Xe displacement: $(\partial\mu_{el}/\partial x) \simeq 0.3$ Debye/bohr. Using a vibrational level spacing of $\hbar\omega = 2.5$ meV, one has a matrix element $x_{01} \simeq 0.15$ bohr. The resulting $|0> \rightarrow |1>$ transition dipole moment is $\mu_{01} \simeq 4.5 \times 10^{-2}$ Debye, which implies an inelastic tunneling fraction of $(\mu_{01}/ea_0)^2 \sim 3 \times 10^{-4}$. The resonance mechanism for inelastic tunneling should also be considered because a comparison of experimental and calculated STM images of Xe [9] indicates that tunneling is due to the density-of-states at the Fermi level arising from the Xe 6s resonance. Following the calculations of Ref. 9, we take the 6s resonance to be located $\simeq 4$ eV above the Fermi level, with a width of $\simeq 1$ eV. One also needs the force resulting from the formation of a transient negative Xe ion. From recent LDF calculations on ion-surface potentials [10], a realistic estimate of the force is about one half of the classical image force. We take the equilibrium distance of Xe from the image plane to be $\simeq 1.6$ Å, which thus yields $F \simeq 0.4$ eV/bohr. Then using $f_{in} \simeq |F \ x_{01}/(E_a - E_F)|^2$, and $x_{01} \simeq 0.15$ bohr, the estimate for the inelastic tunneling fraction via the resonance mechanism is $\sim 4 \times 10^{-4}$. This is quite comparable to the estimate for the dipole mechanism.

In order to estimate the vibrational temperature, one also needs the vibrational relaxation rates. For the low-frequency Xe-metal vibration, one can consider damping via excitation of phonons or electron-hole pairs in the metal. Estimates for the e-h pair mechanism suggest lifetimes of order 10^{-9} s [11]. However, damping by phonon excitation is much faster, as shown below. A good description of the phonon mechanism is provided by an elastic continuum model [12] in which the vibration of frequency ω is regarded as driving waves of frequency ω in the elastic medium of the substrate. These waves radiate energy away from the region containing the adsorbate, thus damping the vibration. In a simplified picture, one can regard each oscillation as a collision with the substrate, with an energy transferred to the substrate of $\Delta E \simeq E(m_a/m_s)$, where E is the energy in the adsorbate vibration, m_a is the adsorbate mass, and m_s is the effective mass of the substrate. This effective mass is simply the mass contained in a volume of $\sim (\lambda/2\pi)^3$, where λ is the wavelength of the sound wave of frequency ω that is driven by the adsorbate vibration. Thus $m_s \sim \rho(c_s/\omega)^3$, where ρ is the mass-density and c_s is the transverse sound speed of the substrate. Since collisions with the substrate occur with frequency $v = \omega/2\pi$, the net rate of damping of the vibrational energy is $dE/dt \sim - (\omega/2\pi)(m_a/m_s)E$. This results in simple exponential decay of the energy in the adsorbate vibration. From the elastic continuum model [12] the overall energy damping rate is $\gamma \simeq 1.6(\omega/2\pi)m_a/[\rho(c_s/\omega)^3]$, which agrees with the heuristic derivation above to within a multiplicative numerical factor. For our purposes it is necessary to describe damping in terms of the discrete energy levels of the adsorbate vibration. One can regard the system as a forced oscillator [13], with transition probabilities that are determined by the magnitude of the average energy transferred to the substrate per cycle, ΔE. When ΔE is small compared to $\hbar\omega$, it takes many cycles to damp the vibrational energy. In this case the decay of a level $v = n$ is primarily by transitions to v = n-1; and the transition probability [13] is proportional to the upper vibrational quantum number, n. It is easy to show that this scaling relation leads to simple exponential decay of the energy in excess of the zero-point energy, i.e. a very close analogy to classical exponential damping. The classical damping rate, γ, is equal to the rate of population decay from the v = 1 level. The higher levels decay at a rate of $n\gamma$, where n is the quantum number. If the energy-transfer per cycle exceeds $\hbar\omega$, it would then be necessary to take into account larger changes in the oscillator state, by using the known transition probabilities for a forced oscillator [13].

Now we apply these results to Xe atoms adsorbed on a nickel substrate. The elastic continuum model yields a vibrational lifetime for the v = 1 level of $\simeq$25 ps, where we have taken $\hbar\omega \simeq$ 2.5 meV [2]. From our earlier work on the detailed solution of the master equations for an atom in a double well [2], we take the inelastic tunneling fraction to be $f_{in} \simeq 6 \times 10^{-4}$, which is comparable to the estimate for the dipole and resonance mechanisms. We take the substrate temperature to be 4 K, as in experiment [3]. The resulting vibrational temperature calculated from Eqn. (1) is shown in Fig. 1 as a function of the tunneling current. One should keep in mind that the vibrational temperature describes the steady-state population of the vibrational levels in the presence of the tunneling current. One can see from Fig. 1 that the Xe-surface vibration can be heated by inelastic tunneling to temperatures that are substantially above the ambient temperature.

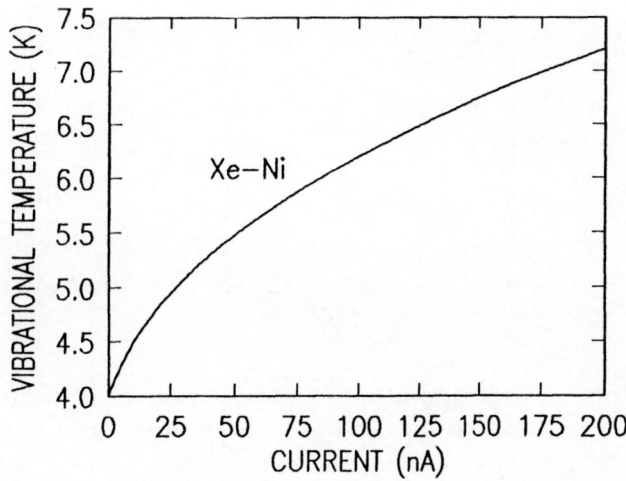

Figure 1. The vibrational temperature is shown as a function of the tunneling current for Xe adsorbed on a nickel surface. The temperature is calculated from Eqn. (1) with $1/\gamma$ = 25 ps, and $f_{in} = 6 \times 10^{-4}$.

As an additional example, consider atomic hydrogen on a silicon surface. The Si-H vibration has a frequency $\simeq 2084$ cm^{-1} which is well above the substrate phonon modes, thus vibrational relaxation is not very efficient. For a hydrogen-terminated Si(111) surface, the lifetime of the Si-H vibration (v = 1) has been directly measured [14] to be $\simeq 10^{-9}$ s at room temperature. One can estimate an inelastic tunneling fraction for the dipole mechanism, using a dipole transition moment from gas-phase data [15]. This results in $f_{in} \simeq 10^{-3}$. With these numerical values and with the assumption that all of the tunneling current goes through a single Si-H bond, the vibrational temperature can be calculated from Eqn. (1). The result is shown in Fig. 2 (solid curve). There is substantial vibrational heating above the 300 K sample temperature for currents as low as ~ 1 nA. Under more conservative assumptions, such as spreading the current over $\simeq 5$ Si-H bonds (dotted curve), the effect of vibrational heating should still be dramatic for currents in excess of 5 nA. Generally speaking, vibrational heating depends strongly on the vibrational lifetime and on the rate of excitation by inelastic tunneling. Significant heating is possible for systems that have rather long vibrational lifetimes, especially for current pulses with peak currents that are substantially greater than typical STM imaging currents.

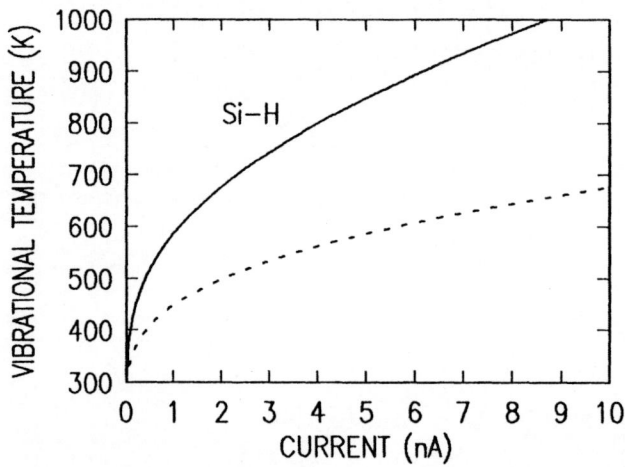

Figure 2. The vibrational temperature is shown as a function of the tunneling current for H atoms adsorbed on silicon. The temperature is calculated from Eqn. (1) with $1/\gamma = 1$ ns, and $f_{in} = 10^{-3}$, assuming that the tunneling current goes through a single Si-H bond (solid curve). The dotted curve shows the calculated temperature if a fraction 0.2 of the total current flows through the Si-H bond.

3. Atom transfer

Local heating by inelastic tunneling has important implications for the manipulation of individual atoms or molecules. For example, heating of the relevant vibrational mode can promote diffusion, desorption, or reaction. Here we consider the effect of vibrational heating on the transfer of an atom between sample and tip. The relevant potential energy surface for atom-transfer can be envisioned as a double-well, with one minimum on the sample, and another minimum on the tip. When the tip-sample separation is small, the energy barrier for atom-transfer is small. When an atom is localized in one of the potential wells, its environment is roughly harmonic, thus the atom moves on a vibrational ladder by taking nearest-neighbor steps. The rate of transfer from the sample to the tip is approximately the probability of finding the atom in the highest state, n^*, localized on the sample side of the barrier, times the rate of excitation to states above the barrier, times the probability that the hot atom will settle onto the tip. This is illustrated in Fig. 3, which shows the quantum-mechanically exact energy levels for a Xe atom in an asymmetric double well [2]. Examination of the corresponding eigenfunctions indicates that the states below the barrier are well-described by harmonic-oscillator wavefunctions that are localized to a particular side of the barrier. Within a harmonic description, the probability of finding the atom in level n^* - see Fig. 3 - is $\simeq (u/d)^{n^*}$, where (u/d) is the ratio of upward to downward transition rates, and where we assume that $(u/d) << 1$. One can express this probability in terms of the energy barrier and the vibrational temperature given by Eqn. (1). The rate of atom transfer can then be written as

$$R_{S \to T} \sim \left[\gamma f_T \frac{\Delta E}{\hbar \omega} \right] e^{-\Delta E/kT_{vib}} , \qquad (2)$$

where ΔE is the energy barrier for transfer from the sample to the tip, kT_{vib} is the vibrational temperature in the presence of the tunneling current, γ is the decay rate of the $v = 1$ level, and f_T is the probability that an atom with an energy above the barrier will decay into the well on the tip.

    Eqn. (2) has the familiar form of a prefactor times a Boltzmann factor. Note that the prefactor is not equal to the vibrational frequency, but instead is proportional to the damping rate. This result is known in the chemical kinetics literature, where the model of a harmonic vibrational ladder with an "absorbing" barrier has been extensively studied [16]. Given the number of assumptions made in the context of the present problem, Eqn. (2) should be regarded as a rough but useful approximation. For quantitative studies, a detailed solution of the master equations for an atom in a double-well is preferred, as described in Ref. 2. With master equations one can account for the anharmonic character of the potentials, atomic tunneling across the barrier, and the possibility of multi-quantum transitions, etc.

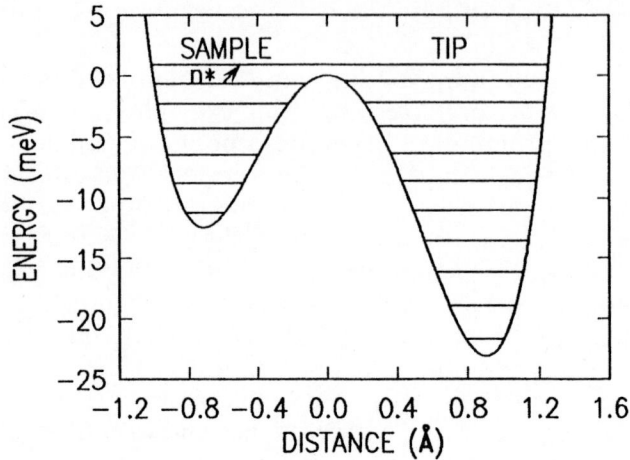

Figure 3. Schematic diagram of atom transfer from a sample to the tip. The energy levels shown are quantum-mechanically exact results for Xe in the double well. The rate of atom transfer is approximately given by the probability to find the atom in state n*, times the rate of excitation to levels above the barrier, times a branching ratio for relaxation into the well on the tip.

4. Application to Xe atoms adsorbed on metals

Systematic studies of atom transfer by the application of voltage pulses were reported by Eigler and co-workers [3] for Xe atoms adsorbed on metal surfaces. Within the context of the model outlined above, the key parameter for atom transfer is the energy barrier, ΔE. Here we briefly consider the barrier for Xe transfer. Xe atoms were observed to spontaneously transfer to the tip at close tip-sample separations. Reversible transfer by voltage pulses was obtained when the tip was displaced outward from the point of spontaneous transfer by roughly 0.1 Å, and lateral displacement of Xe, but not atom transfer, was obtained when the tip was retracted further. From this one can conclude that the relevant energy barrier in the Xe experiments is comparable to the diffusion barrier. The energy barrier can also be obtained by assuming that spontaneous transfer is thermally activated at the ambient 4 K temperature, and then estimating the barrier required to yield a reasonable spontaneous transfer rate. Both of these arguments indicate that the barrier for Xe transfer is in the 10-20 meV range.

Additional information on the interaction of adsorbed Xe with the STM tip can be obtained by using empirical pair-potentials and an atomistic model of the sample and tip. Results [2] are shown in Fig. 4 for the potential energy of Xe as a function of distance from the surface, for several tip-sample separations. A double-well potential is obtained, and the energy barrier decreases sharply as the tip approaches the sample surface. Energy barriers in the 10-20 meV range occur for tip-sample separations of ∼ 4.3 Å, measured from the point of hard-sphere contact.

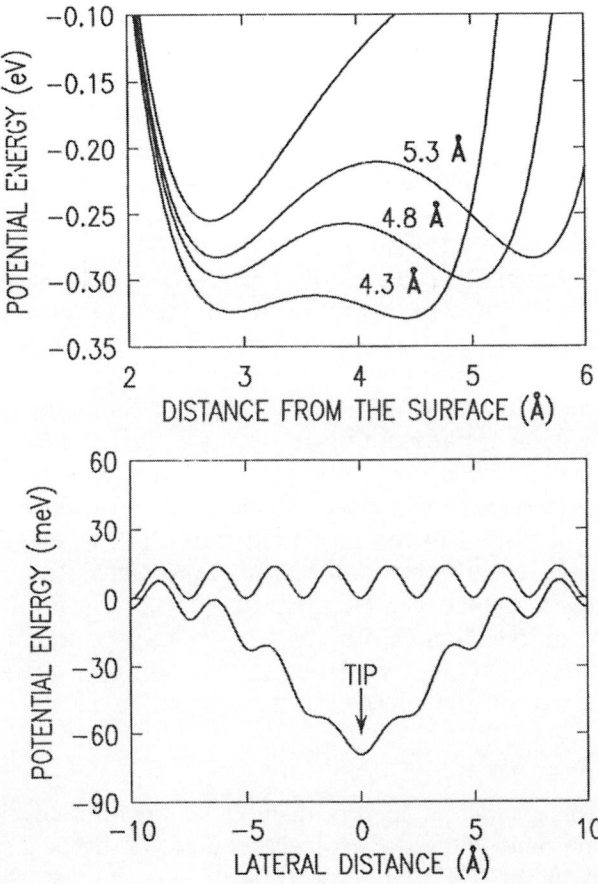

Figure 4. Interaction of an adsorbed Xe atom with the STM tip. The upper panel shows the potential energy as a function of distance from the surface, for several tip-sample separations. The lower panel shows the energy vs lateral distance along a closed-packed direction of the Ni(110) surface. The tip creates a substantial van der Waals well which can be used to slide the atom laterally.

The effect of the tip on the lateral degrees of freedom is also shown in Fig. 4. In the absence of the tip, the Xe atom on Ni(110) experiences a weakly corrugated potential, with diffusion barriers of ~ 15 meV in the closed-packed direction, and ~ 60 meV for motion across the rows. When the tip is lowered into the tunneling position, a strong van der Waals minimum develops which confines the Xe atom to the region directly under the tip, as shown in Fig. 4 for a tip-sample separation of 4.3 Å. This lateral confinement enables one use the STM tip to slide a Xe atom along the surface to a new location, as demonstrated by Eigler and Schweizer [17].

The energy barrier can be affected by the electric field in the tunnel junction and also by the tunneling current. This can result in directionality of the atom transfer process. The most important effect of an applied electric field (for small fields) is due to the adsorption-induced dipole moment, as pointed out by Cerda and co-workers [18]. Adsorption of Xe on a metal surface results in a dipole moment, $\mu \sim 0.2$ D [7]. The sign of the dipole is consistent with a small amount of charge transfer from Xe to the metal. Since the sample and tip surfaces oppose each other, the sign of the dipole moment is reversed upon transfer of Xe to the tip. When a negative bias is applied to the sample, the dipole moment will be aligned with the applied electric field for Xe adsorption on the tip, and counter to the field for adsorption on the sample. Thus the dipole interaction with the applied field favors adsorption on the positively biased electrode. This is consistent with the direction of atom-transfer observed in experiment [3]. From basic considerations of electrostatics, a bias-dependent potential, $\mu(x)V_{bias}/w$, should be added to the zero-field potential, where $\mu(x)$ is the dipole moment (which varies with the position of the Xe nucleus, x), V_{bias} is the sample bias, and w is the distance between effective image planes on the sample and tip. For an atom between two identical slabs, the energy barrier is located at the midpoint, and the dipole moment vanishes at this point by symmetry. Thus the effect of the dipole term on the energy barrier is given by:

$$\Delta E \simeq \Delta E_0 + \mu_{ads}V_{bias}/w \qquad (3)$$

where ΔE_0 is the energy barrier in the absence of an applied electric field, and μ_{ads} is the dipole moment when the atom is located at the adsorption well of interest (either sample or tip). For the experimental system reported by Eigler and co-workers, one can estimate $\mu_{ads} \simeq 0.2$ Debye, and $w \simeq 4$ Å [2]. The resulting barrier lowering is ~ 10 meV * V_{bias}(Volts). A current-dependent force referred to as the "electron wind" [19] can also give rise to directionality. This force arises from momentum transfer from the current carriers to the Xe atom. Our estimates indicate that for a 1 MΩ tunnel junction, the "wind" force is roughly an order-of-magnitude smaller than the force arising from the variation of the dipole moment. Such a wind force would be more important for tunnel junctions with a low impedance.

By using the analytic expressions Eqns. (1)-(3) above, one can fit the experimental Xe transfer-rate [3] with two adjustable parameters: the zero-field barrier height and the inelastic tunneling fraction. The fit to the data is shown in Fig. 5 (solid curve). The best-fit values are $\Delta E_0 = 11.7$ meV for the zero-field energy barrier, and $f_{in} = 6.6 \times 10^{-4}$ for the inelastic tunneling fraction. The other parameters were kept fixed at $\hbar\omega = 2.5$ meV, $1/\gamma = 25$ ps, and $f_\Gamma = 0.5$, with the dipole term given above. The barrier height obtained from the fit is quite reasonable, and the inelastic tunneling fraction is within the theoretically expected range. The results obtained from this simple approach agree very well with a detailed solution of the rate equations for a Xe atom in a double-well potential [2]. The results of Fig. 1 and Fig. 5 make a very strong case that heating of the Xe atom by inelastic tunneling should play an important role in the atom-transfer process. A key experiment for placing additional constraints on the transfer mechanism would be a measurement of the rate of transfer from the sample to the tip for both bias polarities. For small electric fields the dipole term has only a small effect on the energy barrier, thus the transfer rate at small fields should be almost independent of the direction of current flow. In contrast, the transfer rate at higher fields depends strongly on the direction of current flow. The calculated transfer rate for the reversed polarity is shown in Fig. 5 (dotted line).

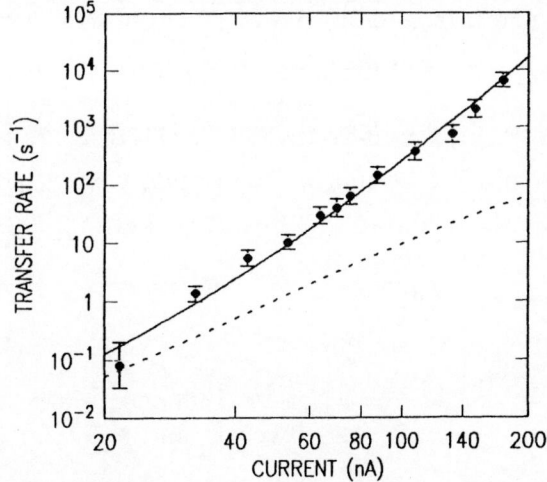

Figure 5. The rate of transfer of a Xe atom from the sample to the tip is shown as a function of the tunneling current. The experimental data are from Eigler and co-workers [3], and the solid curve is the best fit obtained via Eqn. (2). The two parameters determined by the fit are the zero-field energy barrier (11.7 meV) and the inelastic tunneling fraction (6.6×10^{-4}). The dotted curve shows the calculated rate of atom transfer with the direction of current flow reversed.

5. Conclusions

For a harmonic vibrational mode subjected to current flow in an STM, inelastic tunneling leads to a Maxwell-Boltzmann distribution of vibrationally excited levels. The characteristic vibrational temperature is determined by a balance between inelastic tunneling and vibrational damping. The energy pumped into the vibration can be used to promote diffusion, desorption, reaction, or atom transfer to (or from) the STM tip. In the case of weak vibrational damping, one can obtain a simple analytic expression for the rate of atom transfer. Calculations for Xe atoms adsorbed on metals indicate that there should be substantial vibrational heating for tunneling currents > 100 nA. This is consistent with observations of Eigler and co-workers [3]. The experimental transfer rate vs. tunneling current [3] can be accurately fitted by choosing reasonable values for the energy barrier and the inelastic tunneling fraction. These results indicate that vibrational heating by inelastic tunneling can play an important role in STM-based manipulation of individual atoms and molecules.

References

1. P. K. Hansma, pg. 135 in Vibrational Spectroscopy of Molecules on Surfaces, ed. J. T. Yates, Jr. and T. Madey (Plenum, NY, 1987)

2. R. E. Walkup, D. M. Newns, and Ph. Avouris, submitted to Phys. Rev. Lett.

3. D. M. Eigler, C. P. Lutz, and W. E. Rudge, Nature 352 , 600 (1991)

4. B. N. J. Persson and J. E. Demuth, Solid State Comm. 57 , 769 (1986)

5. B. N. J. Persson and A. Baratoff, Phys. Rev. Lett. 59 , 339 (1987)

6. N. G. van Kampen, Stochastic Processes in Physics and Chemistry (North Holland, Amsterdam, 1981)

7. C. P. Flynn and Y. C. Chen, Phys. Rev. Lett. 46 , 447 (1981)

8. N. D. Lang, Phys. Rev. Lett. 46 , 842 (1981)

9. D. M. Eigler, P. S. Weiss, E. K. Schweizer, and N. D. Lang, Phys. Rev. Lett. 66 , 1189 (1991)

10. R. E. Walkup, Ph. Avouris, N. D. Lang, and R. Kawai, Phys. Rev. Lett. 63 , 1972 (1989)

11. B. N. J. Persson, Phys. Rev. B 44 , 3277 (1991)

12. B. N. J. Persson and R. Ryberg, Phys. Rev. B 32 , 3586 (1985)

13. D. Rapp and T. Kassal, Chem. Rev. 69 , 61 (1969)

14. P. Guyot-Sionnest, P. Dumas, Y. J. Chabal, and G. S. Higashi, Phys. Rev. Lett. 64 , 2156 (1990)

15. A. A. Radzig and B. M. Smirnov, Reference Data on Atoms, Molecules, and Ions (Springer-Verlag, Berlin, 1985)

16. E. W. Montroll and K. E. Shuler, Adv. Chem. Phys. 1 , 361 (1958), reprinted in I. Oppenheim, G. H. Weiss, and K. E. Shuler, Stochastic Processes in Chemical Physics, (MIT Press, Cambridge MA, 1977)

17. D. M. Eigler and E. K. Schweizer, Nature 344 , 524 (1990)

18. J. R. Cerda, F. Flores, P. L. de Andres, and P. M. Echenique, to be published

19. See, for example, A. H. Verbruggen, IBM J. Res. Develop. 32 , 93 (1988)

TIP-INDUCED MODIFICATIONS OF ELECTRONIC AND ATOMIC STRUCTURE

S. CIRACI
Department of Physics
Bilkent University
06533 Ankara, Turkey

ABSTRACT. Tip-sample interaction at small separation is strong enough to modify the electronic structure and potential barrier, and to induce strong forces which can move atoms on the surface. Using the results of self-consistent calculations for the interaction energy we studied the states of a single atom (Al and Xe) between tip and sample. In particular, we investigated the perpendicular and lateral motions of the atom, and the resonant tunneling through the states bound to the potential well created by this atom.

1. Introduction

Initially, scanning-tunneling microscopy (STM) was considered as a powerful tool to probe the local "atomic" structure of surfaces. Keeping the tip-sample separation large and hence probing the weak and exponentially decaying coupling of the electronic states between electrodes, STM achieved resolutions on the atomic scale. However, it has been not too late to realize that the real power of STM lies in the tip-sample interaction and several novel effects derived thereof. Early on we pointed out significant coupling between tip and sample by distinguishing *conventional tunneling* from *electronic* and *mechanical contact regimes* in STM [1]. Modification of electronic states of the electrodes due to the tip approaching the sample affects not only the tunneling current, but also creates other effects: For example, the short-range forces become site-dependent and overtake the long-range Van der Waals (VdW) force. Site-dependent effective barrier is induced by the confinement of current transporting states between tip and sample even if the potential barrier is already collapsed [2,3].

Strong tip-sample interaction may lead irreversible (even hysteretic) modification of atomic structure [4,5]. Recently, it has been shown that atoms can be manipulated on the surface [6,7] even they can reversibly be moved between tip and sample providing a switching effect [8]. These controlled modification of materials on the

111

P. Avouris (ed.), Atomic and Nanometer-Scale Modification of Materials: Fundamentals and Applications, 111–119.

atomic scale is expected to bring about several important applications in future.

In this work we present a theoretical study of the tip-sample interaction in STM which leads to the controlled motion of the atom. We first outline the theoretical approach and discuss various aspects of the interaction between a single atom and sample surface. Then, we examine the interaction energy curves of Al and Xe atoms between the Al tip and Al sample to understand the controlled motion of an atom. Finally, considering various features of the interaction energy curves we study perpendicular and lateral motions of the atom between tip and sample, and the resonant tunneling effect created thereof.

2. Single atom-electrode interaction

Interactions on the atomic scale are calculated by using self-consistent field (SCF) pseudopotential method within the local density approximation (LDA) [9]. To represent the wave function in the plane wave basis set our system consisting of an atom interacting with the sample and also two electrodes (tip and sample) interacting with an atom between them are considered in supercells, which are periodically repeated in 3D. As long as the interatom interaction between adjacent supercells are negligible this periodic model can yield accurate values of potential energy $V(\mathbf{r})$, total charge density $\rho(\mathbf{r})$, interaction energy $E_i(\mathbf{r})$ and lateral force (parallel to the surface) $F_{\parallel}(\mathbf{r})$ and perpendicular force $F_{\perp}(\mathbf{r})$ between atom and electrode. Here $\mathbf{r}$ is the position vector relative to a point in the atomic plane of the sample surface, and z is its projection on the axis perpendicular to the same plane. The interaction energy of a single atom with the surface of an electrode $E_i(\mathbf{r})$ is the energy required to take the atom from position $\mathbf{r}$ to infinity. The variation of E_i with the distance z from the surface of the sample is site-dependent for small z, but becomes uncorrugated at large distance. The minimum of $E_i(z)$ corresponds to the binding energy E_b for a given lateral position of the atom. In most of the cases the minimum of $E_i(z)$ corresponding to the top (T) site is only a saddle point.

The corrugation of $E_i(\mathbf{r})$ is relevant for the motion of an atom on the surface. For atoms engaging in strong interaction with the surface and hence causing to massive charge rearrangement the corrugation of $E_i(\mathbf{r})$ can be in the range of 0.5 to 1 eV at the equilibrium distance, but decays to zero at large z. There are, however, pathways along which the corrugation and hence the barrier for an atom to move from one position to the other may be relatively smaller. The corrugation of $E_i(\mathbf{r})$ is, in general, very small for the physisorbed atom and is in the range of ~ 50 meV near the equilibrium distance from the sample. The total force on the atom exerted by the sample has short range (electrostatic) and long-range (VdW) components. The VdW component is small and is overshadowed by the electrostatic force at small z if the interaction involves charge rearrangements as in a chemical bond. The perpendicular component of the short range force is site-dependent since it is equal to $-\partial E_i(\mathbf{r})/\partial z$.

As for the lateral short-range force, it is also site-dependent, but its magnitude is normally smaller than $F_\perp(z)$ [9].

3. States of an atom between two electrodes

The above arguments, which are derived for a single atom interacting only with one electrode (sample) become complicated in the presence of a second electrode (tip) separated by s from the surface of the first one (see the inset in Fig.1). If the separation between two electrodes is large, the interaction energy of the atom with these electrodes $E(s,z)$ can be expressed as the sum of its interaction energy with only one electrode, i.e. $E_{i,1}(z) + E_{i,2}(z)$. When the atom is closer to one of two electrodes

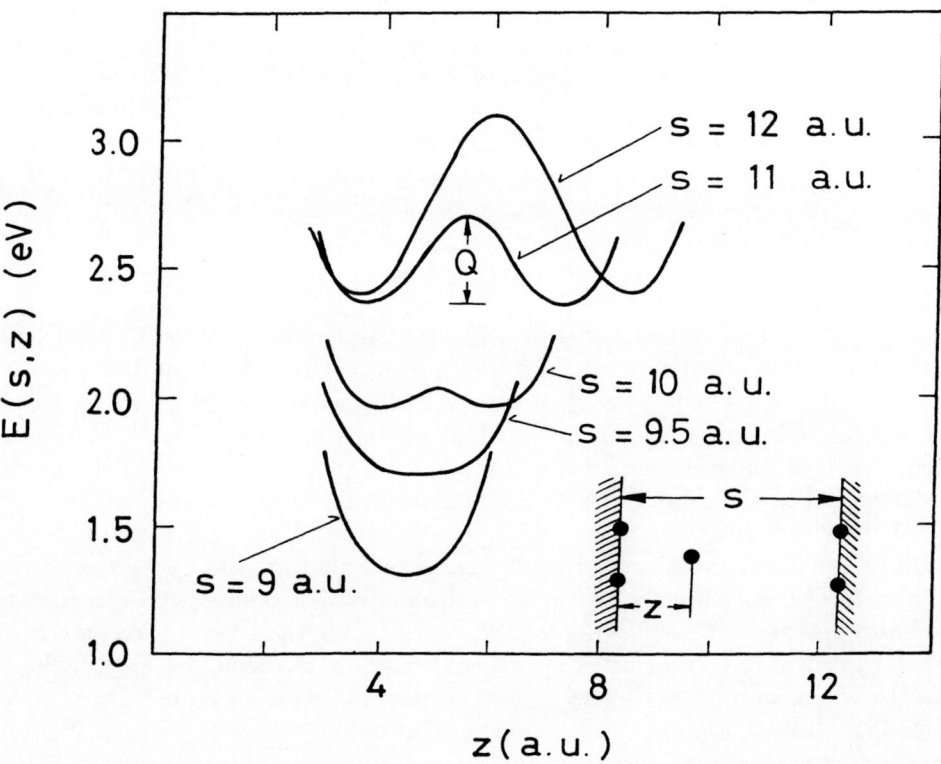

Figure 1. Variation of the interaction energy $E(s,z)$ of an Al atom between two Al(001) slabs with its distance, z, from the left slab and with the separation of Al slabs, s, as described in the inset. The Al atom faces the hollow sites of both slabs.

$E(s,z)$ rises to large values in the repulsive range, but becomes attractive with increasing z and passes through a minimum and rises again to form a double well potential. The barrier Q between two wells is large and approximately equal to the binding energy E_b if s is large. Because interaction with, as well as between both electrodes are included, the actual interaction energy at small s deviates from the superposition of interaction energies $E_{i,1}(z) + E_{i,2}(s-z)$. For small separations we therefore calculated $E(s,z)$ as a function of z in a supercell with lateral (3x3) periodicity of the Al atom facing the hollow (H) site of two Al(001) surfaces [9]. Figure 1 illustrates the evolution of $E(s,z)$ curves. Since the position of Al atom relative to both Al(001) surfaces is identical, the resulting $E(s,z)$ curve has a reflection symmetry with respect to a plane bisecting s. Consequently, Al atom has the same energy in both wells. Since our discussion covers motion of atoms on the surface of one of the electrodes (lateral motion in the xy-plane), as well as between electrodes (perpendicular motion along z-direction) one has to go beyond the $E(s,z)$ curves and consider also the interaction energy surface $E(s,\mathbf{r})$. We note that the surface of $E(s,\mathbf{r})$ varies also with the relative displacement of electrodes. The evolution of the interaction energy curve with s and z as illustrated in Fig. 1, and also with the relative displacement of electrodes display a number of interesting features, which bear on the controlled motion of the atom. In what follows we discuss the perpendicular and lateral motions of the atom with reference to these features.

3.1. PERPENDICULAR MOTION OF THE ATOM

As seen in Fig. 1, the barrier Q (Q_+ and Q_- for an asymmetric double well structure) is high for large s and close to the binding energy E_b of the atom, but decreases rapidly with decreasing separation. For $s=11$ a.u. it is already reduced to a small fraction of E_b, and completely disappeared at $s=9.5$ a.u. transforming the double well structure into a single, shallow minimum. At the separation for which the barrier Q collapses the distance of the Al atom from either electrode is larger than the distance if this atom were bound to a single electrode. Upon further approach of the electrodes towards equilibrium the minimum gets sharper and hence the binding energy gets larger. Apparently, the atom is stable in the middle with considerably larger binding energy with separation significantly larger than twice the corresponding quantities with only one electrode present. This means that the atom gains an additional stabilization energy between two electrodes. Although the $E(s,z)$ curves shown in Fig. 1 are symmetric, the difference between Q_+ and Q_- comes from the asymmetry of the interaction energy and will generally occur owing to several reasons: First of all the tip and sample are usually made of different materials, and the tip is usually chosen to be a hard material like W. Even if the tip and sample consists of, or coated with the same material, their shapes are different. The $E(s,z)$ curve shown in Fig. 1 and the corresponding energy surface $E(s,\mathbf{r})$ becomes also asymmetric when one of the electrodes is laterally shifted. As s of an asymmetric $E(s,z)$ decreases the shallow minimum first disappears by changing into a flat plateau connected to the remaining relatively deeper minimum. Note that the shallow minimum is usually associated with the top site in a chemical bond, and may correspond to a saddle point. The asymmetry imposed on the form of the energy surface by an externally applied electric field will be discussed later in this section.

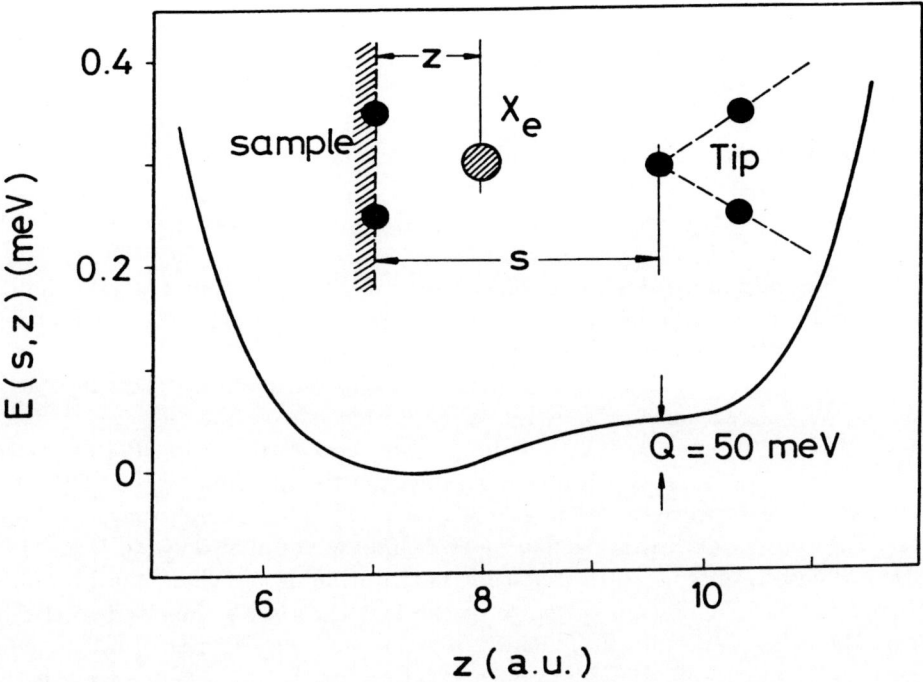

Figure 2. Variation of the interaction energy $E(s, z)$ of a Xe atom between the Al(001) slab (sample) and sharp Al tip. The inset describes the atomic configuration. Q is the barrier calculated for the Xe atom to transfer from the sample surface to the tip.

In the presence of Q the atom can hope from one well to the other at finite temperature with a rate $\nu \exp(-Q_\pm/k_B T)$. The lower is the barrier the higher is the rate in this thermal process. Atom transfer via tunneling through the barrier can in principle also contribute at sufficiently low temperature, but such a process can be meaningful only for very low barriers [10]. In the case of an symmetric interaction energy surface $E(s, \mathbf{r})$ the time spent by an atom in the deeper minimum will be much larger, and the probability of the atom transfer towards deeper minimum will be much larger. This determines the directionality of transfer in the absence of any kind of external agent, and underlies the wetting mechanism. According to this picture atoms of one electrode is transferred and hence wets the other electrode if their binding with the latter is stronger.

External agents can reduce the barrier towards the desired direction either by emphasizing the asymmetry of $E(s, \mathbf{r})$ or by exciting the atom and hence by raising its energy. For instance as in the field desorption [11] the quasi symmetric positions of energy minima can be momentarily modified in such a way that the probability of the atom transfer to one side will be enhanced. An external field of ~ 1 V/Å can achieve this since it can easily penetrate not only the apex, but also the top layer of a metal sample [11]. In some case a negatively (positively) charged atom move against (along) the direction of the field and hence leads to a reversible transfer [6,12]. An atom bound to one of the well in the double well structure becomes vibrationally excited (or heated) by the high current and moves along the direction of the field inducing this current level [8,13,14].

As a specific example to the reversible perpendicular motion of the atom let us consider the reversible transfer of Xe atom between Ni(110) sample and tungsten STM tip [8]. To simulate this experiment we carried out SCF pseudopotential calculations of Xe atom between Al(100) surface and Al tip having a single atom at the apex [15]. In the resulting interaction energy curve illustrated in Fig. 2 only the effect of weak chemical bond between the metal surface and Xe atom is taken into account, but the VdW interaction is not included. For this reason the minimum of $E(s, z)$ occurs at the H-site. For $s = 17$ a.u., which is comparable to the spacing estimated by Eigler et al. [8]. We found $Q_+ \simeq 50$ meV at the sample site, but $Q_\sim 0$ at the tip site implying that Xe atom becomes unstable at the tip site. Increasing s is, however, resulted in a double well structure having two shallow minima um with $Q_+ - Q_- \simeq 40$ meV [15]. These results indicate that the contribution of the chemical interaction to the physisorption is significant, in spite of the fact that it was omitted in the earlier studies. By using parabolic fit to the potential well we estimate phonon energies $\hbar\omega_+ \simeq 3$ meV, which can allow phonon excitation for the adsorbed Xe atom.

3.2. LATERAL MOTION OF THE ATOM

The lateral component of the forces on the atom exerted by the tip and sample are responsible from the lateral motion. Initially, in the presence of the weak tip interaction, the atom (say Xe) is bound to the sample and hence occupy the sample side of the double well structure. Normally, the tip approaching the atom changes the double well into a single, flat minimum as illustrated in Fig. 1. This is the appropriate situation to drag or to push the atom on the surface. Lateral motion of the atom (or lateral motion of the minimum of $E(s, \mathbf{r})$ correlated with the displacement of the tip) depends not only on the material and structural parameters of the electrodes but also on the character of the moving atom. The limitations in the lateral motion can be better understood by the analysis of the lateral forces acting on the atom. By assuming the additivity of forces we consider the lateral forces exerted by the sample

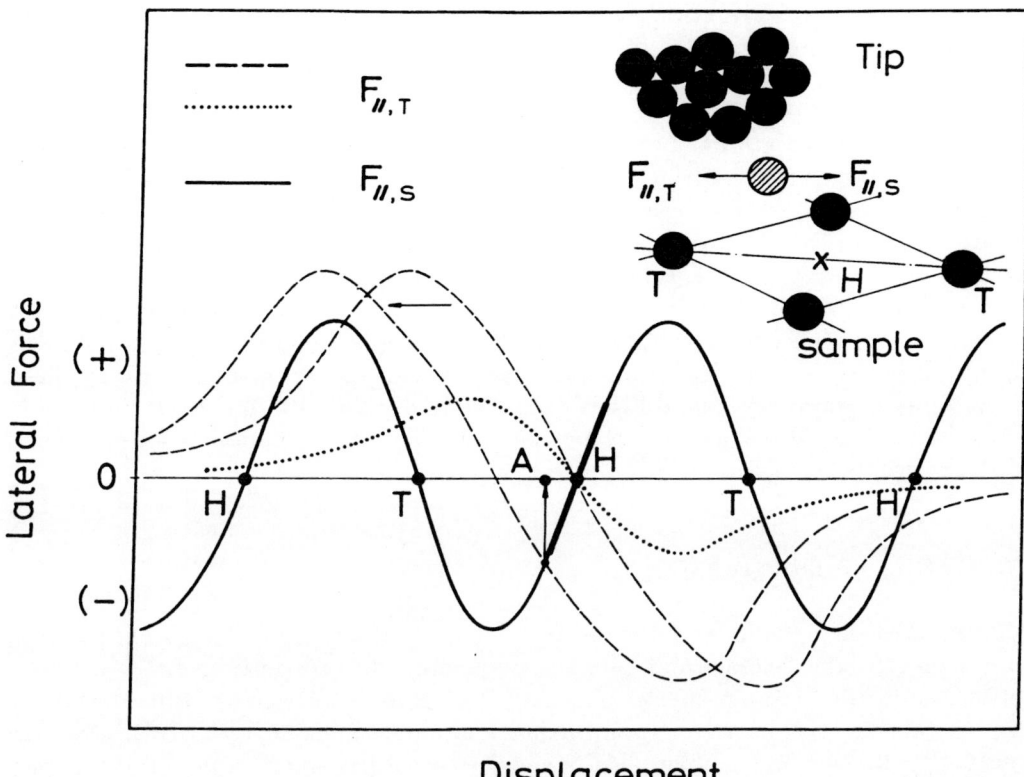

Figure 3. Schematic description for the variation of the lateral forces on an atom moving on the diagonal connecting the hollow (H) and top (T) sites. $F_{\|,S}$ and $F_{\|,T}$ are the lateral forces exerted by the sample and tip, respectively, as described by the inset. The position A of the atom is set by the intersection of the $F_{\|,S}$ curve with the attractive $F_{\|,T}$ curve which is displaced with the tip.

$F_{\|,S}$ and by the tip $F_{\|,T}$, separately. The addition of the force between the atom and sample to that between the atom and tip (which are calculated independently) is justified if the proximity of the tip does not induce massive charge rearrangement. This is the case for an inert gas atom between sample and tip. The variation of $F_{\|,S}$ with the position of the atom is periodic as schematically described in Fig. 3. The force $F_{\|,T}$ is directed always towards the apex of the tip in the attractive range, but decays to zero far away from the tip. In a slow motion of the tip in the attractive range, the position of the atom between tip and sample surface is the intersection of $F_{\parallel,S}$ curve with the displaced $F_{\parallel,T}$ curve. The form of $F_{\parallel,T}$ determines whether the atom moves with the moving tip or it oscillates around its initial equilibrium position. The former situation occurs when $F_{\parallel,T}$ is significant, i.e. the amplitude of the $F_{\parallel,T}$ curve exceeds that of the $F_{\parallel,S}$ curve. As the atom moves with the tip it can periodically flick back and fort around the apex of the tip. Depending upon the slope of the $F_{\parallel,T}$ curve, this motion occurs either continuously or discontinuously. In the discontinuous motion the atom jumps from back to front executing a stick-slip motion. If the $F_{\parallel,T}$ force is weak, the atom restores its original position on the sample surface after the tip passed over it. These arguments point to the fact that the lateral motion of the atom requires significant lateral force exerted by the tip. If the corrugation of $E(s, \mathbf{r})$ is large as in the case of atoms engaging in strong chemisorption bond with the surface of the sample the lateral motion of the atom requires strong forces. This can be achieved only by small tip-sample separation which produces strongly repulsive lateral forces on the atom. While the atom is pushed by these strongly repulsive forces, instabilities at the saddle points of $E_{(s}, \mathbf{r})$ or irreversible deformations can take place. On the other hand the corrugation of $E(s, \mathbf{r})$ is weak for the physisorbed atom, and its controlled motion can easily be achieved.

3.3. RESONANT TUNNELING

SCF calculations [3] show that the electronic potential barrier collapses for $s \simeq 9\text{-}10$ a.u.. However, the lateral confinement of the current transporting states between tip and sample induces an effective barrier. Therefore, electrons continue to tunnel through this effective barrier. Upon further approach of the tip the effective barrier eventually collapses and a channel is opened between tip and sample. This way the character of the transport changes from tunneling to ballistic regime, and the value of conductance becomes close to the quantum conductance $2e^2/h$. An atom between tip and sample causes the (effective) potential barrier to lower locally so that a situation reminiscent of the resonant tunneling double well structure can occur. The states of the atom between two electrodes, which are subject to a shift, can be viewed as the states bound to this well. If one of these empty states coincides with the Fermi level of the electrodes a resonant tunneling condition is realized and hence the value of conductance can jump to $\sim 2e^2/h$ before the first conduction channel is opened. This gives rise to a negative differential resistance condition.

A similar situation was found earlier for a local widening of the constriction between two 2D electron gas reservoirs [16], where a resonance peak appeared on the plateau before the threshold of a new channel. The resonance condition was also found for scattering by an impurity in a 1D mesoscopic channel [16]. Negative differential resistance was observed by Lyo and Avouris [6] for an STM tip over the specific sites of the boron exposed Si(111) surface.

4. Conclusions

It is shown that the interaction energy surface of an atom between two electrodes as a function of the relative position of electrodes is essential to understand the controlled motion of this atom. The form of this surface can be influenced by an external agent such as electric field and tunneling electrons scattered from the atom. The corrugation of the interaction energy surface for a chemisorbed atom is large. The lateral motion of this atom requires strongly repulsive force induced by the tip. On the other hand, the potential barrier to move a physisorbed atom is usually small. The controlled lateral motion of this atom can be achieved at separations without causing irreversible damage either to the sample or to the tip.

References

[1] Ciraci, S. and Batra, I. P. (1987) Phys. Rev. **B 36** 6194; Tekman, E. and Ciraci, S. (1989) Phys. Rev. **B 40** 10286; For a discussion of recent work in this area, see for example: Ciraci, S. (1990) *Tip-Surface Interactions*, in R. J. Behm, N. Garcia, and H. Rohrer (eds), Basic Concepts and Applications of Scanning Microscopy and Related Techniques, Kluwer Academic Publishers, Amsterdam, pp. 119.
[2] Lang, N. D. (1987) Phys. Rev. **B 36** 8173; *ibid* (1988) **B 37** 10395.
[3] Ciraci, S., Baratoff, A. and Batra I. P. (1990) **B 41** 2793; (1990) **B 42** 7168.
[4] Pethica, J. B. and Sutton, A. P. (1988) J. Vac. Sci. Technol. **A 6** 2490; (1990) J. Phys. Condensed Matter 2 5317.
[5] Landman, U., Luedtke, W. D., Burnham, N. A. and Colton, R. J. (1990) Science **248** 454.
[6] Lyo, I. W. and Avouris, P. (1989) Science **245** 1369; (1991) Science **253** 173.
[7] Eigler, D. M. and Schweizer, E. K. (1990) Nature **344** 524.
[8] Eigler, D. M., Lutz C. P. and Rudge, W. E. (1991) Nature **352** 600; Science **253** 173.
[9] Ciraci, S., Tekman, E., Baratoff, A., and Batra, I. P. (1992) B46 xxxx.
[10] Gomer, R. (1986) IBM J. Res. Develop. **30** 426.
[11] Kreuzer H. J., Wang, L. C. and Lang, N. D. (1992) Phys. Rev. **B 45** 12050.
[12] Mamin, H. J. Guethner, P. H. and Rugar, D. (1990) Phys. Rev. Lett. **65** 2418.
[13] Persson, B. N. J. and Demuth, J. E. (1986) Solid State Commun. **57** 769.
[14] Walkup, R. E., Newns, D. M. and Avouris, Ph. (paper in this proceeding).
[15] Ciraci, S. and Baratoff, A. (to be published).
[16] Tekman, E. and Ciraci, S. (1989) Phys. Rev. **B 40** 8559; Tekman, E. and Ciraci, S. (1990) Phys. Rev. **B 42** 9098.

FIELD EMISSION FROM SINGLE-ATOM PROTRUSION TIPS: ELECTRON SPECTROSCOPY AND LOCAL HEATING

Vu Thien Binh [a], S.T. Purcell [a], G. Gardet [a], and N. Garcia [b]

[a] Département de Physique des Matériaux, UA CNRS
Université Claude Bernard Lyon 1. 69622. Villeurbanne. France
[b] Departamento de Fisica de la Materia Condensada,
Universidad Autonoma de Madrid. 28049. Madrid. Spain

ABSTRACT. The total energy distributions of field-emitted electrons from a single-atom on top of nano-scale protrusions show the existence of a band-structure at the tip apex. The energy exchange at the tip apex during electron emission must take into account this band-structure and this can lead to an important heating of the nano-scale cathode even at very low emission currents.

1. Introduction

Well-defined field emission tips ending with a single-atom are of current interest for producing self-collimated, coherent electron beams [1,2]. A knowledge of the total energy distributions (TED's) of the field-emitted electrons is needed for an understanding of the physics of field emission (FE) from one atom including the electronic structure, the tunneling barrier and the energy exchanges during emisssion. TED's from single-atom protrusion tips may help to explain, for example, the high value of the tunneling barrier found in our former work [3]. They are also important for the detailed analysis of scanning tunneling spectroscopy (STS) because tunneling characteristics vary with the electronic structure of both the tip apex and the sample surface [4]. The results may also help to answer questions about the local heating during STM deposition process [5,6] and the stability and emission properties of single atom tips for its use as field electron sources of high brightness in applications such as electron microscopy [7,8,9].

2. Experimental set-up

The single-atom protrusion tips were obtained in ultra-high vacuum ($\sim 5 \times 10^{-11}$ Torr), in-situ, after thermal sharpening and cleaning at ~3200 K, and by the combined action of field and temperature using the field-surface-melting process [3]. The tip used in this study was a W<111> single-crystal, spot-welded to a loop which allowed control of the tip temperature by Joule heating. The protrusion was formed along the <111> axis of the macroscopic substrate tip. A schematic of the apparatus used for measurement of the TED's is shown in fig. 1. A FE screen was placed at ~3 cm from the tip with a probe hole of diameter 0.8 mm in its center. An electron energy analyzer was placed behind the probe hole. We used a hemispherical energy analyzer (VG-Clam 2) in which we had modified the entrance lenses to adapt it for the measurement of TED's. The tip was held on a mechanical

P. Avouris (ed.), Atomic and Nanometer-Scale Modification of Materials: Fundamentals and Applications, 121–131.
© 1993 Kluwer Academic Publishers.

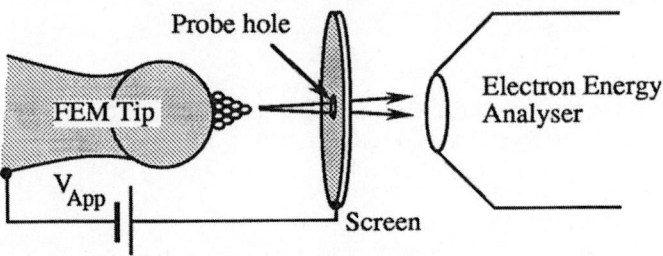

Figure 1. Schematic drawing of the experimental FEES setup.

movement that allowed the choice of the analyzed region of the tip. The cleanness of the macroscopic tip was controlled by FE Microscopy (FEM). FEM observations of the protrusion tip showed only FE from the protrusion zone, which created a FE spot of < 2 mm on the screen. No other pattern out of this spot is observed over a large range of V_{App}, in agreement with the field confinement behavior for a protruding geometry [10]. FE currents were determined by a combination of total current measurements and counting rates in the analyzer.

3. Field Electron Emission Spectroscopy

The measure of TED's is termed Field Emission Electron Spectroscopy (FEES). FEES has been used previously to study the electronic states of individual surface atoms and molecules chemisorbed on the metallic surfaces of tips [11]. For macroscopic tips, with radius greater than 100 nm, the area sampled by FEES has a diameter of more than 20 Å [12]. Recently, the controlled formation of nanoscale protrusions on the tops of macroscopic tips has been obtained by the field-surface-melting-mechanism [3]. These protrusions are fabricated by the application of field and temperature under ultra-high vacuum (UHV) conditions. They have been shown by field ion microscopy to have a conical shape with 2-3 nm height ending in one atom. The sharply pointed geometry of the protrusions strongly enhances the applied field over the topmost atom [13] which then permits all of the field emission (FE) to come from only one atom. The emitted electrons are then self-collimated to form a beam of ~4° and the Field Emission Microscopy (FEM) pattern of such a single-atom tip consists only of a single spot [3]. FEES measurements from such a spot then allow the determination of the TED of the electrons coming from one atom. In this article we describe FEES studies of both macroscopic and single-atom tips formed on a W<111> single crystal.

3.1. EXPERIMENTAL PROCEDURE AND RESULTS

The experimental procedure was as follows: (i) *in situ* fabrication of a clean macroscopic tip of less than 100 nm radius; (ii) FEES measurements of this tip centered on the (111) region; (iii) fabrication of a single-atom protrusion on top of this tip; (iv) FEES study from this protrusion; (v) destruction of this protrusion by a controlled heating of the tip and FEES measurements of the resulting macroscopic tip at the (111) region. All measurements were performed at room temperature.

The TED spectra recorded for macroscopic tips during steps (ii) and (v) had the same shape, were located at the same Fermi level (E_F), and showed the well-known behaviors for clean W<111> tips [14]. Two example spectra, measured after the destruction of the protrusion, for different values of the applied voltage are presented in fig.2. Basically, one strong peak was observed with a sharp edge at E_F for any V_{App}. An increase of V_{App} did not change the position of the Fermi edge, only causing a broadening of the peak on the low energy side. To fit the spectra we have used the classic equation for the tunneling current from a free electron metal which was developed by Young [15]. Excellent agreement between the experimental data and theory was found, as is shown in fig.2.

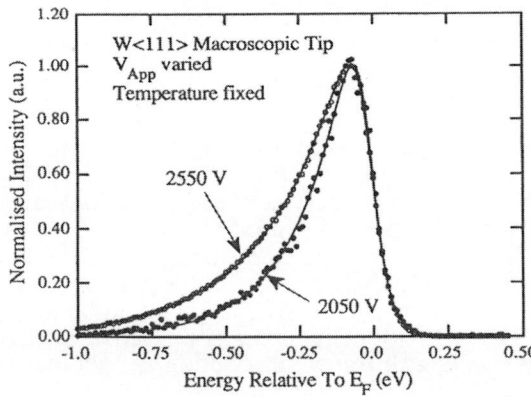

Figure 2. TED Spectra of a macroscopic tip for two V_{App} showing a broadening of the low energy side.

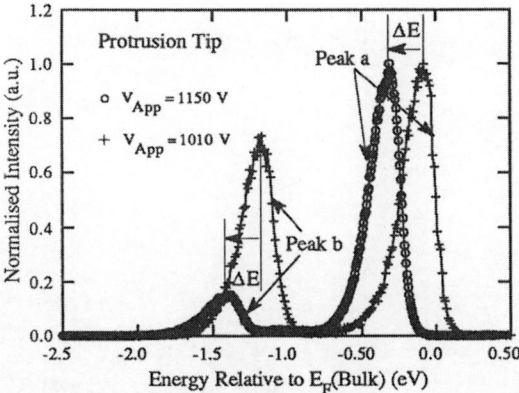

Figure 3. TED spectra of single-atom protrusion tip for two different V_{App} with the corresponding shift.

The fits are represented by the solid curves over the experimental points. The value of E_F for macroscopic tips was constant and was then taken as the reference level for all the spectra. The measured FWHM's for the spectra in fig.2 were 0.23 eV and 0.30 eV for 2050 V and 2550 V respectively, a variation which is a consequence of the increase of the slope of the tunnel barrier. The TED spectra were recorded for different V_{App} after formation of a protrusion on top of the macroscopic tip. Characteristic examples of the observed spectra are shown in figs. 3. Four salient features, that are not present for macroscopic tips, can be discerned in the experimental observations:

- (1) The spectra are composed solely of well-separated peaks. In the presented example of fig. 3. there are two peaks, labelled (a) and (b). Peak (a) is near E_F and peak (b) ~1.1 eV lower in energy. In general, the number of peaks and their relative intensities depends on the protrusion geometry and on the V_{App}.

- (2) A linear shift of the whole spectrum is observed as a function of V_{App}. This is illustrated in fig. 4 where we plot the positions of the peak maxima as a function of V_{App}. All the data fall on parallel lines, with slopes of 1.65±0.02 meV per applied Volt, showing that the separation between the peaks remains constant. The total shift of the peaks for the range of V_{App} in this experiment was ~0.7 eV. Note that no shift was detectable for macroscopic tips and for similar changes of V_{App}.

- (3) None of these peaks could be fitted satisfactorily by the Young equation [15]. For the same FWHM, the spectra from the macroscopic tip have wider tails on the low energy sides and sharper maxima than the peaks from the protrusion.

- (4) The shapes of the peaks vary little with V_{App}. The FWHM's remain ~0.24 eV for all values of V_{App}.

The shifts and the intensity variations of the peaks were reversible; they could be varied reproducibly by changing V_{App}. For clean single-atom protrusion tips, stable FEM patterns and FEES spectra with constant FE intensities were observed over a period of many hours. Presence of adsorption from the background gases on top of the protrusion was easily characterized by large instabilities of the emission current and strong modifications of the TED's. The adsorption could be removed by application of field and temperature after which the emission properties were again stable. The FEES spectra could be the same as before the cleaning or show changes in the positions, number and relative intensities of peaks for the same V_{App}, probably reflecting a different structure in the geometry of the protrusion.

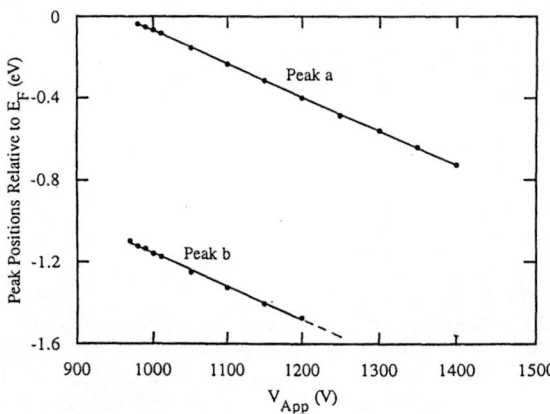

Figure 4. Positions of the peaks of fig. 3 versus V_{App}. The slope of the lines are 1.65 ± 0.2 meV/V

3.2. DISCUSSION

Single-atom TED spectra show well-separated peaks and also that the electrons do not tunnel directly from the Fermi level to the vacuum. These strongly suggest the presence of a localized band structure or more complex energy broadening of the atomic levels. The peaks in the TED spectra are then explained by FE resonant tunneling only through these bands. To take into account the band structure, we multiplied the equation for free electron metal by a Lorentzian distribution. We then found very good agreement between the experimental TED spectra and the fits. This is illustrated in fig.3 where the fits are shown as the solid curves over the experimental points. This result supports the hypothesis of a local band structure on the topmost atom of the protrusion which in turn explains the peaks in the TED spectra and the constant band-gap observed experimentally. Therefore, the tunneling equations through single atoms must have a different functional form than the standard F-N models for a plane [15] or for non-planar metal emitters [16].

Resonant tunneling through atomic energy levels of adsorbed atoms, which have been broadened due to interaction with the underneath surface, has first been introduced by Duke and Alferieff [17] and later developed more fully by Gadzuk [18]. This was used to explain the small bumps added to the energy distributions of the clean macroscopic tips observed in earlier FEES experiments [19] with chemisorbed atoms on metallic surfaces. It must be emphasized that the presence of chemisorbed atoms in these experiments only slightly modified the standard peak of a clean macroscopic tip, in contrast to the spectra from the protrusions which consist solely of well-separated peaks. This latter behavior could have its origin in the atomic size and shape of single-atom protrusion tips, and in particular on the reduced coordination number of the atoms that constitute the apex compared to a single atom on the surface.

The shifts of the peaks also run counter to a metallic behavior of the topmost atom. The linearity of the shifts versus V_{App} (fig.4) means linear shifts versus applied field F at the cathode surface because $F = \beta V_{App}$. β is a geometrical factor that depends on the tip shape [12]. We propose to explain this shift and its linearity versus the applied voltage by a charge confinement in the region of the topmost atom, which implies a field penetration into the tip.

The charge confinement and the penetration of the field can be estimated by the Thomas-Fermi model of screening [20,12]. To our knowledge, there are no calculations which give the screening for a conical protrusion geometry. Therefore, to estimate field penetration x_0 for the protrusion we use, as a first approximation, the expression for the potential of the electric field penetration into a flat subsurface region:

$$V_F = x_0 F \exp(-\frac{r}{x_0}) \tag{1}$$

where r is the distance from the surface to a position within the cathode. Thus the energy of the emitted electrons varies linearly with V_{App} as:

$$\Delta E = e x_0 \beta V_{App}. \tag{2}$$

by taking into account the relation between F and V_{App}. e is the electron charge.

Applying eq. (2) to our experimental results of $\Delta E / V_{App} = 1.65$ meV/V and taking β as 5 to 10×10^6 m^{-1} for a protrusion of 2 to 3 nm height [10], gives x_0 of 2-3 Å for the single-atom protrusion. This value should be compared to the screening length of a metal surface.which is less than 0.5 Å [20] and also with the estimation of field shift with single adsorbed Ba which is 1.3 to 1.7 Å [19]. It is also roughly the dimension of an atom and this strongly supports the idea that the observed peaks in the TED spectra are related to localized levels at the topmost atom. This estimate of x_0 is also in agreement with the fact that we can field evaporate the protrusion atom by atom [3].

To illustrate the above hypothesis, a model for tunneling from the single atom protrusion tip is presented in fig.5, which shows the one-dimensional potential energy diagram in the presence of an applied field. The band structure, the field penetration, the band-shifting and the TED for field emission are all depicted. Many of these features are similar to a semiconductor, except they pertain only to the topmost atom of the protrusion.

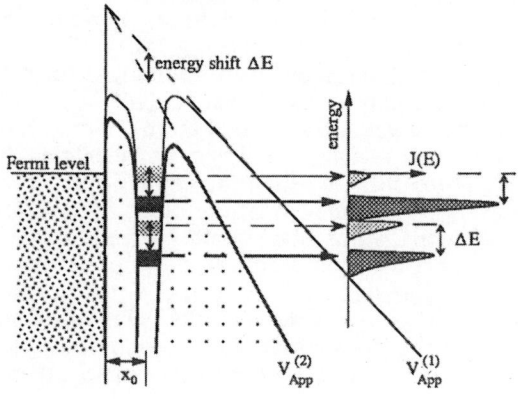

Figure 5. Simplified model of FE from single-atom tips for two V_{App}. The lightly-shaded bands signify the positions of two bands for $V_{App}^{(1)}$ and the two darkly-shaded bands are the same bands after they were shifted by increasing V_{App}.

4. Atomic scale temperature measurement

4.1. NOTTINGHAM EFFECT

During the field electron emission (FE) process, energy exchanges take place between the emitted electrons and the cathode surface. These exchange processes, or so-called Nottingham effect [21], are important in determining the local temperature at the emitter surface. In macroscopic metallic tips, the electrons which replace the emitted electrons are assumed to come from near the bulk Fermi energy (E_F). When all the energy states above E_F are empty, all emitted electrons have less energy than E_F. The average energy of the emitted electrons is then less than that of the replacement electrons and the energy exchange, ΔE_1, then necessarily produces heating of the cathode. When energy states above E_F become populated, the average energy of replacement electrons is lower than that of emitted electrons from these levels, and the energy exchange, ΔE_2, tends to cool the cathode. The heating or cooling of the cathode is governed by the balance between $n_1.\Delta E_1$ and $n_2.\Delta E_2$, where n_1 and n_2 are the replacement rates for each process. For macroscopic tips the temperature increases caused by FE are negligible for FE current < 1 μA [22,23].

It was shown in the previous section that the TED's of single-atom protrusion tips are drastically changed from those of macroscopic tips. Referring to (Fig.5), we see that the standard Nottingham effect mentioned above (process 1) still occurs within each band, but there are now two additional energy exchange mechanisms. The replacement electrons can now come from the higher energy levels of Fermi sea of the underneath substrate to fill the localized levels in each band that are emptied by emission (process 2) and also from the upper localized band levels to the lower (process 3). In both processes 2 & 3, energy is lost by the replacement electrons. The amount depends strongly on the number and position of the bands with respect to E_F, which also means on the protrusion geometry and the V_{App}. Moreover, since the energy exchange per emitted electron of processes 2 & 3 could have values of the order of eV, this can lead to a larger increase of the temperature at the single-atom apex compared to the standard Nottingham effect.

To check the above idea, we have set up experimental procedures to measure the temperature increases at the apex during the FE through single-atom protrusion tips. The experimental problem in the studies of such temperature increases was how to measure the local temperature at the apex of a single-atom protrusion tip during FE. This temperature may be very different from the temperature of the whole tip because of the very small emitting area. It is necessary to have a local probe of the temperature giving atomic-scale resolution. Our determination of the local temperatures was based on two effects. The first effect is that a repetitive, back-and-forth motion of a single atom between neighboring atomic sites can exist at the atomic-scale apex of a protrusion tip whose frequency is temperature dependant. The second effect is that the shape of the TED's from the protrusion tips depends on temperature. Using procedures based on these effects, we have determined the local temperatures at the apex of a nanoprotrusion for electron emissions in the 10^{-13} to 10^{-9} A range, with the whole tip at room temperature.

4.2. EXPERIMENTAL RESULTS FOR LOW CURRENTS: "FLIP-FLOP" METHOD

The first procedure we have used is based on the fluctuation method introduced by Gomer [24]. The emission current fluctuates due to the back and forth movement, or "flip-flop", of the adsorbed atom between two neighboring sites. The flip-flop frequency is directly related to the local temperature. Thus flip-flop frequency can be used to determine the temperature changes during FE. Based on this idea we have adapted the flip-flop method to measure the temperature at the apex of the single-atom protrusion tips. In the case of Gomer the current fluctuations had to be treated statistically because they were due to the random movement of a large number of atoms. For single-atom protrusion tips, the FE comes from the topmost atom at the apex and the flip-flop then causes changes in the FE characteristics which are related only to the movement of an individual atom. During the flip-flop, the total FE current switches between two discrete values [2] which remain constant and depend on the atomic configuration of the protrusion. Their difference is typically 5 to 10 % of the total current. Moreover, each current level is associated with a

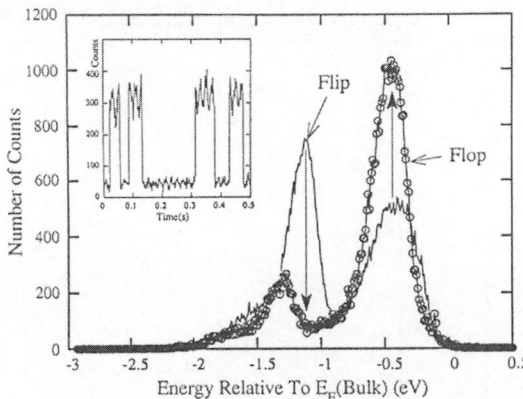

Figure 6. Effect of a flip-flop process on a TED from a single atom tip. Inset represents the change in counts for one energy.

particular TED. An example of two TED's measured during the two states of a flip-flop are shown in Fig. 6. The number of peaks and their relative positions are preserved during the flip-flop but the TED's shift as a whole and the relative peak intensities change. The switching between the two spectra is repetitive as long as the flip-flop continues. A consequence of the TED changes is that if the emission current is measured for one chosen energy during the flip-flop process, its variation can be very large even if the total current variation is only of a few percent. This phenomena allows very easy detection of the flip-flop even for total FE currents from the single-atom tips in the range of 10^{-13} A. This is illustrated by the inset in Fig.6 on which we show the variation of the number of counts at a fixed energy during a flip-flop process. This technique is then used to measure the flip-flop frequency.

We have used the variation of the frequency of the flip-flop of an adsorbed atom coming from the residual gas to determine the temperature at the apex of single-atom protrusion tips. The experimental procedure for the determination of the temperature is the following:

1) The first step was the calibration of the heating loop. The local temperature of the FE area of a macroscopic tip could be deduced from the experimental TED spectra, and in particular for the high energy side, by fitting them to the tunneling equation given by Young [15]. We have checked the validity of this method for two fixed temperatures (liquid nitrogen and room temperature) and for different temperatures in the range of 1000 K to 1600 K with an optical micro-pyrometer. The temperatures from the TED's agreed to the known temperatures to better than 5%. This agreement allowed us then to calibrate the heating loop current with the local temperature at the apex of a macroscopic tip for the range from 150 K to 1600 K.

2) A clean protrusion tip was then fabricated from the macroscopic tip and we waited until a flip-flop process appears, due to an adsorbed atom on the apex of the protrusion. The flip-flop frequency was then measured versus the temperature of the tip as controlled by the heating loop, at fixed V_{App}. It followed an Arrhenius function and the deduced activation energies for several flip-flop processes (~0.39 eV) were in rough agreement with former results given by Gomer for O/W (~0,55 eV). This agreement supported the validity for the calibration of the variation of the flip-flop frequency with the local temperature.

3) The last step for this experiment is the measurement of the FE current at fixed energy as a function of time of a stable flip-flop and for different values of V_{App}. We observed an increase of the flip-flop frequency with increasing V_{App} (fig.7a) and by using the calibration values of steps 1 & 2 the variation of the local temperature at the apex of the protrusion is then deduced as a function of V_{App} (fig.7b). It is of ~ 30 K for V_{App} increase from 950 V to 1070 V, which corresponds to measured total currents from the protrusion apex respectively of ~$3x10^{-13}$ to ~$9x10^{-12}$ A.

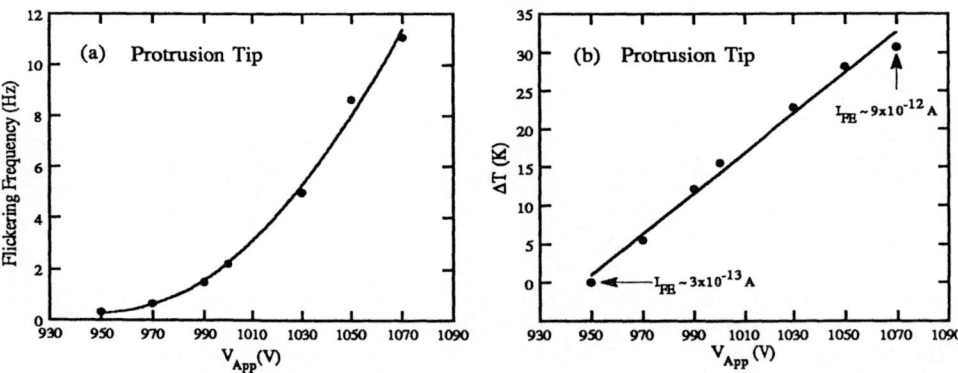

Figure 7. (a) Frequency of flip-flop vs V_{App}. (b) Temperature variation induced by FE from single-atom tip

4.3. EXPERIMENTAL RESULTS FOR HIGH CURRENTS: TED METHOD

A too large temperature causes an instability of the adsorbed atom on the protrusion. Thus, for larger temperature increases, we use another method which is based on the fitting of the experimental TED spectra by an equation derived from the classic tunneling equation [15]. This method has been set up from the following experimental observation: the shape of the peaks in the TED from single atom protrusion tips is temperature dependant. In fig.8(a) we show the shape variation of the spectrum of a single-atom protrusion tip obtained for one value of V_{App} and two different controlled temperatures. It shows a broadening of the high energy edge of the spectrum with the temperature. The temperatures found from the TED's are in agreement with the known values from the calibrated heating loop. We have used this phenomenon to measure the local temperature at the apex of single-atom protrusion tips for different V_{App}. In fig.8(b) we show that, concomitant to the shift of the spectra which is characteristic of FE from single-atom protrusion tip, there is a broadening of the high energy side of the spectra as a function of V_{App}. This indicates an increase of the temperature at the apex due to FE. The local temperatures at the apex of the protrusion tip found by fitting the spectra for varying V_{App} are shown in Fig 9. For the range of applied

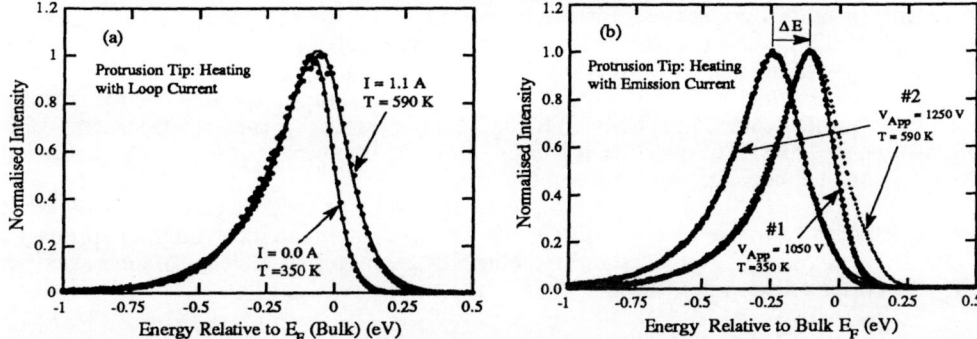

Figure 8. (a) Spectra from protrusion tip for room and 590 K. The higher temperature is created by loop heating current. (b) Spectra from the same protrusion tip for different applied voltages and different emission currents. The spectra at higher voltage, #2, shifts to lower energy due to field penetration. We have shifted it numerically by ΔE to the position of the lower voltage peak to show the broadening of the high energy side of the spectra that is caused by the temperature increase.

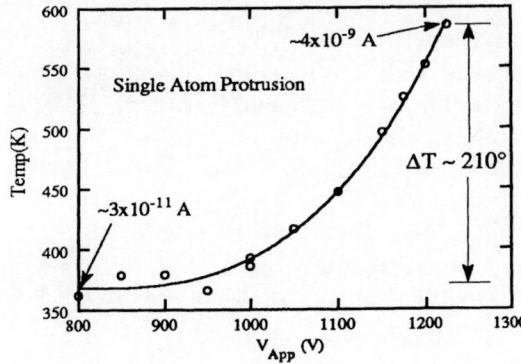

Figure 9. Changes in temperature versus V_{App} for single-atom tip

voltages of the experiment (800 V - 1220 V), we determined the temperature variations as a function of V_{App}. The corresponding FE current increase is from 3×10^{-11} to 4×10^{-9} A. The temperature increase can reach a value of ~210 K at 1220 V. The value of 210 K in the increase of the apex temperature is not a limiting value. For this experiment we have not increased V_{App} more in order to preserve the same protrusion geometry. For higher FE currents the temperature increase is indeed larger. The protrusion becomes unstable and it can be destroyed by a local melting.

5. Conclusions

We have shown here that the field emission through single atom tips has a the band-structure TED spectra. The spectra are field and temperature dependent. The existence of the peaks in the TED's is important for quantitative interpretations of scanning tunneling spectroscopy at the atomic level because the hypothesis that the electronic structure of the scanning tip is the same as a macroscopic tip is no longer valid in general. The band structure can also be the source of large energy exchanges between the emitted electrons and the cathode surface. If the tunneling current produces only minor increases in temperature for macroscopic tips, this effect cannot be ignored in the case of single-atom emitting tips. Such large temperature increases have now to be considered in experiments like FEM of adsorbed species (atoms or clusters) or STM configuration for tunneling or nano-deposition. Furthermore, we have shown that local temperature can be measured of an area with atomic scale. This is possible because the conical protrusion geometry of the nano-tips confines the field over the top-most atom of its apex. They also show that one can monitor the atomic switching between two sites.

In summary, these studies show that atomic-scale tips have specific characteristics which cannot be deduced from macroscopic studies. New theoretical and experimental approaches will have to be developed to understand the complex behavior of the electron emission and thermal effects revealed in this work.

ACKNOWLEDGEMENT. This research was sponsored by an EEC "Science" project and by French (MRT n° 90.S.0229) and Spanish scientific institutions. The authors would like to thank V. Semet his contribution.

REFERENCES

[1] N. Garcia and H. Rohrer, J. Phys. Cond. Matt. **1** (1989) 3737.
[2] J.J. Saenz, N. Garcia, Vu Thien Binh and H. De Raedt, in "Scanning Tunneling Microscopy and Related Methods" Eds. R.J. Behm, N. Garcia and H. Rohrer, NATO-ASI Series E Vol. 184. Kluwer Acad. Publ., The Netherlands, 1990.
[3] Vu Thien Binh and N. Garcia, J. de Phys. I **1** (1991) 605; and STM'91 Interlaken, Ultramicroscopy, in press (1992).
[4] R.J. Hamer, R.M. Tromp and J.E. Demuth, Phys. Rev. Lett. **56** (1986) 1972.
[5] P.F. Marella and R.F. Pease, Appl. Phys. Lett. 55 (1989) 2366.
[6] Y.Z. Li, L. Vazquez, R. Piner and R.P. Andres, Appl. Phys. Lett. 54 (1989) 1424; 55 (1989) 2367.

[7] T.H.P. Chang, D.P. Kern, M.A. McCord and L.P. Muray, J. Vac. Sci. Technol. B **9** (1990) 438; and J.C.H. Spence and W. Qian, Phys. Rev. B **45** (1992) .

[8] P. Grivet, "Electron Optics", Pergamon Press Ltd., Oxford, (1972).

[9] A.L. Blelock, M.G. Walls and A. Howie, EUREM 92, to be published.Grenada Spain.

[10] D. Atlan, G. Gardet, Vu Thien Binh, N. Garcia and J.J. Saenz; Ultramicroscopy STM'91 in press (1992).

[11] for a review see J.W. Gadzuk and E.W. Plummer, Rev. Mod. Phys. **45** (1973) 487.

[12] R. Gomer, "Field Emission and Field Ionization", Harvard Monographs in Applied Science 9, Cambridge, USA (1961).

[13] D.J. Rose, J. Appl. Phys. **27** (1956) 215; and ref. [10].

[14] C.E. Kuyatt and E.W. Plummer, Rev. Sci. Instru. **43** (1972) 108.

[15] R.D. Young, Phys. Rev. **113** (1959) 110.

[16] Jun He, P.H. Cutler and N.M. Miskovsky, Appl. Phys. Lett **59** (1991) 12.

[17] C.B. Duke and M.E. Alferieff, J. Chem. Phys. **46**, 923 (1967).

[18] J.W. Gadzuk, Phys. Rev. B **1** (1970) 2110.

[19] E.W. Plummer and R.D. Young, Phys. Rev. B **1** (1970) 2088.

[20] C. Kittel, Introduction to Solid State Physics (Wiley, New York, 1968).

[21] W.B. Nottingham, Phys. Rev. **59**, 908 (1941).

[22] L.W. Swanson, L.C. Crouser and F.M. Charbonnier, Phys. Rev. **151**, 327 (1966)

[23] L.W. Swanson and A.E. Bell, in Adv.Electr. Electron Phys. Vol. 32 ,pp. 194, (1973), Ed L. Marton, Acad. Press New York & London. (for review on energy exchange during FE)

[24] R. Gomer, review article on the flip-flop "Surface mobilities on solid materials" pp 7-62, Ed. Vu Thien Binh, NATO-ASI Series B Vol. 86 (1983) Plenum Press, New York.

FOUR-POINT RESISTANCE MEASUREMENTS OF WIRES WRITTEN WITH A SCANNING TUNNELING MICROSCOPE

A. L. DE LOZANNE, E. E. EHRICHS, and W. F. SMITH
Department of Physics
The University of Texas
Austin, TX 78712-1081, USA

ABSTRACT. We review briefly our work on using the Scanning Tunneling Microscope (STM) to break down organometallic precursor gases to form metallic lines. We are particularly interested in the transport properties of wires smaller than 100 nm. This requires four-probe measurements of the wires and therefore a good method for locating the probes on the substrate and for bringing the STM tip to the probes. We have built a UHV STM inside a commercial Scanning Electron Microscope in order to accomplish this. This instrument is briefly described here.

1. Introduction and Background

The STM is now becoming popular for the modification of surfaces down to the atomic scale[1,2]. The interest ranges from the basic mechanism of the process itself, to the fabrication of interesting structures without regard for the process. We are interested in the electronic transport properties of metallic wires narrower than 100 nm, which show novel effects like universal conductance fluctuations, magnetofingerprints, and flux quantization in rings[3]. The STM is for us a tool that can fabricate these wires with enough flexibility and at a reasonable cost.

One classification scheme for STM nanofabrication is to make a distinction between techniques that add atoms of a different kind to the surface from those that do not. Most of the techniques involving mechanical indentation[4-5], electrical pulses[6-9], surface melting[10], localized etching[11,12], and field evaporation[13,14] or desorption fall within the latter classification: a rearrangement or removal of substrate atoms. The former classification, namely adding atoms of a different kind, includes transfer of atoms from the tip to the sample by field emission/evaporation[15] and the method we describe here. The beautiful work of Eigler et al.[16] falls in between these two categories: the atoms are different from the substrate, but they are randomly adsorbed all over the surface; the STM moves them around to make atomic scale structures.

Our fabrication method is essentially a very localized version of Chemical Vapor Deposition (CVD), one of the workhorses of the semiconductor industry. This is one of the strenghts of this method, because the CVD industry has developed a precursor gas for most elements of interest. Furthermore, one can envision fabricating more complex structures by switching precursor gases in-situ to deposit different materials, in a similar way to what has been demonstrated with laser beams[17]. While standard CVD and laser CVD use heat to break the organometallic precursor gas, we use the electrons from the STM tip, with energies ranging from 5 to 40 eV.

P. Avouris (ed.), Atomic and Nanometer-Scale Modification of Materials: Fundamentals and Applications, 133–137.
© 1993 *Kluwer Academic Publishers.*

In our early work demonstrating the feasibility of this technique [18-21] we found that the fabricated structures had a large content of carbon, at least 40%[20]. This has also been observed by others using our technique[22,23]. In fact, our smallest structures were made with no gas at all: the surface contamination can be used to form structures down to 10nm in size. This work, while interesting, served only to motivate the design and construction of the STM/SEM briefly described below.

2. Description of the STM/SEM

The main features of our STM/SEM are the following:
1) Commercial SEM (JEOL Model JSM 820), with base pressure of 1×10^{-6} Torr (diffusion pumped)
2) STM base pressure of 1×10^{-10} Torr (differentially pumped with an ion pump)
3) Loadlock for tips and samples.
4) Homemade precision UHV tweezer [24]
5) Port available for in-situ Auger (future purchase)
6) Homemade UHV backscattered electron detector
7) Eight translational degrees of freedom:
   - 3 for standard STM scanning (20 µm range)
   - 3 for coarse tip-sample position (2 mm - 2cm range)
   - 2 for placing the tip-sample junction under the SEM beam

This instrument, shown in Fig. 1, has been described in more detail in our previous publications [24-26]. The best demonstration that all of the essential elements of this complex instrument work is to bring the tip to a unique location on the substrate under the guidance of the SEM, as shown in Fig. 2.

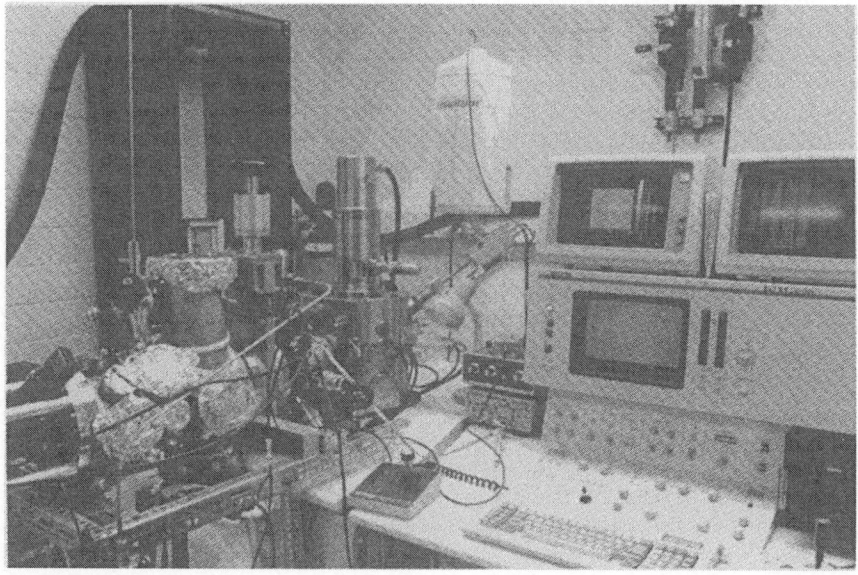

Figure 1 Photograph of the UHV STM chamber (left) plugged into the commercial SEM (right). The two monitors on top are for STM display and control.

3. Experimental Results

Figure 3 shows a line written using nickel carbonyl as the organic precursor. Auger measurements of larger deposits have shown that these deposits have at least 95% nickel. The pattern consists of a vertical wire, with voltage probes attached diagonally near the top and bottom. The left voltage probe appears discontinuous in the micrograph, however it was in fact electrically continuous, allowing a genuine four-probe measurement of the wire, as shown in fig. 4. The general behavior is qualitatively similar to that previously observed by a two-probe measurement [21] for larger and more contaminated aluminum deposits. As the temperature is lowered from room temperature, the resistance increases sharply. The observed characteristic temperature for the resistance rise is the same as for a blank test sample, indicating that the rise is due to freeze-out of conduction through the Si substrate. However, from the symmetry of the four-probe behavior with respect to the current path (fig. 3), it would appear that much of the conduction through the substrate must be through the area directly under the nickel deposit. It may be that the process of writing creates a more highly doped region of Si directly under the nickel deposits. A more likely explanation is that there are one or more points in the sample where the nickel is very thin or narrow. At high temperatures, the current flows through the Si in these small regions. As the temperature is lowered, the Si freezes out, and the current flows through the Ni.

The inset to fig. 3 shows a magnified view of the behavior below 230 K. After the initial rise, the resistance decreases slowly as the temperature is lowered, with a temperature coefficient of resistance of 150 ppm/K. This is as expected for a very thin disordered metal film[27]. Below 50 K, there is a slight rise in the resistance. This may be due to freeze-out of small semiconducting regions in the sample or to the temperature dependence of the resistance of a parallel channel of nickel silicide (similar rises in resistance are observed in 6 nm-thick films of iron or cobalt silicide on silicon [28]). A more recent sample with similar dimensions shows the same qualitative behavior seen in Fig. 4.

The dimensions of the wire in Fig. 3 have been determined with an atomic force microscope: length=3.7μm, width=190 nm, and thickness=4nm. The resistance of the wire implies a resistivity of

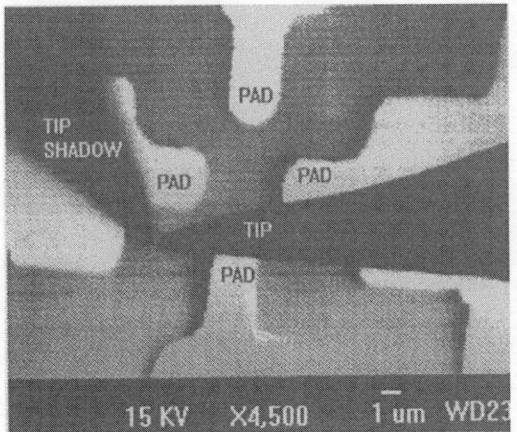

Figure 2. SEM micrograph of the STM tip positioned over a unique location in the sample. This location shows four contact pads for the transport measurements described below.

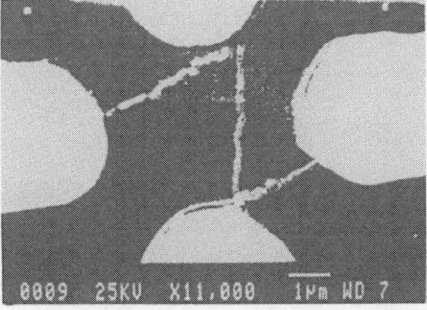

Figure 3. SEM image of lines connecting the contact pads of Fig. 1. The vertical line is the "sample", while the two diagonal lines are voltage probes. All lines are about 170 nm wide and 5 nm thick; the scale bar at the bottom is one micron long. The poor quality of the image is due to the fact that these lines are too thin for the sensitivity of the commercial SEM.

136

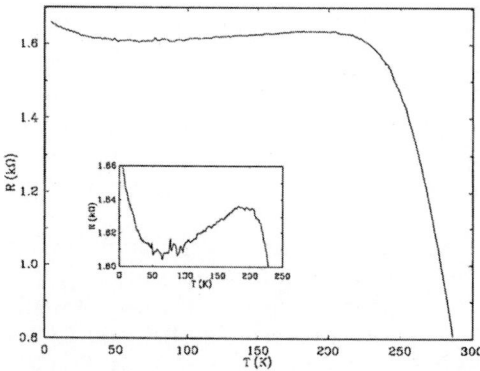

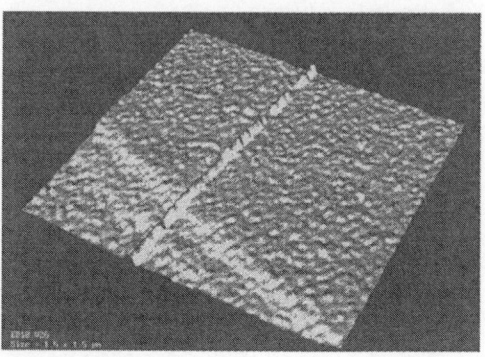

Figure 4. Four point resistance as a function of temperature for the sample shown in Fig. 3. The inset shows a close-up of the metallic-like region. The calculated room temperature resistivity is 34 μΩ-cm.

Figure 5. AFM image of a nanoscopic nickel line, only 35 nm wide and 4 nm thick. The image is 1.5μm on a side.

about 34 μΩ-cm at 300 K. Clearly, the resistivity of this wire is much less than the previous lowest figure obtained with this technique, 0.01 Ω -cm [23]. Also, our estimate of the resistivity is consistent with Auger analysis [1] of larger deposits, which indicated very high (95%) Ni content.

The narrowest line that we have obtained in the STM/SEM system is 35 nm wide and 4 nm thick, as shown in Fig. 5. While this line was quite uniform and longer than 2 μm, unfortunately it did not reach the contact pads, so that a resistance measurement was not possible. In the future we will reduce the distance between the contact pads to have a higher probability of obtaining a continuous line between the pads. In our initial work demonstrating this technique we obtained lines as narrow as 10 nm [20] by just using the contamination present on a silicon surface as a "precursor". We therefore hope to make 10-nm lines with very high purity in our UHV STM/SEM and to measure their magnetoresistance in the near future.

4. Conclusions

We have used a combined UHV STM/SEM to write well-defined nickel wires connected to electrical probes, with linewidths down to 140 nm, and disconnected wires down to 35 nm. We wrote wires connecting four pre-existing contact pads, and made four-probe resistance measurements as a function of temperature. Our best estimate of the resistivity of the wires is 34 μΩ-cm at room temperature. This is the first four-point resistance measurement of a structure made with STM.

5. Acknowledgements

This work is supported by the National Science Foundation (DMR-8553305), the Defense Advanced Research Projects Agency, the Welch Foundation, and a Hertz Foundation fellowship (E. Ehrichs). We are grateful to Jesse Martinez for performing the contact-pad depositions.

6. References

[1] See, for example, W. F. Smith, E. E. Ehrichs, and A. L. de Lozanne, in Nanostructures and

Mesoscopic Systems M.A. Reed and W.P. Kirk, eds. (Academic Press, 1991) pp85-94, and references therein.

[2] See, for example,"Surface Modification with a Scanning Proximity Probe Microscope" by U. Staufer in Scanning Tunneling Micrsocopy II: Further Applications and Related Scanning Techniques, R. Wiesendanger and H.-J. Güntherodt, editors. (Springer Series in Surface Science, Vol. 28, pp 273-300; Springer-Verlag, New York, 1992).

[3] R.R. Webb, these Proceedings.

[4] D. W. Abraham, H. J. Mamin, E. Ganz, and J. Clark, IBM J. Res. Develop. 30, 492 (1986).

[5] R.C. Jaklevic and L. Ellie, Phys. Rev. Lett. 60, 120 (1988)

[6] Y. Z. Li, L. Vazquez, R. Piner, R. P. Andres, and R. Reifenberger, Appl. Phys. Lett. 54, 1424 (1989)

[7] J. P. Rabe and S. Buchholz, Appl. Phys. Lett. 58, 702 (1991)

[9] G. C. Wetzel and S. E. McBride, Proc. Conf. on Scanned Microscopies, Jan. 6-11, 1991, Santa Barbara CA, K. Wickramasinghe and F. A. McDonald, eds.

[10] U. Staufer, L. Scandella, and R. Wiesendanger, Z. Phys. B77, 281 (1989); and U. Staufer, R. Wiesendanger, L. Eng, L. Rosenthaler, H.R. Hidber, H.-J. Güntherodt, and N. Garcia, J. Vac. Sci. Technol. A6, 537 (1988)

[11] E.E. Ehrichs and A.L. de Lozanne, J. Vac. Sci. Technol. A8, 571 (1990)

[12] O. E. Husser, D. H. Craston, and A. J. Bard, J. Electrochem. Soc. 136, 3222 (1989)

[13] I.-W. Lyo and Ph. Avouris, Science 253, 173 (1991); and these Proceedings.

[14] M. Aono, these Proceedings.

[15] H. J. Mamin, P. H. Guethner, and D. Rugar, Phys. Rev. Lett. 65, 2418 (1990); and H. J. Mamin, P. H. Guethner, S. Chiang, and D. Rugar, J. Vac. Sci. Technol. B9, 1398 (1991).

[16] D. M. Eigler and E. K. Schweizer, Nature 344, 524 (1990)

[17] B.M. McWilliams, I.P. Herman, F. Mitlitsky, R.A. Hyde, and L.L. Wood, Appl. Phys. Let. 43, 946 (1983).

[18] R. M. Silver, E. E. Ehrichs, and A. L. de Lozanne, Appl. Phys. Lett. 51, 247 (1987)

[19] E. E. Ehrichs, R. M. Silver, and A. L. de Lozanne, J. Vac. Sci. Technol. A6, 540 (1988)

[20] E. E. Ehrichs, S. Yoon, and A. L. de Lozanne, Appl. Phys. Lett. 53, 2287 (1988)

[21] E. E. Ehrichs and A. L. de Lozanne, Proc. Int. Symp. on Nanostructure Physics and Fabrication, College Station, TX, March 13-15, 1989, M. A. Reed and W. P. Kirk, eds. (Academic Press 1989) pp. 441-446.

[22] D. D. Awschalom, M. A. McCord, and G. Grinstein, Phys. Rev. Lett. 65, 783 (1990)

[23] M. A. McCord, D. P. Kern, and T. H. P. Chang, J. Vac. Sci. Technol. B6, 1877 (1988)

[24] W. F. Smith, E. E. Ehrichs, and A. L. de Lozanne, J. Vac. Sci. Technol. A10, 576 (1992)

[25] E. E. Ehrichs, W. F. Smith, and A. L. de Lozanne, J. Vac. Sci. Technol. A9, 1381 (1991)

[26] E. E. Ehrichs, W. F. Smith, and A. L. de Lozanne, STM-91 Proceedings (J. Ultramicroscopy), to appear.

[27] See, for example, Experimental Techniques in Condensed Matter Physics at Low Temperatures, edited by R. C. Richardson and E. N. Smith, Addison-Wesley, Redwood City CA, 1988, p. 201.

[28] R. Miranda, private communication.

HIGH RESOLUTION PATTERNING WITH THE STM

C.R.K. Marrian[1], E.A. Dobisz[1] & J.M. Calvert[2]
Naval Research Laboratory
Washington DC 20375, USA.

[1] *Code 6864, Electronics Science and Technology Division.*
[2] *Code 6090, Center for Bio/Molecular Science and Engineering.*

ABSTRACT. The use of low energy electrons for e-beam lithography in radiation sensitive materials is of interest because the detrimental effects of electron scattering can be eliminated. The scanning tunneling microscope (STM) is a convenient way to produce such electrons in a spatially localized beam. The STM has proved to be a valuable probe of resist materials and a viable method of lithography for high resolution patterning of semiconductor substrates. Thin layers of commercially available resist materials have been patterned and the patterns transferred into a substrate with a reactive ion etch. The absence of electron scattering is evidenced by the elimination of proximity effects and an improvement in resolution over that obtained with a tightly focussed 50 kV e-beam in identically prepared and processed resist films. Preliminary studies are also reported of STM lithography of ultrathin resists formed by self assembly techniques. The use of such thin films promises to increase further the resolution capability of STM based lithography. Issues related to the implementation of a low voltage lithography tool are discussed. A single tip instrument built with a fast STM like scanner appears viable for small scale nanolithography whereas a multiple tip approach is required for applications where throughput is an issue.

1. Introduction

The physical phenomena apparent at sub-100 nm dimensions have led to numerous concepts for nanoelectronics devices, such as resonant tunneling structures, quantum confinement and inter-ference based devices, single electron transistors and Coulomb blockade switches. In a few cases working prototype devices have been fabricated with 100 nm scale geometries, but operation at liquid He temperatures is required for device demonstration. Higher temperature operation should be possible for smaller scale geometries. However, fabrication procedures for 100 nm geometries are presently extremely difficult and time consuming and operational devices are only produced with very low yields. Radically different approaches to nanofabrication are needed to achieve the necessary dimensional control, reliability and reproducibility.

A steered beam lithography will be an essential part of the technology for ultrahigh resolution nanofabrication. In the case of e-beam lithography, the understanding and correction of proximity effects are the major obstacles to increased resolution. Proximity effects are the result of resist exposure away from the point of impact of the primary electron beam. This exposure is due to

P. Avouris (ed.), Atomic and Nanometer-Scale Modification of Materials: Fundamentals and Applications, 139–148.
© 1993 *Kluwer Academic Publishers.*

primary electrons forward scattered in the resist layer, electrons backscattered from the substrate, and secondary electrons generated by the primary and scattered electrons. Even in current e-beam lithography systems, resolution is limited by proximity effects as opposed to the size of the focussed beam. The conventional approach to overcoming proximity effects is to increase the primary beam energy and create a more diffuse 'fog' of backscattered electrons for which corrections can be made. This technique has worked in some applications but it is not clear that correction can be made for a broad range of resists and substrates at a resolution below 100 nm. An alternative strategy is to decrease the primary beam energy so the scattered electrons responsible for proximity effects are spatially localized. The logical extension of this approach is to reduce the electron energy to the threshold required to induce the resist chemistry. However, creation of such a focussed low energy electron beam is difficult. The scanning tunneling microscope (STM) provides a technique to spatially confine the electron beam.

In the STM, a sharp tip is maintained very close to a surface by controlling the tip-sample separation with a simple servo loop. The servo controls the position of the tip to maintain a constant tip-sample current. In most applications, the STM is operated with a voltage bias in the range -4 to +4 V between tip and sample, so the tip-sample current is determined by electron tunneling through the electronic barrier between tip and sample. As the tip is scanned laterally, the servo loop adjusts the tip height to track variations in the surface electronic structure. With suitable sample preparation, atomic resolution imaging can be achieved. At higher tip-sample biases, the tip-sample current is determined by field emission. For a given tip-sample current, the tip-sample separation is greater in the field emission mode than in the tunneling mode. However, the servo loop is still able to track variations in the surface topography for imaging (with nanometer resolution) and lithography.

Low voltage electron beam techniques hold considerable promise as an approach to nanometer scale fabrication. The ability to manipulate materials on the atomic scale has been demonstrated with proximal probes such as the scanning tunneling microscope (STM). The STM can be operated to produce a spatially confined, low energy (5-50 eV) electron beam which has been shown capable of material modification on the nanometer scale. Examples of STM induced material modification are described in this volume and elsewhere [1]. This paper summarizes recent studies of resist materials with the STM at NRL [2-4].

2. Experimentation

The STM head, obtained commercially from W.A. Technology, is mounted in a stainless steel chamber with ion and turbo pumps. The STM is driven by custom built electronics and in-house developed software. Vacuum operation is necessary to avoid electrical breakdown between tip and sample at the high tip-sample voltages used. Typically, lithography is performed at pressures of about 10^{-7} torr. Samples are introduced under dry N_2 and the chamber can be evacuated in about 15 min with the turbo pump which is then valved off and turned off for the lithographic patterning.

Lithographic exposure has been performed with an elevated bias between tip and sample (typically -10 to -50 V) and moving the tip laterally (at a constant speed). The STM servo loop is operated to maintain a constant tip-sample current. By appropriate lateral movement of the tip with the STM scanner patterns can be defined in the resist material. As operation is in the constant current mode, the exposure dose applied to the resist can be easily monitored and varied. The STM tip is biased negatively with respect to the sample so electrons pass from the tip through the resist and into the substrate.

Operation of the STM requires a conductive substrate, so care must be taken to remove non-conductive oxides from semiconductor surfaces. Highly doped p-type Si was etched in dilute HF to remove the native oxide and to passivate the surface with hydrogen. GaAs was etched in a 7:1:1 $H_2SO_4 : H_2O_2 : H_2O$ solution, rinsed in water and coated with 10 nm of Si to stabilize the surface and improve resist adhesion. Films of the e-beam resist SAL-601 from Shipley were spin coated onto the samples. Resist thickness was nominally 50 nm and verified with a surface profilometer. Following exposure, the resist was baked (107 °C for 7 min) and developed (12 min in MF-322) as recommended by the manufacturers [5]. The GaAs samples were etched in a small custom built reactive ion etch (RIE) system with BCl_3. Etching was performed at an RF power of 80 W, a pressure of ~1 mtorr and a flow rate of 6 sccm. Samples were coated with 10 nm of Au or Au-Pd for inspection in a scanning electron microscope (SEM).

3. Results and Discussion

3.1 LOW VOLTAGE LITHOGRAPHY IN POLYMERIC RESISTS

Successful resist exposure and patterning of GaAs has been achieved at tip-sample voltages up to -35 V. The smallest voltage at which resist exposure has been observed is -12 V. With our present experimental set-up, 3 pA is the minimum tip-sample current possible and 1 μm/s the maximum lateral tip velocity. Even the minimum line dose, 30 nC/cm, is sufficient to expose the resist. An example of STM lithography of SAL-601 is shown in figure 1, which is an SEM micrograph of an etched GaAs sample at a tilt of 45 °. The micrograph has not been expanded vertically to correct for the tilt. The developed resist was used to mask for a RIE of the substrate to a depth of approximately 100 nm. The pattern was written at -25 V with a line dose of 200 nC/cm. The beam was 'blanked' between the non-contiguous parts of the pattern by retracting the tip before moving it laterally to the start of the next element of the pattern (i.e. letter). It would clearly be faster if blanking could be achieved without retracting the tip by adjusting the operation of the STM so that the tip could be moved laterally without exposing the resist. Some success with this approach has been achieved with resist that was less sensitive than the SAL-601 used in the experiments described here [4].

The finest resolution STM lithography has been performed in a 30 nm film of resist on Si as shown in figure 2. Here 23 nm wide lines of resist written at -15 V are shown. This a only few nm less than has been written in a 50 nm film. The issue of resolution is discussed further in the next section. Also shown in the figure are lines written at -25 and -35 V illustrating the ability to control the feature size with the tip-sample bias. Results from a series of lithographic exposures are summarized in figure 3 where feature size is plotted as a function of tip-sample voltage for lithography performed at an exposure of 200 nC/cm. The points lie close to a line passing through the origin which can be explained by considering STM operation in the field emission mode. As the tip-sample bias is increased the action of the STM servo is to retract the tip to maintain the constant tip-sample current. Field emission from the tip is a function of the electric field at the tip as given by the Fowler-Nordheim equation. To first order the field at the tip is proportional to the field between tip and sample. As a result, the tip-sample separation will be proportional to tip-sample bias for a given tip-sample current. As the electrons flowing from the tip diverge, the diameter of the beam impinging on the sample will be proportional to the tip-sample separation and hence bias.

Figure 1. Non-contiguous pattern written with the STM in SAL-601 and etched into GaAs with a BCl₃ reactive ion etch.

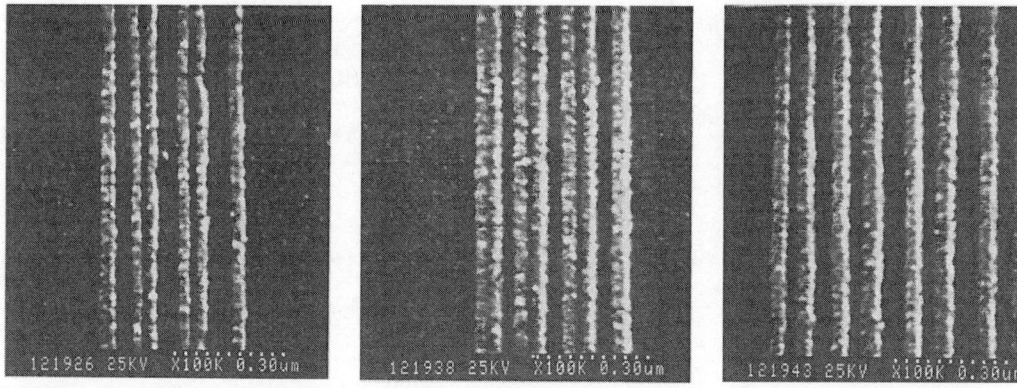

Figure 2. SEM micrographs of STM lithography in a 30 nm film of SAL-601. The patterns were written at -15 V (left), -25 V (center) and -35 V (right).

3.2 COMPARISON TO HIGH VOLTAGE LITHOGRAPHY

Comparisons have been made between the STM lithography and identically processed resist films exposed with a tightly focussed 10 nm 50 kV e-beam in a JEOL JBX-5DII. We have shown that there is a significant advantage in using low voltages for e-beam lithography. Smaller feature sizes can be written with the STM than with the 50 kV e-beam. In a 50 nm film of SAL-601 on silicon 27 nm lines have been written at -15 V. This is three times smaller than has been defined with a 10 nm, 50 kV e-beam in an identically prepared and developed 50 nm resist film. Most importantly, proximity effects are eliminated in STM lithography. Figure 4 compares identically prepared and processed samples of SAL-601 on silicon. On the left, STM written 40 nm lines on 55 nm spacing are clearly resolved. Isolated lines written under the same conditions (-25 V, 200 nC/cm) are of the same size. In contrast, on the right 'lines' written on 100 nm centers with the 10 nm, 50 kV are not resolved. In the absence of proximity effects, such a grating should be resolved as isolated 75 nm lines can be written [3].

By using the STM as a low voltage lithographic tool, we have shown that the polymeric resist has a intrinsic resolution superior to that observed with conventional high voltage lithography. This has provided valuable insights into the mech-

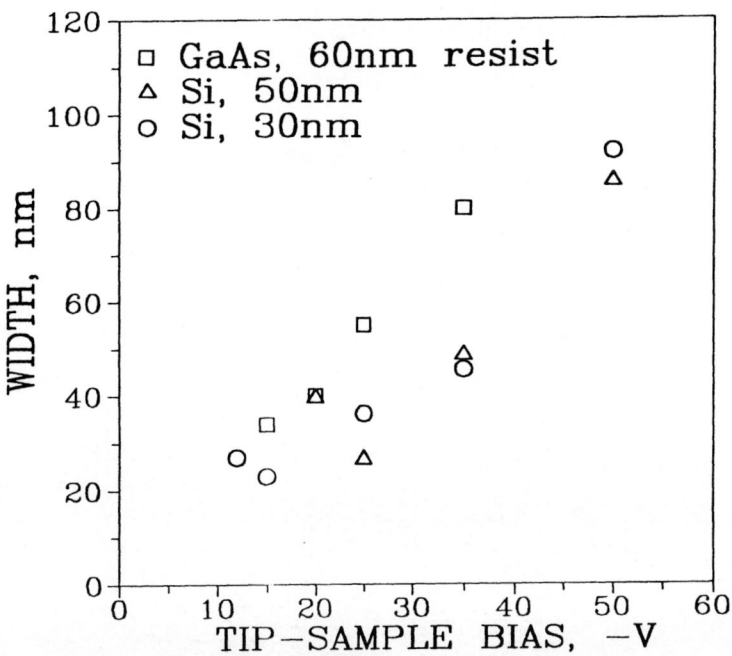

Figure 3. Measurements of feature size in developed SAL-601 as a function of STM writing voltage.

anisms defining the resist resolution. For example, electron exposure of SAL-601 releases a catalyst which promotes crosslinking during a post exposure bake. Thermally activated diffusion during the bake was believed to be responsible for significant increase in minimum feature size. The STM lithography has provided an upper bound to the catalyst diffusion length of about 12 nm (half the observed minimum feature size). Similar minimum feature sizes have been achieved with STM lithography in PMMA [1]. However, the same resolution in PMMA can be achieved with a tightly focussed 50 kV e-beam. Thus, in contrast to SAL-601, no benefits in terms of resolution improvement are realized by going to ultralow exposure voltages with PMMA. Our current interpretation is that SAL-601 is more sensitive to low energy electrons than PMMA which causes the resolution to be degraded by low energy scattered and secondary electrons created by a high voltage primary e-beam. These results have been extremely valuable in our studies of proximity effect reduction in high voltage e-beam lithography [6].

3.3 ULTRATHIN CHEMISORBED FILMS

As has been shown above, low voltage e-beam lithography in 30 to 50 nm films of polymeric resist materials has been achieved using the STM as a source of electrons [1-4]. Over the range -15 to -50 V, minimum feature size scales with tip-sample voltage. The lowest voltage at which exposure has been possible is -12 V. Below this voltage the STM tip has to push into the resist to maintain the tip-sample current. To achieve resolution on the 10 nm scale using the STM in the

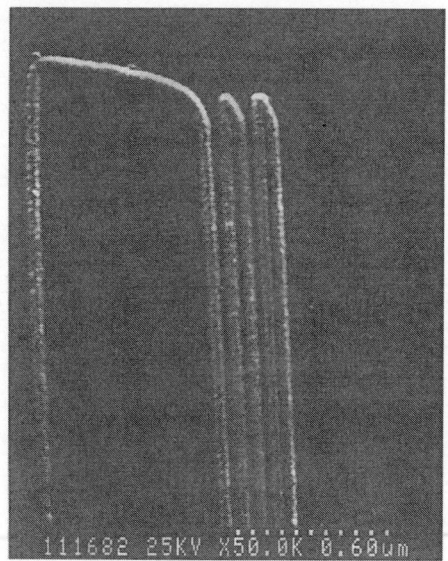 

Figure 4. Comparison of STM (left) and 50 kV (right lithography in identically processed 50 nm films of SAL-601 on a bulk Si substrate.

constant current mode, the exposure must be performed at lower voltages. However, thinner (~10 nm) resist films are necessary. It is not practical to use polymeric resists at these thicknesses as uniform and continuous films cannot be spin cast. However, molecular self assembly (SA) is a technique whereby homogeneous radiation sensitive layers of this thickness range can be formed.

Molecular self-assembly is a chemical process in which molecules are spontaneously organized to form larger structures. Chemisorbed films are a class of self-assembling materials in which precursor molecules from solution or vapor phase react at interfaces to produce layers that are chemically bonded to those surfaces. SA film precursor molecules are of the general type: G-S-R, where G is a surface reactive group at one end of the molecule, R is a functional group that extends away from the surface, and S is an optional spacer group. With a trichlorosilyl ($-SiCl_3$) group as G and an organic R group, these molecules are referred to as organosilanes. The R group can be chosen to provide essentially any desired chemical characteristic such as radiation sensitivity.

Substrates to be coated with organosilane self-assembled monolayer (SAM) films are cleaned or oxidized to produce a sufficient density of surface hydroxyl (-OH) groups. The surface is then exposed to the organosilane precursor (usually in dilute solution or vapor phase, although spin coating can be used) for several minutes to allow the chemisorption reaction (siloxane bond formation) to occur, for example:

$$\text{R-SiCl}_3 + \text{HO-(substrate)} \longrightarrow \text{R-Si-O-(substrate)} + \text{HCl}.$$

Organosilane precursors can be used to form chemisorbed SAM films on a wide variety of surfaces, unlike an approach such as organothiol adsorption which is restricted primarily to Au. Formation of SAM films on the surfaces of semiconductors (Si, poly-Si, Ge), dielectrics (SiO_2,

Al_2O_3, Si_3N_4, SiC), metals (Au, Pt, W, Al, In-doped SnO_2), polymers (epoxy, novolak, polyvinyl-phenol), plastics (ABS, polycarbonate, polysulfone), and diamond have all been demonstrated at NRL [10-12].

Surfaces treated with organosilane SAM films have been patterned by exposure to deep UV, soft x-ray, and focussed ion beams [10-13]. A variety of surface photochemical and beam-induced transformations have been effected by irradiation of organosilane SAMs including bond cleavage, molecular elimination, rearrangement, and oxidation.

The patterned organosilane SA layers have been demonstrated to serve as a template for subsequent selective attachment and buildup of a variety of materials. For example, irradiation of films with ligating R groups with an affinity for a metal complex catalyst damages the ligating groups and prevents a Pd-based catalyst from binding to the film. Treatment of the catalyzed surface, followed by immersion in an electroless (EL) Ni plating bath, results in selective deposition of Ni in the unexposed areas. The thickness of the metal film is controllable by manipulating the plating bath conditions. The metal films are of high quality and can be used as conductive paths or as masks for energetic plasma etches.

An example of STM patterning of an ultrathin organosilane film and EL plating is shown in figure 5. An SEM micrograph of a Ni film (dark regions) selectively deposited on the unirradiated regions of a Au substrate is shown in the figure. The unplated (light) regions were defined by the STM with a tip-sample voltage of -20 V and current of 10 pA. Successful patterning was achieved at tip biases down to -10 V. Inspection of the unplated regions in the

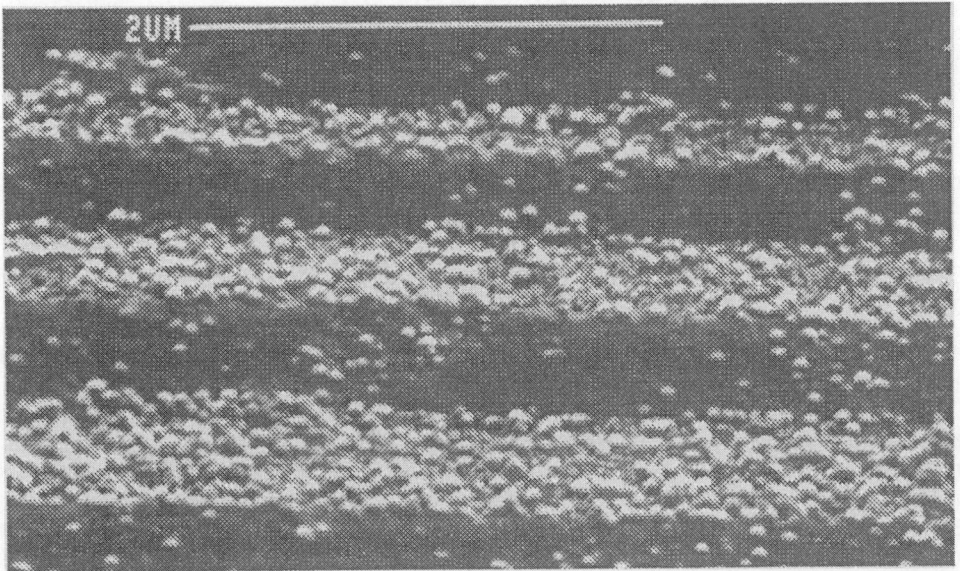

Figure 5. SEM micrograph showing selective metallization of a STM (-20 V, 10 pA) patterned SA film: the brighter regions are the unplated Au substrate and the darker regions are the EL deposited Ni film.

figure reveal that they are considerably rougher than the unexposed areas. In addition to destroying the ligating functionality of the R groups, the action of the STM patterning has damaged the underlying Au. Indeed this is not unexpected as -20 V is significantly higher than the threshold voltage observed by Mamin et al [14] for the creation of Au mounds on various substrates with the STM. However, this substrate damage does not occur with Si substrates used for our lithographic studies. These results demonstrate the potential of lithographic patterning of SA films capable of pattern transfer through an plasma etch. The matching of such a resist technology to the capabilities of the STM is crucial for the development of an ultra high resolution low voltage lithographic tool.

3.4 LITHOGRAPHIC SPEED

A key limitation of beam based techniques is that they are serial in nature and thus inherently slow. The issue of lithographic speed must be addressed to develop a technologically viable STM based lithography system. Speed gains can be obtained in two ways. First, the speed of the STM can be increased through improved design although new resist technology then becomes critical. Second, parallel operation of multiple tips is possible.

The STM tip can provide such a high current that the time required for resist exposure time is not the speed limiting factor. In effect, the lithographic speed of the STM is determined by the maximum lateral velocity of the tip. In practice this is limited by the resonant frequency and/or the resonance damping of the scanner used to move the STM tip [7]. To obtain the fastest transient response the resonance should be as high a frequency as possible. Recently, with piezoelectric slabs with resonant frequencies in the MHz range, STMs have been built which can scan at close to 1 mm/s [8]. In contrast, the pattern in figure 1 was written at a speed close to 0.5 μm/s with a current of 3 pA. The same exposure dose would require 6 nA when the tip is moved laterally at 1 mm/s. Such a current is well within the capabilities of conventional STM tips.

Resist exposure time is the speed limiting factor for nanolithography with our JEOL JBX-5DII used with its smallest probe sizes. The time required to expose a 100 μm line of 30 nm width has been estimated. For the STM, a lateral tip velocity of 1 mm/s is assumed. To write the line, a single pass at a tip-sample voltage chosen to give a 30 nm feature size would take approximately 0.1 s. For the JEOL a 30 nm linewidth requires a high resolution resist such as PMMA which has a low sensitivity of 2 nC/cm. A single pass line with a beam current of 100 pA would take 0.2 s. Thus at these dimensions, an STM based lithographic tool would give a comparable writing speed.

Further increases in speed could be realized by the parallel operation of multiple STM tips, with each tip having its own piezoelectric scanner and servo loop. For example, an array of about 10 tips operating in parallel is feasible with conventional STM technology. Greater numbers of parallel STMs would require microfabrication techniques similar to that demonstrated in Japan and at Stanford University [9]. Table 1 gives an estimate of the time required to write a 'mask' requiring 30 nm minimum feature sizes over a 4 cm^2 area with a 50% fill factor. It has been assumed that 50% of the writing would be performed at -15 V (30 nm effective spot size) and 50% at -50 V (100 nm spot).

TABLE 1. 'Mask' writing times for different numbers of tips

Number of tips	1	10	1000
Time (hrs)	5000	500	5

Thus, ~1000 tips are required for this approach to full scale nanolithography. Such a number is well within the capabilities of present microfabrication technology.

4. Summary

Our results have demonstrated that the STM can be used for lithography of resist materials and nanofabrication. The resist patterns are sufficiently robust to act as a mask for a reactive ion etch with BCl_3. The results with the resist SAL-601 further demonstrate that there are specific advantages in using an STM (i.e. low voltage) approach to e-beam lithography in that the resist resolution is enhanced and proximity effects are absent. The combination of the STM approach to lithography and self assembled resist films has been shown to be viable and has the potential to improve upon the ~25 nm resolution routinely obtained with thin films of polymeric resists. Significant challenges exist but there seems every reason to believe that they can be overcome so proximal probe lithography can be considered viable outside the research laboratory.

5. Acknowledgements

The authors would like to thank the following for valuable interactions and discussions: J. Murday, R. Colton, C. Dulcey, T. Koloski, J. Griffith, J. Dagata and M. Peckerar.

6. References

1) For example, M.A. McCord and R.F.W. Pease, **J. Vac. Sci. Tech.** B4, 86 (1986) & J.A. Dagata, J. Schneir, H.H. Haray, C.J. Evans, M.T. Postek and J. Bennett, **Appl. Phys. Lett.** 56, 2003 (1990).
2) E.A. Dobisz and C.R.K. Marrian, **J. Vac. Sci. Tech.** B9, 3024 (1991).
3) E.A. Dobisz and C.R.K. Marrian, **Appl. Phys. Lett.** 22, 2526 (1991).
4) C.R.K. Marrian, E.A. Dobisz, and R.J. Colton in **Scanned Probe Microscopies**, K. Wickramasinghe, ed., AIP Press, 241, 408 (1992).
5) Processing specified in "Microposit SAL 600 e-Beam Process", Shipley Corporation.
6) E.A. Dobisz, C.R.K. Marrian, L. Shirey and M. Ancona, **J. Vac. Sci. Tech.** B10, to be published Nov/Dec 1992.
7) D.P. DiLella, J.H. Wandass, R.J. Colton and C.R.K. Marrian, **Rev. Sci. Instrum.** 60, 997 (1989).
8) H.J. Mamin and D. Rugar, **J. Ultramicroscopy**, in press.
9) T.R. Albrecht, S. Akamine, M.J. Zdeblick and C.F. Quate, **J. Vac. Sci. Tech.** A8, 317 (1990).

10) J.M. Schnur, M.C. Peckerar, C.R.K. Marrian, P.E. Schoen, J.M. Calvert, and J.H. Georger, **US Patent** 5,077,085 (1991) and **US Patent** 5,079,600 (1992).

11) J.M. Calvert, C.S. Dulcey, J.H. Georger, M.C. Peckerar, J.M. Schnur, P.E. Schoen, G.S. Calabrese, and P. Sricharoenchaikit, **Solid State Technology** 34 (10), 77 (1991).

12) J.M. Calvert, P.E. Pehrsson, C.S. Dulcey, and M.C. Peckerar, **Materials Research Society Proceedings**, 260, in press.

13) J.M. Calvert, T. Koloski, W.J. Dressick, C.S. Dulcey, M.C. Peckerar, F. Cerrina, J. Taylor, D. Suh, O. Wood, A. MacDowell and R.D. Souza to be presented at the SPIE conference "3 Beams for Manufacturing III", 2/28/93, San Jose CA.

14) H.J. Mamin, S. Chiang, H. Birk, P.H. Guenther and D. Rugar, **J. Vac. Sci. Tech.** B9,1398 (1991).

AFM DATA STORAGE USING THERMOMECHANICAL WRITING

H.J. MAMIN AND D. RUGAR
IBM Research Division
Almaden Research Center
650 Harry Rd.
San Jose, CA 95120

R. ERLANDSSON
Ragnar Erlandsson
Department of Physics and Measurement Technology
Linköping University
S-58 83 Linköping
Sweden

ABSTRACT. We report preliminary results on using a using an AFM tip to do reading and writing on a spinning sample for possible data storage applications. The writing is performed using a new thermo-mechanical technique on a polymer substrate.

We have recently developed a new AFM-based indentation technique as a candidate writing mechanism for high density data storage. [1] In this technique, a pulsed infrared laser is focused on the tip and used to heat it. The tip is in contact with a plastic substrate, typically polymethyl methacrylate (PMMA). In the contact region, the heat from the tip softens the substrate above its glass transition temperature (about 120 °C), and the local pressure then creates a pit. This thermomechanical approach to topographic surface modification offers certain advantages over a purely mechanical approach. In the latter case, the contact force from the tip is used to cause plastic deformation or machining of the substrate. [2-4] A mechanical approach is somewhat limited in speed and dynamic range. Since a weak cantilever is used, the peak force that can be applied is limited, as is the rate at which the load can be changed. When heating the tip, however, writing can be performed without varying the load. In addition, because the technique utilizes a material phase transition, it should lead to a sharp threshold for writing, as well as give relatively high stability marks.

The size of the pits is determined by the tip shape, the loading force, and the laser parameters. We have used commercially available Si_3N_4 levers with integrated pyramidal tips. The pit size can be varied from roughly 1 μm down to several hundred angstroms across (FWHM). Typical pits are roughly 100 nm across and 10 nm deep. Typical loading force is 10^{-7} N; with a 30mW write laser, pits have been made with pulses as short as 0.3 μsec.

P. Avouris (ed.), Atomic and Nanometer-Scale Modification of Materials: Fundamentals and Applications, 149–151.
© 1993 *Kluwer Academic Publishers.*

To investigate issues such as recovery time for writing, reading speed, and wear, it is necessary to achieve fairly high relative velocity between tip and sample. We achieved this by incorporating our AFM onto a rotating sample. The sample was mounted onto a precision air bearing spindle and rotated at 0.5-1 Hz, giving a linear velocity up to roughly 60 mm/sec at a 10 mm radius. On this set-up we found that reading and writing could be achieved at frequencies up to at least 100 kHz. Figure 1 shows an AFM micrograph of two series of such marks written on a sample spinning with a linear velocity of 25 mm/sec. The narrower track is roughly 150 nm wide, and the spacing between marks is 250 nm. These marks are considerably smaller than those used in conventional optical recording, as is shown by the comparison in Fig. 1. When writing was attempted at frequencies higher than 100 kHz, there appeared to be incomplete cooling of the tip, leading to indistinct pits. When reading was attempted at higher frequencies, higher order modes were excited in the cantilever, leading to a degraded signal-to-noise ratio. Smaller, higher frequency cantilevers should allow for significantly higher read rates.

We have also performed a number of wear studies. In one such study, the tip remained on a previously written track for greater than 8 hours, and the signal showed no degradation over time. AFM micrographs of the same track also confirmed that at this load, no groove was formed by the tip. At higher loads, above 1 μN, some evidence of wear was observed.

As a demonstration, we have used this technique to record a few tracks of audio information. This was done by using standard FM techniques to vary the spacing between laser pulses in an analog fashion. A 50 kHz carrier frequency was generated, and the frequency modulated according to the audio input. This signal was then fed into the laser pulser. To read back, the AFM output signal was mixed up to 88 MHz and then demodulated with the use of a FM radio. A few seconds of voice information were recorded in this way. Although crude, this is one of the first examples of storing non-pictorial information with an AFM- or STM- based device.

[1] Mamin, H.J. and Rugar, D. (1992) 'Thermomechanical writing with an atomic force microscope tip,' Appl. Phys. Lett. 61, 1003-1005.

[2] Jung, T.A., Moser, A., Hug, H.J., Brodbeck, D., Hofer, R., Hidber, H.R. and Schwarz, U.D. (1992) 'The atomic force microscope used as a powerful tool for machining surfaces,' Ultramicroscopy 42-44, 1446-1451.

[3] Jin, X. and Unertl, W.N. (1992) 'Submicrometer modification of polymer surfaces with a surface force microscope,' Appl. Phys. Lett. 61, 657-659.

[4] Kim, Y. and Lieber, C.M. (1992) 'Machining oxide thin films with an atomic force microscope: pattern and object formation on the nanometer scale,' Science 257, 375-377.

1 μm

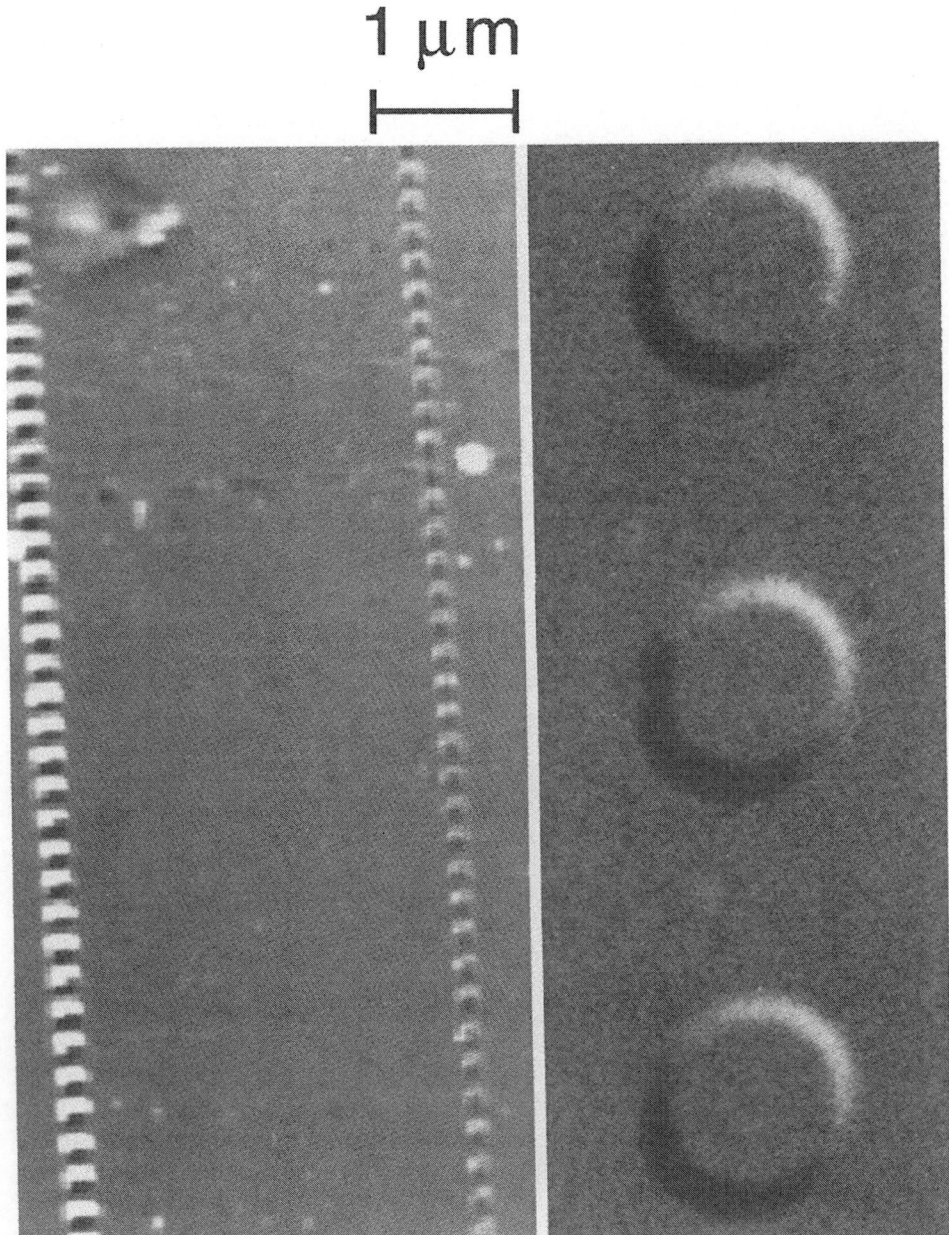

Figure 1. AFM micrograph (left) of two tracks written with the AFM tip on a rotating PMMA sample. On the right is a Lorentz micrograph of typical marks written thermo-magnetically into a standard magneto-optic recording medium. The considerably difference in density is readily apparent. (Images supplied courtesy of J.A. Logan and R. Geiss.)

BEEM: A PROBE OF NANOSCALE MODIFICATION

H. D. HALLEN
Physics Division
AT&T Bell Laboratories
600 Mountain Avenue
Murray Hill, NJ 07974
USA

R. A. BUHRMAN
School of Applied and Engineering Physics
212 Clark Hall
Cornell University
Ithaca, New York 14853
USA

ABSTRACT. Hot electrons injected by a scanning tunneling microscope (STM) tip with a few volts tunneling bias scatter and modify a gold film not only at the top surface of the film, but throughout the film and at the inner or interfacial surface. The ballistic electron emission microscopy (BEEM) measurement technique is a powerful method with which one can probe such modifications. In particular, the production of adatoms and their subsequent coagulation into atomic terraces on the inner surface of the gold is demonstrated. Quantitative measurements of the adatom production rate are in good agreement with that predicted by a model including bond breaking by the hot electrons. The stability of the created structures is shown to be related to the physical properties of the gold film.

1. Introduction

Traditionally the Scanning tunneling microscope (STM) has been used as a surface sensitive instrument, with little concern for the fate of the electrons once they have entered the sample. Recently, with the innovation of ballistic electron emission microscopy (BEEM) [1], the power of STM-related techniques for the study of buried interfaces has become apparent. We use the BEEM technique as a tool to study the subsurface modifications of a thin metallic film. Hot electrons injected by the STM tip are used to produce the modifications. This paper will focus on modification of a gold film deposited on silicon. A carbon-based passivation layer lies between the film and substrate.

The types of hot electron induced effects can be grouped into two categories: those that occur at a surface and those in the bulk of the film. Examples of each type are seen. By surface we refer to the outside of a grain: the top surface is the side of the grain scanned by the STM; the inner surface is opposite, i.e. at the interface of the gold and silicon; the grain boundaries are lateral surfaces which can be important during grain growth. At surfaces, the hot electrons induce the formation of adatom-vacancy pairs. The adatom is a gold atom which has moved out onto a surface. It may combine with other adatoms to form a terrace or diffuse to a sink such as a step edge, vacancy or grain boundary. The vacancy diffuses into the film. We have observed[2-4] terrace growth on the inner surface, grain growth on the lateral surfaces, and mound growth on the top surface of the films. The stability of the structure is found to depend strongly on the properties of the gold film.

The hot electrons can scatter from vacancies once they are in the bulk of the film. This can result in an enhanced, non-thermal motion of the vacancies. We have observed[2-4] the creation of large areas of defect-filled film. Such areas strongly scatter even lower energy electrons. The system begins to react to the induced changes even while still under hot

153

P. Avouris (ed.), Atomic and Nanometer-Scale Modification of Materials: Fundamentals and Applications, 153–164.
© 1993 *Kluwer Academic Publishers.*

electron bombardment, and continues after the stress is removed. The reaction of the system can be a measure of the stability of the structures which were created, but also can reflect further development of the structures. One example of the latter occurs when several layers of terraces have been grown on the inner surface of the gold film, bringing the gold into contact with the silicon. As has been well documented[5], silicon will diffuse into a gold film evaporated into intimate contact with it, provided that the terrace is large enough. The resulting gold-silicon alloy scatters all electrons strongly. The resultant structure is very stable. A benefit of the BEEM technique is that the stability of bulk and subsurface interfacial structures can be observed in addition to changes of the film topography. Correlations between topography and subsurface properties can be used to understand the mechanism by which mounds on metal surfaces are formed, and what parameters are important in their decay. We believe that the BEEM technique will also be of aid in understanding other systems.

2. Experimental Technique

The sample preparation has been described elsewhere[3]. In brief, a few monolayer thick carbon-based passivation layer is deposited on a prepared silicon surface, followed by evaporation of a 150 Å thick gold film - the system under investigation. The choice of sample geometry is governed by the BEEM measurement technique. The gold film must be thin enough so that most of the electrons pass through it without scattering[4]. The gold/silicon interface provides a Schottky barrier to filter the electron distribution so that only those electrons within the proper transverse wave vector and energy ranges pass into the silicon. The BEEM current is that current which is collected in the silicon after having passed over the Schottky barrier.

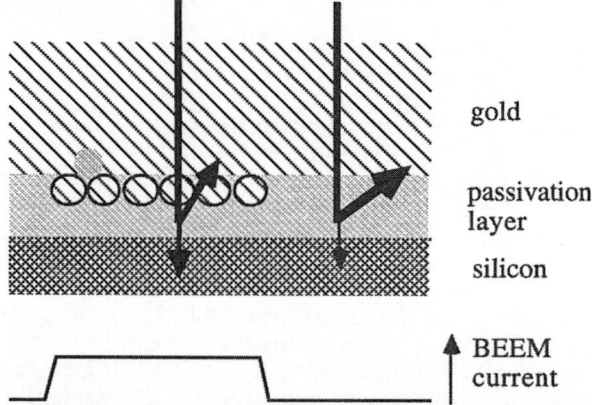

gold

passivation layer

silicon

BEEM current

Figure 1. A schematic drawing which illustrates how an inner surface terrace can increase the BEEM current due to a high scattering rate in the passivation layer material which it displaces. The BEEM electrons pass through a thinner layer of passivation material where a terrace exists. The electrons which are scattered back, in addition to many which cross the passivation layer but are not able to surmount the Schottky barrier, are eventually collected to maintain the tunnel current.

To understand how the technique can be used to study terrace growth on the inner surface of the gold film, consider Figure 1. The passivation layer scatters electrons strongly and to a good approximation, independent of the electron's energy[3]. A variation in the thickness of this passivation layer caused by the presence of a terrace layer induces a large change in the magnitude of the BEEM current. Therefore a measure of the BEEM current in this system acts as a sensitive detector of single atom high terraces on the inner gold surface. By scanning the tip laterally over the top gold surface while maintaining a constant tunnel current, one obtains concurrently a standard constant current STM image by monitoring the tip height displacement, and a BEEM image. One would expect to find regions of higher BEEM current while the tip scanned over a terrace region, and a lower but non zero current elsewhere. An experimental image is shown in Figure 2. Figure 2(a) shows the STM topograph taken simultaneously with the BEEM image in (b). The constant 1 nA current STM image shows that the grain size is typically a few hundred angstroms. All of the whitish areas in the BEEM image are regions where inner surface terraces were produced in the following manner with the parameters for each given in the figure caption: the STM is assumed to be tunneling to the sample with the sample to tip bias at the level used for imaging (typically ~1.4 V) which was chosen to be below any thresholds for modifying the system. The tip is then scanned to the point where stressing is to be done. The sample to tip bias is swept to the stressing level at a rate slow enough that the feedback loop is able to maintain the constant current state. The voltage is held at this higher level for a specified amount of time - always with the feedback maintaining constant current. The tunnel bias is then reduced to the imaging value at the same rate as it was increased. An area scan measuring both the topography and a BEEM image is then taken. Subsequent images of the same region are often taken to observe the stability of the structures.

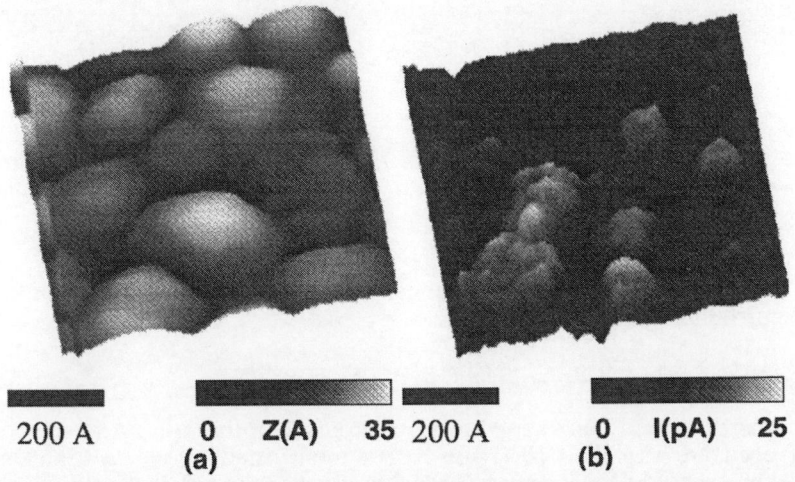

200 A   0   Z(A)   35   200 A   0   I(pA)   25

(a)   (b)

Figure 2.   1.   Gray scale 1.4 V constant 1 nA current STM (left) and corresponding BEEM (right) images illustrate enhancement type modifications of a sample. The images are 800 Å square. (a) shows the STM topograph which did not visibly change as a result of the modifications. (b) is the BEEM image where all the whitish areas were individually created by stressing with the STM current. The BEEM image before any stressing was uniformly gray. Clockwise from the two largest (just touching, in the lower left) the modifications were created with a 2.5 V for 3.8 sec, an voltage sweep 0.4->2.88 V in 5.7 sec, 2.1 V for 6.7 sec, 2.0 V for 6.5 sec, 2.25 V for 3.3 sec, and the at the bottom center 2.25 V for 7.0 sec.

3. Qualitative Results

3.1 TERRACE GROWTH - INCREASES IN THE BEEM CURRENT

Consider the image shown in Figure 2. The STM topograph figure 2(a) exhibits the gold grain structure and did not change noticeably during the production of the inner surface terraces seen in the BEEM image of figure 2(b). Recall that the terraces are on the inner surface of the gold so are not observed in the STM image. The resolution of the BEEM image is not as high as the STM image because it is quite far (~150 Å) from the tip. The resolution is better than one would naively expect at this distance, however, due to the constraints on transverse momentum imposed on electrons before they are allowed to cross the Schottky interface[1]. One can tell that the BEEM resolution is quite high by looking at the BEEM image profile near a inner surface terrace edge. The resolution can also be measured while creating small terraces at the inner surface as will be described below. The measured resolution for this system is ~10-20 Å.

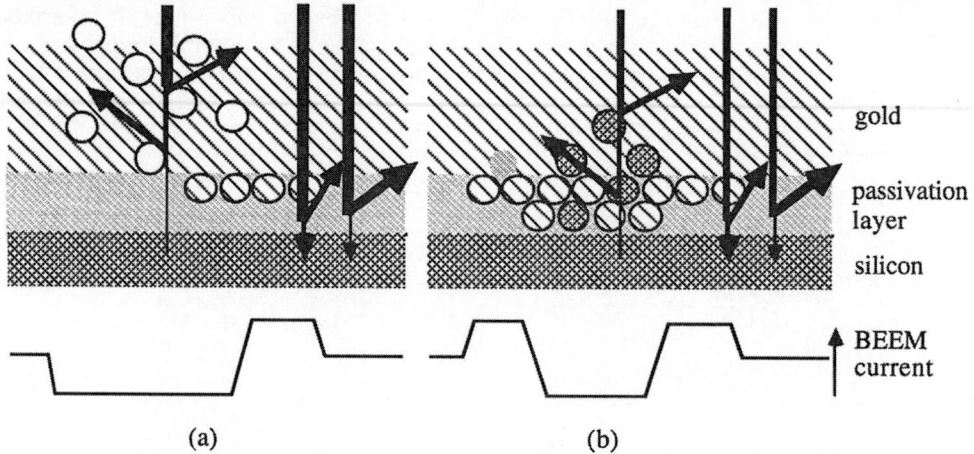

gold

passivation
layer

silicon

BEEM
current

(a)   (b)

Figure 3.   Sketches of mechanisms which can result in the local decrease of the BEEM current from surrounding values.

3.2 MECHANISMS FOR DECREASES IN THE BEEM CURRENT

The mechanisms which can result in the decrease of the BEEM current from the surrounding area are sketched in figure 3. As mentioned above, a high density of vacancies within the gold film can strongly scatter electrons. Such is the situation depicted in figure 3(a). The inelastic scattering of hot (> ~2 eV above the gold Fermi level) electrons can stimulate the motion of the vacancies. Lower energy electrons, such as those used to image the modification results, where a typical bias of 1.4 V is used, do not substantially alter the structure but probe it. The imaging electrons do not scatter as strongly from the vacancies so are less sensitive to vacancies in the film. Thus a very high density of vacancies is required before a noticeable structure will be seen in an image taken at 1.4 V. Such a high density of defects in a small area can be expected to be unstable. Indeed, defect structures of this type will often disappear within a time scale of minutes accompanied by changes in the topography of the gold film as seen in the STM image. Examples of such defects have been previously shown[3,4]. These observations underline the reversibility of the vacancy motion process.

Another method through which the BEEM current can be reduced in a local region is given in figure 3(b). This effect depends on the nature of the gold/silicon system and results when a second layer of gold grows on the inner surface which brings the gold into intimate contact with the silicon. A much higher rate of adatom/vacancy production is required to nucleate this second terrace due to the high loss rate of adatoms to grow the first terrace. The loss rate is higher since the first terrace is much closer to the production site than the nearest adatom sink was to the production site while nucleating the first terrace. The narrow ring of enhanced BEEM current shown in figure 3(b) over the region at the edges of the modification where only one terrace exists is an identifying feature of this type of modification. The upper left hand part of figure 4 shows an region containing such structures. The BEEM image of figure 4(b) shows the ring of enhancement surrounding a region of immeasurably small BEEM current. The existence of a region in which the BEEM current is enhanced implies that the density of silicon scattering centers in the gold falls rapidly with distance into the gold. Otherwise one would expect to find larger regions with reduced BEEM current and a less reproducible width to the enhanced ring. Of course the strongly reduced BEEM current in the degraded regions then implies that silicon within gold is a very strong scatterer of electrons within the energy range used for imaging. Note that this type of modification is irreversible at room temperature by the second law of thermodynamics.

500 A   0   Z(A)   80   500 A   0   I(pA)   25
          (a)   (b)

Figure 4.   1.   Gray scale 1.4 V constant 1 nA current STM (left) and corresponding BEEM (right) images illustrate various modifications of a sample. The images are 1800 Å square. (a) shows the STM topograph which exhibited some grain growth as a result of the intermixing modifications. (b) is the BEEM image which was uniformly gray before stressing with the STM current. The top, left and center intermixing modifications were created with 3.1 V and 3 nA for 5.4 sec, 0.1 nA for 5.4 sec, and 3 nA for 2.7 sec, respectively. The creation of the lines near the bottom and right is described in the text.

3.3 DRAWING LINES

At the bottom and right hand sides of figure 4 are some terraces which appear as lines. These are literally the first attempts at making any structures besides dots with the technique. First the line near the bottom was made by positioning the tip at the left end, raising the tip bias to 2.3 V@1 nA tunnel current, and moving 700 Å to the right at 230 Å/sec and back to the starting position at the same rate. When no increase in BEEM current under the tip was observed, the tip was swept again to the right at 47 Å/sec, leaving the line in the figure. The lines at the right of figure 4 were formed by moving to the right hand end of the upper line, where the tip bias was increased to 2.3 V@1 nA tunnel current. The line segments were then swept in order -- the last one is vertical in the image right at the edge of the figure. The horizontal lines are 350 Å long and the vertical lines 400 Å long, all swept at ~30 Å/sec.

In principal, arbitrarily shaped and sized structures can be created. The limitation is thermodynamic stability of the structures, i.e. they must be large enough that surface energy does not cause them to evaporate, and diffusion rates of the terrace atoms on the surface and vacancies within the film. Stability issues will be discussed at the end of section 4.

4. Quantitative Results and Discussion

Before embarking on a quantitative analysis of the data, we must first remark about its generality. We have found that the quantitative behavior of the modification properties, including threshold levels, rates and stability, depend on the nature of the gold film under study and the interface condition. This is presumably due to differing diffusion rates of vacancies and inner surface adatoms, in addition to other extended defects in the system which interrupt vacancy motion. A striking example of interface preparation dependence is given when the carbon-based passivation layer is replaced by 1-2 monolayers of SiO_2, which is grown in an ultra high vacuum (UHV) chamber following silicon cleaning. The oxide thickness is estimated[3] from the ratio of the substrate to oxide silicon 2p photoelectron peaks measured by x-ray photoemission. The oxide scatters electrons very strongly so the background BEEM current levels are reduced by a factor >20 from those with a few monolayer thick carbon passivation layer. The modification properties also differ. To within the signal to noise of the images, we have seen no evidence for an enhancement of the BEEM current following stressing of an oxide-passivated sample. One explanation for this difference is that the carbon-based passivation layers are easily compressed or incorporated into the inner surface gold terraces as they grow, but that the oxide is much harder to displace, prohibiting growth on the inner surface of the gold. To insure that such systematic errors do not effect the data presented in this section, all data in this section, except where explicitly noted, was taken on a particular carbon-layer-passivated sample within a few microns of each other and a few days of each other.

For the hot electron scattering model for gold adatom formation one expects the number of adatoms created to depend on the number of injected electrons. Indeed for the modification parameters studied here the effect of longer stressing time was found to be redundant with that of higher current over the range of tunnel currents (0.1-10 nA) used. The relevant parameter is the electron dose given as the product of the current times the stressing time. This is not a surprising result since for a 1 nA tunnel current, the average time between electron arrivals is $~10^{-10}$ sec for a region the size of the BEEM resolution. This time is much longer than electron relaxation times and phonon periods. Thus one would expect the electron system to have settled into an equilibrium or metastable state before the next electron arrives.

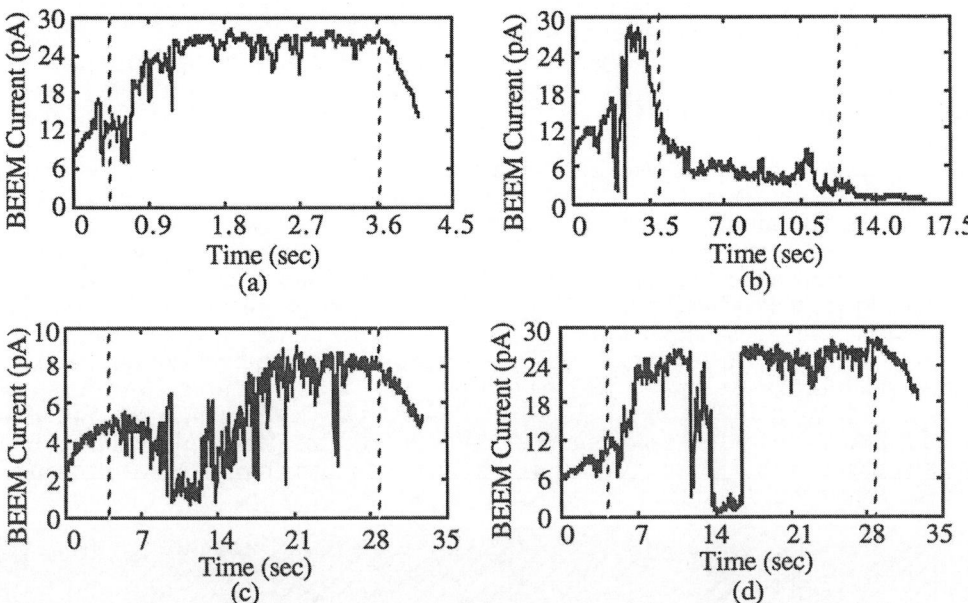

Figure 5. Plots of the BEEM current taken in real-time during modifications of the sample. The sample-tip bias was increased from the imaging value of 1.4 V linearly to the stressing value, at which it was held constant for a period of time, then decreased linearly back to the imaging value. The dotted lines indicate the times at which the bias reached and left the highest stressing value. The stressing voltages and tunnel current values for the plots are: (a) 2.25 V, 1 nA, (b) 3.1 V, 1 nA, (c) 2.1 V, 0.3 nA, and (d) 1.9 V, 1 nA. Note that the time axes differ between the plots.

4.1 REAL TIME OBSERVATIONS -- TERRACE NUCLEATION

The modification which occurs due to the stressing with hot electrons as described in section 2 can be followed in real time by monitoring the BEEM current. Experimental data are shown in figure 5(a-d). Note that there are several distinct phases which occur while the voltage is held constant. Before describing the properties of each phase in detail, it is important to recall that the BEEM current is only sensitive to what is occurring in a region the size of the BEEM resolution (~10-20 Å in diameter) at the interface. We have seen that the modification process often creates structures much larger than this. The real time BEEM current plot will reflect only what is occurring at the region underneath the tip (usually the center of the region unless the tip is nearly over a grain boundary). Following along figure 5, one finds that a latency period during which the BEEM current does not change follows the ramp to the stressing bias, then the current rises, saturates, and decreases if the bias is held at a high enough level. The time scales for the various phases depends on the stressing bias. There are noisy periods in the plots, especially (c) and (d). These are expected as will be discussed below and are the result of inelastic scattering as the interface through which the BEEM electrons pass is modified.

160

The latency period is best described by the picture in figure 6(a). We know that hot electrons at these energies stimulate the production of adatoms at the inner surface of the gold. When no terrace exists, however, the adatom will tend to diffuse away, inhibiting the formation of a terrace. It is found that the latency time for a given dose is either close to zero or depends on voltage in a roughly exponential manner as $\exp[(-6.6 \pm 1.7)$ V]. The cases when the latency time is near zero occur when the new modification is within 10-30 Å of a previous modification. In these cases, the nearby terrace presumably prevents the adatoms from freely diffusing.

When the BEEM current begins to rise, some adatoms must be localized at the inner surface underneath the tip. Figure 6(b) depicts the situation. A small terrace has formed. The evaporation rate from such a small terrace is expected to be very high due to its thermodynamic instability[6]. Thus, the initial growth rate of the terrace will be much slower than the growth rate when the terrace is larger. This is simply checked by comparing the time it takes for the real-time BEEM plot to saturate compared to the time it would take to grow a region the size of the BEEM resolution using the rates measured for larger terraces. One finds that the initial growth is ~10 times slower. A measure of this rate is given by the time it takes for the BEEM current to reach half its saturated value, a quantity which behaves $\sim \exp[(-6.8 \pm 0.4)$ V] up to ~2.5 V stressing bias. The functional form of the current rise is approximately proportional to the square root of time. Figure 5(b) decreases before reaching a saturation level. This decrease is exponential in time as is expected if a layer of material with a short mean free path was growing in thickness. This material is silicon intermixed with gold.

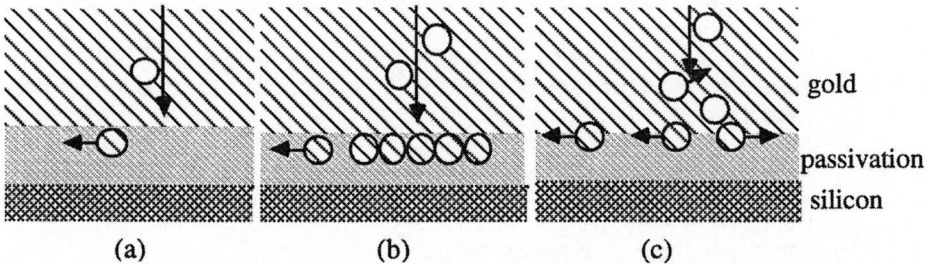

| (a) | (b) | (c) |

Figure 6. Schematic drawings of the state of the interface at various stages of the terrace growth process. (a) During the latency period, the adatoms which are created diffuse away. (b) Eventually some adatoms coagulate to form a small terrace, but the terrace is not thermodynamically stable since the curvature is very large. The evaporation rate of atoms is high so the terrace grows slowly. The small terrace allows the BEEM current to increase somewhat. This is the situation as the BEEM current rises in the plots of figure 5. (c) Vacancies are formed in the bulk of the film at all stages. Hot electrons scatter from the vacancies if they are in the path of the incident beam as is shown here during the initial growth phase.

4.2   VACANCIES

Note that vacancies are produced with adatoms. A hot electron can scatter from a vacancy similarly to the way it scatters at the inner surface, so one would expect that whenever a vacancy gets into the (resolution-sized) region in the path of the incident electrons, it would decrease the BEEM current. A cartoon of such an event is shown in figure 6(c). Eventually the vacancy is driven out of the path of incident electrons. Bursts of such decreases are seen at all parts of the real-time BEEM plot. For example one is found near the end of the latency period in figure 5(c) and in the saturated region of figure 5(d).

After the real-time BEEM current plot saturates, the terrace is larger than the resolution of the BEEM technique. This provides a convenient way to measure the BEEM resolution. The stressing is stopped as soon as the BEEM current is found to saturate. A BEEM image shows the size of the structure. Subsequent images reveal if the terrace is evaporating or if it retains its size. This is the origin of the ~10-20 Å BEEM resolution quoted above. After the sample-tip bias has returned to the imaging value, one can compare it to the starting value to ascertain whether the region directly under the tip has had a net increase or decrease in transmittance. The results of the stressings shown in figure 5(a,c,d) was a net increase. Subsequent area scans showed terrace formation. The stressing shown in figure 5(b) caused an intermixing modification.

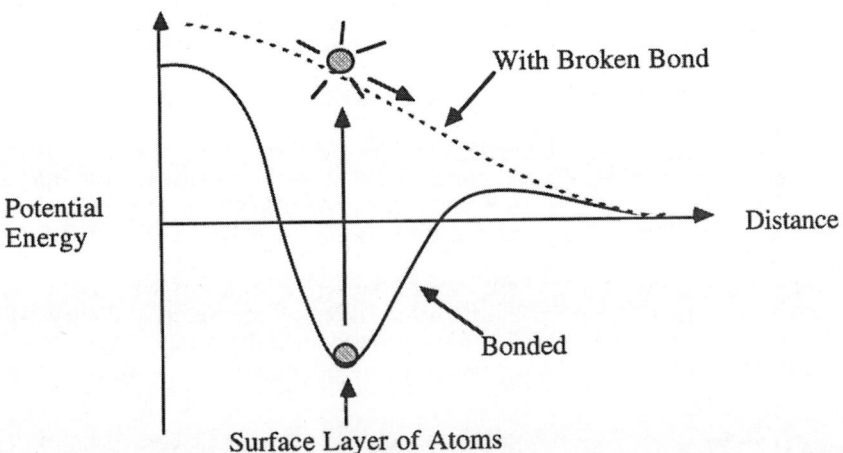

Figure 7.   When a hot electron inelastically scatters from a gold atom, it may break a bond causing the atom to become unstable in its present location. A schematic potential energy vs. position is shown indicating how an atom can be accelerated onto a surface by a potential energy gradient if the bond remains broken long enough.

4.3   ADATOM PRODUCTION

Terrace growth rate can be measured from a BEEM image (e.g. figure 2) as a function of electron dose and bias voltage. The initial increase of the area of the terrace is linear with the dose as is expected if the growth is limited by the adatom production rate. The size of the terrace eventually saturates, probably due to annihilation of adatoms with vacancies from within the grain or loss of adatoms to nearby grain boundaries. The initial terrace growth rates can be converted into a lower limit on the adatom production rate. It is a limit since some adatoms are lost through annihilation with vacancies. The adatom production rate depends upon voltage as $\exp(5.9 \text{ V})$ at lower voltages, and begins to saturate ~2.5 V. It is not surprising that the voltage dependence of the latency time, the rate of BEEM current rise and the adatom production rate all share the same voltage dependence. They all depend on the rate at which adatoms are produced.

So far we have attributed the adatom-vacancy production to 'inelastic scattering of hot electrons.' A model has been proposed for this process[4]. It is similar in nature to electron stimulated desorption (ESD), which has recently been reviewed in [7]. The difference from

ESD is that the electron energies are much lower in the present investigation, and the ions are not ejected from the surface. The similarities are that an injected electron breaks a bond of an atom in the sample. This makes the atom unstable in the lattice -- it undergoes a Franck-Condon transition. A picture is given in figure 7. The atom was originally sitting in a potential well formed by the bonding to its neighbors. When the bond is broken, it shifts to the upper potential energy curve which is shown for a surface atom. If the bond remains broken for a long enough period of time for the atom to be accelerated by the potential energy gradient and move out onto the surface, an adatom-vacancy pair will have been created. Typically a bond will not remain broken for such a long time unless it has been stabilized by a lattice distortion. Instead, the hole (broken bond) will jump from one atomic site to another through the lattice until either it is localized or decays. Atoms near surfaces or defects in the bulk (e.g. vacancies) do not have as high a coordination number so generally have softer bonds. These bonds can therefore deform more easily to localize the hole than others. Thus a hole created in the bulk of the film can move to a defect or surface and get localized. It must be fairly close, however, or it will decay on the way. We hope to be able to measure such decay lengths and localization probabilities in the future using this BEEM technique with an appropriate range of samples. Some discussion of hole localization is given in [7].

A model to predict the voltage dependence of the adatom production rate was given in [4]. The model used an approximate calculation of the hole formation rate at finite temperatures for a system with a gaussian broadened threshold energy. A broadened threshold for hole creation is not unrealistic: the bonds involved are near defects so are strained and in a variety of environments. A good fit to the latency time, half-time for BEEM current rise and the area growth rate of the terraces, all of which depend on the adatom production rate, is found for a broadened threshold centered at 2.5 eV with a full width of 0.4 eV. The model predicts a nearly exponential voltage dependence below threshold, with the exponential factor depending on temperature and the width of the threshold distribution.

4.4   STABILITY

The stability of the structures created within the film or on any of the surfaces is strongly tied to the nature of the film itself. Most of the data discussed here were from samples evaporated at a fairly high rate in high vacuum. The inner surface structures are quite stable on such a film. The terraces of figure 2 and the lines and other modifications in figure 4 stay close to their original form for at least hours (as long as we observed them). Vacancy structures within the films do anneal at least somewhat on the minutes time scale, then are either gone or stabilize. The top surface structures viewed in the STM scan change some as the vacancy structures anneal. If a top surface mound is formed on such a sample, it will usually remain with only minor changes.

At another extreme, we have grown samples very slowly (0.1 Å/sec) in ultra high vacuum. These samples will anneal as they grow so presumably will have fewer internal defects in addition to being cleaner. Generally speaking all the structures decay on the minutes time scale. The inner surface terraces decay very quickly -- vacancy diffusion in the film is unimpeded so the vacancies can move to the inner surface where they can annihilate terrace atoms. The top surface structures change greatly. In general they do not return to their initial configuration (observe with STM topography) but vary considerably as the internal vacancy structures (observed with BEEM) anneal. Another feature of this type of sample is that we are not able to grow two layers of terraces on the inner surface, so we do not produce any intermixing type modifications. The terrace annihilation rate is very high so we are not able to produce inner surface adatoms at a high enough rate to nucleate a second terrace.

A final word about stability involves that of the sample structure itself. If one uses hydrogen, chlorine, hydrocarbons, or nothing for the passivation layer, the lifetime is a few days, a few months, around a year or more, or zero, respectively. The lifetime is defined as the time before a significant portion of the sample has had the passivation layer break down and gold-silicon intermixing occur. This problem could be overcome by using multiple types of terrace layers.

5. Summary

We have shown that hot electrons injected by an STM tip scatter and modify a gold film throughout its volume and at all its surfaces. The BEEM measurement technique is found to be a powerful method for probing such buried structures. With an appropriate choice of passivation layer between the gold film and silicon substrate, small single atom high terraces can be studied. This is important as it allows observation of hot electron effects at high spatial resolution without the effects of a high electric field from an STM tip or other tip-surface interactions. The growth of buried-surface terrace structures is observed in real-time. One can measure the effects of the thermodynamic instability of the initially very small terraces by comparing the initial growth rate to the growth rate of larger terraces. The growth rate is related to the rate of adatom production, which can be determined quantitatively as a function of voltage from such measurements. Vacancy formation accompanies adatom formation. The structures formed by the vacancies can effect the BEEM images, which then allows a glimpse into their evolution. The BEEM image is correlated with an STM image to relate vacancy structure changes and topographic changes. The mechanisms of mound formation and stability can be addressed. The stability of the created structures is shown to be related to the physical properties of the gold film.

6. Acknowledgments

We wish to acknowledge the contributions of Tony Huang and Andres Fernandez and constructive discussions with David Peale, Andres Fernandez, Tony Huang, Wilson Ho, John Silcox, and Dan Ralph. We thank Bill Kaiser, Carl Kukkonen and the Jet Propulsion Lab for helping us to get started in STM. During part of the time this work was done, HDH was supported by an IBM Graduate Fellowship. Research support was provided by the Office of Naval Research and the Semiconductor Research Corporation. Additional support was provided by the National Science Foundation through use of the National Nanofabrication Facility and the facilities and equipment of the Cornell Materials Science Center.

7. References

[1]. Kaiser, W. J. and Bell, L. D. (1988) 'Direct investigation of subsurface interface electronic structure by ballistic-electron-emission microscopy', Phys. Rev. Lett. **60**, 1406-1409; and Bell, L. D. and Kaiser, W. J. (1988) 'Observation of interface band structure by ballistic-electron-emission microscopy', Phys. Rev. Lett. **61**, 2368-2371.
[2]. Hallen, H. D., et al..(1990) 'Gold-silicon interface modification studies', *Proceedings of the Fifth International Conference on Scanning Tunneling Microscopy/Spectroscopy*,

164

(Baltimore, July 23-27, 1990) in (1991) J. Vac. Sci. Technol. **B9**, 585-589; and (1990) 'Ballistic electron emission microscopy of metal-semiconductor interfaces',*Proceedings of the Ballistic Electron Emission Microscopy Workshop 1990* , Jet Propulsion Laboratory, California Institute of Technology, Pasadena, CA, March 9, 1990.

[3]. Hallen, H. D. (1991), Ph.D. thesis, 'Ballistic electron emission microscopy studies of gold-silicon interfaces', Cornell University.

[4]. Hallen, H. D. and Buhrman, R. A. (submitted) 'Hot electron induced atomic motion and structural change at the gold-silicon interface'.

[5]. For example Brillson, L. J., Katnani, A. D., Kelly, M., and Margaritondo, G., (1984), J. Vac. Sci. Technol. A **2**, 551.

[6]. Peale, D. and Cooper, B. H., (private communication); and Peale, D.(1992), Ph.D. thesis, 'Diffusion and mass flow dynamics on the gold (111) surface observed by scanning tunneling microscopy', Cornell University.

[7]. Ramsier, R. D. and Yates, J. T., Jr.(1991) 'Electron-stimulated desorption: principals and applications', Surface Science Reports **12**, 243-378.

NANOSCALE FASHIONING OF MATERIALS

R. FABIAN PEASE
Stanford University
CA 94305-4055 , U.S.A.

ABSTRACT. Today's lithographic technology is capable of exposing 0.5μm features over 1 cm^2 at a rate of 1cm^2/s with a 100nm tolerance on feature size (L) and feature placement (overlay). It appears that such optical technology can be extended to L< 0.2μm, maybe even to L=100nm. Going into the'nanometer range' below 100 nm however will probably require radically new technology that will be very expensive to introduce and therefore will have to be significantly better than 100 nm features; 25 nm feature size should be the goal of such 'nanoscale technology'. Merely achieving such resolution has been possible for many years. Achieving such resolution with adequate throughput but without a precise overlay capability may be just possible with special techniques such as contact X-ray lithography. Even finer features (5nm) might be possible for periodic patterns with special techniques such as the use of self assembling structures (e.g.two-dimensional protein crystals). But achieving 'full service' nanolithography, that is, L = 25 nm with overlay and feature size tolerance of 5 nm and with cost effective throughput (1cm^2/s) will be very challenging.

1.Introduction

The 20-year old trend towards finer features in silicon integrated circuits (SIC) is well known. The minimum feature size, L, of an SIC in 1972 was about 8μm. In 1982 the value of L had dropped to 2μm and in 1992 to 0.5μm. At this rate the value of L in 2002 will be 0.125μm. Although the imminent demise of optical lithography has been predicted throughout the last two decades, optical lithography is still the only serious wafer exposure technology. Furthermore values of L=125 nm have recently been demonstrated with an optical projection system (fig. 1) [1] and so it is likely that optical technology will continue to be the chosen technique for the next decade. This extraordinary extension of the 'old' technology has been made possible by a number of factors. Perhaps the most significant has been that the introduction of very high contrast photoresists has enabled us to compensate for the loss in contrast of the optical image as the imaged spatial frequencies have increased. The recent introduction of phase-shifting masks and other optical exotica has been another factor in stretching the resolution of optical techniques below the wavelength of the exposing radiation. However not the most ardent optical bigot is predicting that optics will be used below 100nm (because of the lack of sources and of materials that are mechanically rigid and transparent at wavelengths below 150 nm) and so a sub-optical technology will have to be introduced if we are to maintain the rate of reduction in L well into the next decade. Any such new technology will involve enormous costs in terms of displacing a huge infrastructure of optical lithography technology and so the new technology will have to show significant promise for going substantially below 100 nm; a value of L=25 nm seems to be one good goal.

There is another reason for picking L=25 nm as a goal which is that this represents about the useful resolution limit of polymeric resists. Such resists enjoy a sensivity that is very

P. Avouris (ed.), Atomic and Nanometer-Scale Modification of Materials: Fundamentals and Applications, 165–178.
© 1993 *Kluwer Academic Publishers.*

 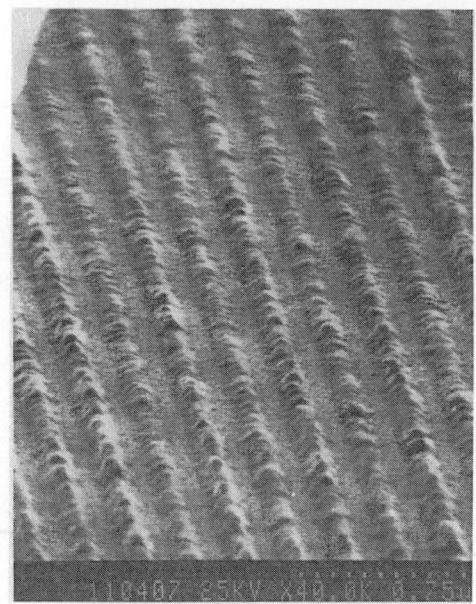

Fig. 1. Features 150 nm wide (left) and 125 nm wide (right) replicated into commercial photoresist using deep ultra violet (248 nm) optics. Photograph courtesy of R. Hsieh [1].

much higher than that of nearly all inorganic resists with resolution much better than 25 nm. However the region between 10 nm and 2 nm is very intriguing for two reasons:

1. It appears to be the most difficult to pattern for periodic patterns (as evidenced by the lack of soft X-ray gratings in this size range). Periodic structures finer than this occur naturally, and hence epitaxial techniques can be used to from finer periodic structures.

2. Quantum transport phenomena in semiconductors become appreciable at practical temperatures (>77K).

Although the main motive for pursuing nanoscale patterning is the fabrication of new functional structures and devices we should remember that we have been fashioning materials on this scale for many years but in the vertical dimension only. Thus a secondary motive is to develop further understanding of the nm-scale patterning that is already in use in the vertical dimension.

Below we discuss some of the issues and techniques for patterning materials in the size range 25 nm and below.

2.Lithographic Parameters

We refer to 'lithography' as the patterning (usually by exposure followed by development) of a resist film; the term 'pattern transfer' is used to convert the resist pattern into a functional material pattern. In this section we discuss the values of lithographic parameters as minimum feature size goes below 100nm. The parameters discussed are **resolution, throughput, overlay tolerance and feature size tolerance.** Other parameters include **defect density** and **cost** . The last is obviously key in any business and includes not only the cost of purchasing and installing but also operating cost and uptime. Both defect density and cost are obviously very difficult to predict and we will restrict our discussion to how the other paramters might affect these two.

The first parameter we discuss is 'resolution', or minimum reliably attained feature size, L. Frequently this is the **only** parameter discussed, but the absurdity of assuming resolution to be the only parameter of note is to consider that optical contact printing can attain a resolution of 0.5μm over a large field of view with a tool that costs less than $50,000 whereas the tools actually used, projection steppers, now cost about $2,000,000. The extra cost must go into providing satisfactory values of parameters other than resolution. We shall return to those other parmeters below. Nonetheless resolution is the one parameter that will force us to abandon optical techniques as we pattern finer features.

Attaining a value of L=25nm has been possible for at least 12 years by exposing a high resolution resist, poly (methyl methacrylate)(PMMA) with a focused, pencil, scanning electron beam. With a specialised exposure reaction a value of L=10nm was similarly demonstrated in a very limited sense as long ago as 1963 and atomic manipulation is now well in hand (see other papers in this conference proceedings). So resolution per se in the sub-optical range does not appear to be the limiting parameter.

In fig. 2 we show a page of a famous text patterned in PMMA by a scanning electron beam. The linewidth is nominally 23 nm. The area of the text is about $33\mu m^2$ and it took about 1 minute to write.This may not seem excessive until we work out that the time to write 1 cm^2 would be about 2 X 10^8 s; i.e. eight orders of magnitude too slow for SIC manufacture. Even allowing for improvements to this technology, eight is too many orders of magnitude to fight by simply pressing on an existing technology.

So we should add **throughput** to the list of relevant parameters and look for a technology that offers both L=25 nm and a throughput of 1 cm^2/s. A mask replication process seems attractive, if not mandatory, because of the needed massive parallelism of

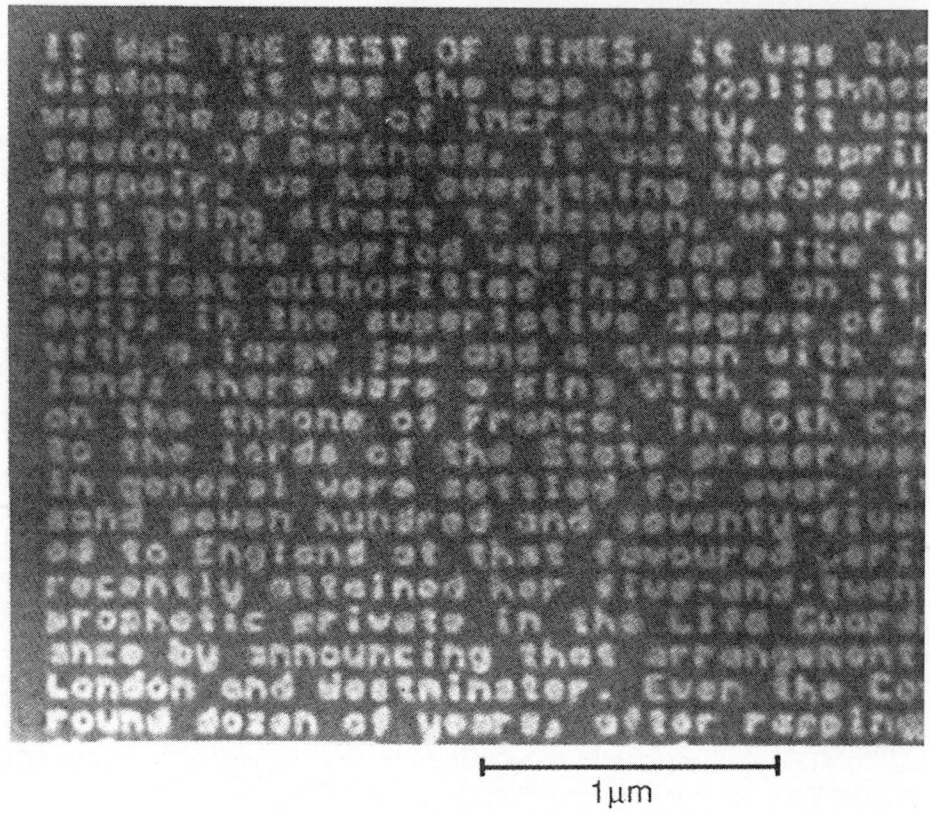

1 µm

Fig. 2. Transmission electron micrograph of a portion of the opening page of 'A Tale of Two Cities' written using direct-write electron beam exposure of poly (methylmethacrylate). The minimum line width is 23nm to give 12 characters per micron (i.e. reduced 25,000X from normal size to win the second Feynman prize). Photograph courtesy of T. H. Newman [2].

the exposure process to allow throughput; we shall return later to the issue of how we might avoid the need to make a mask. For now, assuming we use a mask replication process, the maximum area we can expose at any one time (the exposure field) is set either by the field of view of the replicating process (e.g. the lens system in a stepper) or by the largest area over which there is negligible distortion of the mask pattern relative to the structure on the wafer so that, given accurate alignment, we can achieve accurate overlay over the exposure field.

In today's practice the value of tolerance on overlay (with existing structure on the wafer), Δx, is about $L/5$. This value represents the total budget of error components such as those arising from distortion of the mask pattern, misalignment of the mask, and distortion of the wafer. How these are combined is a matter of some debate but, no matter how the errors are combined, in practice no one component should be larger than half of the total; i.e. $L/10$. A similar situation exists for errors in feature size, ΔL.

Maintaining overlay tolerance $\Delta x = L/10$ as L shrinks to 25 nm looks to be an extraordinary challenge. Assuming an exposure field of 25 mm diameter the above tolerance represents a strain of 10^{-7}. Given that the elastic modulus for silicon and for common mask materials is about 10^6 atmospheres this suggests that a stress of only 0.1 atmospheres is sufficient to cause unacceptable distortion. Such a stress is all too easily found in most mounting schemes and the sag between the three supports in kinematic mounting has been reported to be a serious problem even today in high precision mask making equipment [3]. In terms of thermal expansion directly the relevant coefficient for silicon is about 3ppm/K and this implies a temperature control of 0.03 K while delivering at least several mW/cm^2 over the exposure field. Indirectly the problem is even worse because the frame of the stepper or mask maker is usually made of material with even higher expansion coefficients making even more serious the resulting movement of the workpiece as a result of temperature variations of the frame. Thus it is likely that a radical new approach to achieving $L/10$ overlay tolerance will be needed before L approaches 25 nm.

An analogous problem arises with **critical dimension (CD) control**. Here again present practice is to require a tolerance of $\Delta L = L/10$ on any one contribution (e.g. the resist patterning at the mask making stage); i.e. $\Delta L = 2.5$ nm for the case we are considering. As we pointed out above, our case was chosen because it is close to the resolution limit of polymeric resist so, almost by definition, achieving a tolerance of $L/10$ will be difficult. Achieving this might be easier in thinner resists but the present conventional wisdom is to maintain present resist thicknesses of more than 500 nm because of the brutal pattern transfer processes believed to be needed to achieve anisotropic etching.

Thus if we follow the conventional wisdom we will need to pattern resist at the limit of its resolution not only with a tolerance of 10% of that limit but also with an aspect of ratio of 20:1. A similar situation pertains to the pattern transfer process itself because of the desire, for interconnects, to keep conductance per square high and capacitance per unit area low. Thus in the case of CD control radical new approaches will be needed before L approaches 25 nm. However before we discuss those let's consider how the problem might be eased if we relaxed some of the nominal requirements.

3. Interim measures

If we relaxed the throughput requirement then we can of course go to slower serial processes such a scanning electron beam lithography and this process is well known and widely used for delineating structures, devices and even circuits with features well below 100 nm [4][5][6]. Because of serial nature of the process alignment is relatively straightforward and so overlay tolerance is usually much less than the value of L. Because of the limited area of those circuits built with L<=100 nm it has not been necessary to employ alignment within the chip area. Achieving a linewidth control of L/10 for L<100 nm might well become challenging unless the resists or the imaging layer on the resist is suitably thin (<=L). Thus those applications not requiring the thicknesses required by the makers of SIC's might be able to maintain CD errors significantly less than L.

If we relaxed the overlay requirement then X-ray proximity printing (XRL) becomes attractive. The resolution of proximity printing is nominally $L = \sqrt{\lambda g}$ where λ is the X-ray wavelength and g is the proximity gap width. Very recent studies [7][8] indicate that this is a conservative measure and values of L=50 nm have been reported for λ =1.34 nm and gap 2.7 μm (fig. 3). The choice of l and g affords a rich study in tradeoffs.

The dilemma in the choice of g is straightforward; a small value leads to higher resolution and improved CD control but also greater cost (in providing a well controlled gap) and higher defect density. Present practice in an experimental manufacturing facility is to employ a value of g >= 25 μm for L= 0.5 μm [9].

The choice of value of λ is more involved. As seen above a lower value leads to higher resolution (which is obviously the primary reason for moving from UV to X-ray). A further advantage of a lower value is that the mask substrate can be thicker and more rigid. This is a fundamental difficulty of XRL and any other sub-optical replication process; there is no equivalent to glass, a material that is mechanically rigid yet transparent to sub optical radiation (and as we saw above in the discussion on overlay we need a very rigid substrate). In the case of XRL a membrane is used with obvious problems in the area of dimensional stability. A higher value of λ allows us to use a thicker membrane with less penalty in terms of stability but a thicker absorber is also needed to achieve adequate attenuation (by exactly the same argument). Since this absorber layer is patterned with arbitrary coverage any residual stress in the absorber layer will become increasingly serious as the layer becomes thicker. Similarly a higher value of λ will result in a higher penetration of the resist and so a higher dose (in terms of incident energy per unit area) is needed. During the last twenty years these tradeoffs have been extensively discussed in the literature [10][11].

There are arguments that can be made for improved CD control for XRL for largely because of this high penetration of the resist by the incident radiation and also because of the relatively high contrast of the aerial image compared with optical lithography near the latter's limit of resolution(). However since the contrast of the any aerial image falls, by definition, to an unnacceptably low value at the limit of resolution it is hard to distinguish the comparison of resolution from that of image modulation. The attainable CD control of the X-ray absorber on the mask is also a concern. Present-day masks employ absorbers 500 nm thick for replicating patterns with values of L=500nm. Scaling the thickness down as the value of L drops tenfold does not appear practical unless the value of λ is increased to maintain adequate attenuation and this in turn may lead to undesired attenuation in the nominally clear regions of the mask. Thus it is difficult to say ahead of time how XRL will fare in terms of CD control.

Thus we can regard XRL proximity printing as combining 50-nm resolution with potentially adequate throuput (1 cm^2/s).

If we are prepared to restrict our patterns to, for example a periodic grating or lattice of points then there are other techniques that can be used. Such patterns have been made by

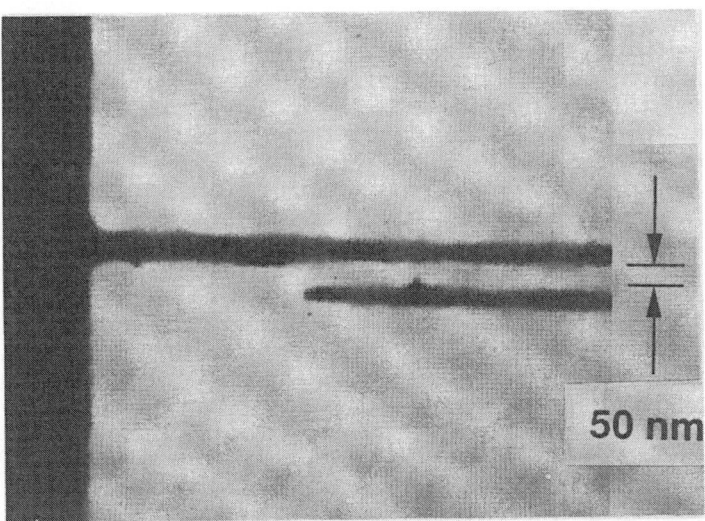

Fig. 3. Example of fine (50 nm) features replicated with X-ray proximity printing. Wavelength 1.34nm and gap 2.72 μm (photograph courtesy of H. I. Smith).

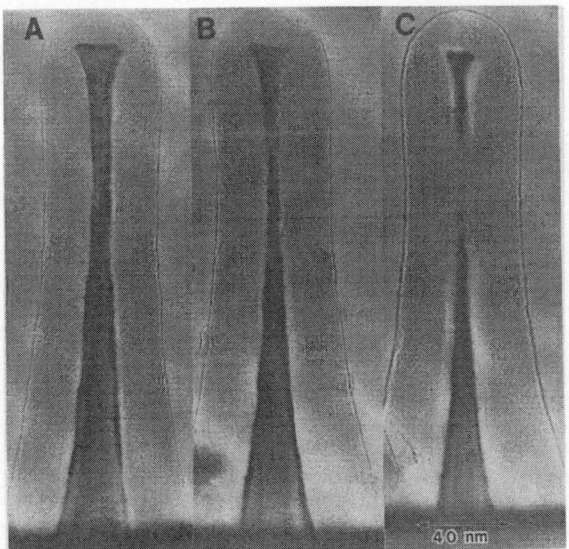

Fig. 4. Three transmission electron micrographs of the same silicon nano-pillar at three stages of oxidation. The silicon core is clearly visible inside the oxide cover. Photograph courtesy of H. Liu and F. Ponce [16]

interfering laser beams and have attained values of L=50 nm (half-pitch) [12] but this seems to be about the limiting resolution A little-used technique is 'natural lithography' in which small spheres are rafted together on the surface of the workpiece and so form *in situ* a periodic resist pattern [13]. The work done to date has employed rather large spheres (800 nm diameter) but there seems no reason why smaller spheres should not be used; polystyrene spheres of 20nm can be obtained commercially.

 Although the techniques described in this section are quite restricted, the practicality of the techniques for achieving nm-scale structures (compared with the non-existence of any full-service nm-scale patterning process) suggests that it might be profitable to look for suitable applications. One possibility that we have been considering is the patterning of silicon into fine (<30 nm diameter) pillars in an attempt to bring about silicon electroluminescent devices [14][15][16]. Another possibility might be for controllably dispersed magnetic recording material to achieve higher packing density. The luminescent silicon pillars is an example of how developing a lateral nm-scale patterning capability is paying off in today's technology is illustrated in fig. 4 which shows some silicon pillars that have been formed by electron beam lithography of overlying chromium followed by reactive ion etching to obtain pillars about 35 nm wide. To obtain narrower pillars the structure was oxidised and then etched. By developing a special specimen handling procedure it was possible follow the progress of the oxidation in a given pillar (fig. 4); we found that the oxidation process did not proceed as expected and hence these results are now being employed to help develop an improved oxidation model for incorporation into existing process emulators [17]. Thus, even before we employ lateral nm-patterning directly we can apply the technology to help understand today's processes which already involve vertical patterning on the nm-scale.

4. Radical new approaches

Here we describe a few ideas that might help address some of the problems of achieving 'full service' lithography in the range 5 <L< 100nm. They do not offer a complete set of solutions but hopefully these descriptions may help others come up with complementary ideas.

4.1 ADAPTIVE ALIGNMENT: ONE APPROACH TO NM-PRECISION OVERLAY

Traditionally we achieve accurate overlay by making very precisely the mask for each level and relying on the unpredictable distortion of the wafer being within our overlay tolerance over the workpiece field. The difficulties of this approach in the nm regime were pointed out above especially for larger workpieces (e.g. IC chips 20mm on the side). However such an approach is fundamentally more than is needed. For IC's (and most other applications) it is only necessary that the different levels overlay each other locally to, say, L/10; e.g. if a chip pattern is nominally 20,000μm on the side it is of no concern if it is 20,001 or even 20,100 μm on the side as long as any two levels are overlaid to the specified tolerance which might be 0.05μm today and 0.005μm for L=25 nm. The question is: how can we take advantage of this degree of freedom?

 One way is to adopt an exposure strategy termed two-dimensional scanning (fig. 5); only a small area (e.g. localized to, say, about 1mm^2) is illuminated and exposed at any one time. Thus it is only necessary that the mask pattern distortion (relative to the structure on the workpiece) be within the overlay tolerance over this small area (which we term the 'exposure field'). As the workpiece and mask are scanned through the illuminated area of

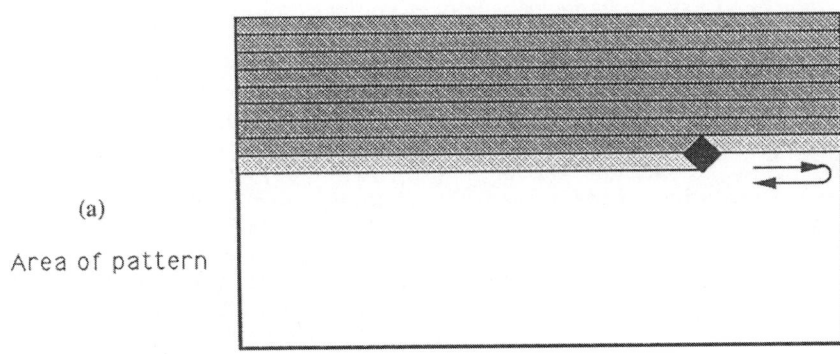

(a)

Area of pattern

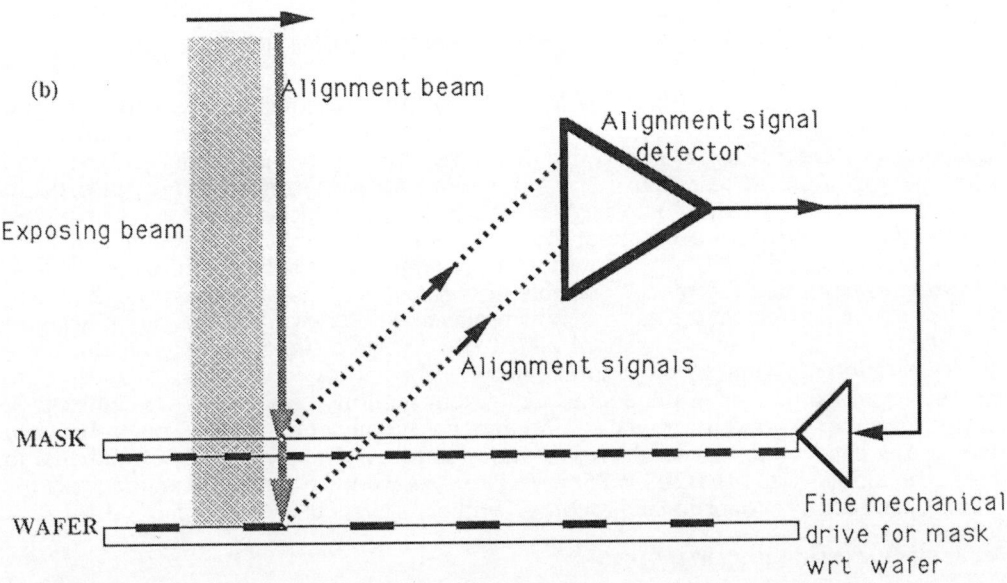

(b)

Alignment beam

Alignment signal detector

Exposing beam

Alignment signals

MASK

WAFER

Fine mechanical drive for mask wrt wafer

Fig. 5 Principle of two-dimensional Scanning. The pattern is built up
by scanning the exposing beam over the workpiece as shown in (a).
This can be accomplished either by directly writing with a close packed array
of pencil beams or by scanning the illumination over the mask and wafer (b).
Alignment can be continously maintained by detecting signals from wafer and
mask.

the exposure field the mask can be displaced relative to the workpiece to maintain precise overlay. Thus it is necessary that alignment signals originating from the mask and wafer be continuously detected and processed to provide input for the vernier drive to bring about the necessary displacement to maintain alignment. It is also necessary that the distortion have no appreciable high spatial frequency (wavelength smaller than the exposure field) components although low spatial frequency components may run to several μm. Given these two conditions we can, in principle, track, rather than correct, the distortion of the mask relative to the workpiece. Finding a means for continuously generating the alignment signals from the workpiece without exposing the workpiece is one problem that needs to be addressed. However lest the reader believe that the adaptive alignment approach is far fetched there is one reported example which is the electron beam proximity printer poineered at IBM Germany [18]. In that system the distortion of the (stencil) mask is first measured in the exposure tool and a two-dimensional power series function fitted to the distortion and stored. This stored function is then used to control the inclination of the exposing beam and so effectively displacing the mask relative to the wafer during the subsequent, two-dimensionally scanned, exposing process. The system was designed for features no smaller than 500 nm; this lower limit was probably set by the difficulties of making a stencil mask.

4.2   ARRAYED STM TIPS

Since STM lithography was first reported in 1985 [19] there has been a wide range of wrtiting mechanisms reported [20][21]. However all have in common certain drawbacks. The throughput is even slower than direct-write electron-beam lithography and the tip to workpiece gap must be less than L, about 10nm for L=25 nm, and this leads to tip crashes. Solutions to the small gap problem are being pursued along the lines of adding focusing fields [22][23]; no practically attractive method of doing this has yet been demonstrated but neither has any 'showstopper' emerged. One solution to the throughput problem is to have an array of tips and this now looks very much like the 2D scanning described above except that the mask is now replaced with an array of tips that is scanned across the target. If we have an array of 200 X 200 = 40,000 tips in a square lattice array with a separation of 5 μm , then the square array will be 1mm on the side and so we have the same exposure field as in the adaptive aligment section above. Thus we can employ adaptive alignment with this arrayed stm approach. Of course a 5μm separation between adjacent tips means that we will have to rotate the array slightly (by 1/200 radians) to achieve an address size of 5/200 = 25nm (fig. 6). As we have more tips in the array the scanned speed can be proportionately reduced without sacrificing throughput.

4.3   QUANTUM LITHOGRAPHY

We are all familiar with the ease of drawing good-looking diagrams with a computer-aided-drafting tool. The essence of the tool is that it will only allow feature edges that are straight, or at least smooth, and that these edges only occur along certain directions (e.g. parallel to the cartesian axes) and at certain co-ordinate values .It would clearly be an advantage if we could devise a lithography technique that enjoys such properties since most IC design rules already require such spatial quantization of feature edge positions (and directions). Although we have not devised a practical approach for such quantum lithography on the wafer, we have demonstrated the technique for a 5X reticle (albeit at rather coarse design rules, L= 500nm)[24] (fig. 7). The chromium film is first patternéd with a square array of trenches 250 nm wide that divide the workpiece into a square array of tiles each 2.5μm on the side; such a process might be carried out with an expensive,

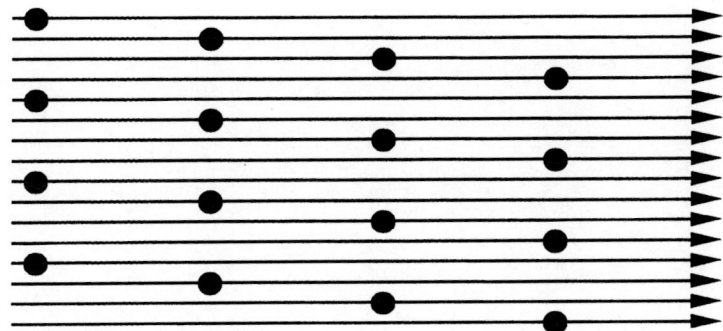

Fig. 6. To obtain a close-spaced line exposure a two-dimensional array can be either skewed (as shown) or simply rotated.

Fig. 7. Example of 'quantum lithography' in which feature edges can only occur at certain locations. Left picture shows the 5X reticle pattern in which the 2.5μm chromium tiles are visible and are separated by 0.5 μm trenches which define the allowed locations of feature edges. In the replicated (5X reduced) pattern with 0.5μm lines (right) the trenches (0.1μm wide) are not resolved. Photograph courtesy of N. Maluf [24].

176

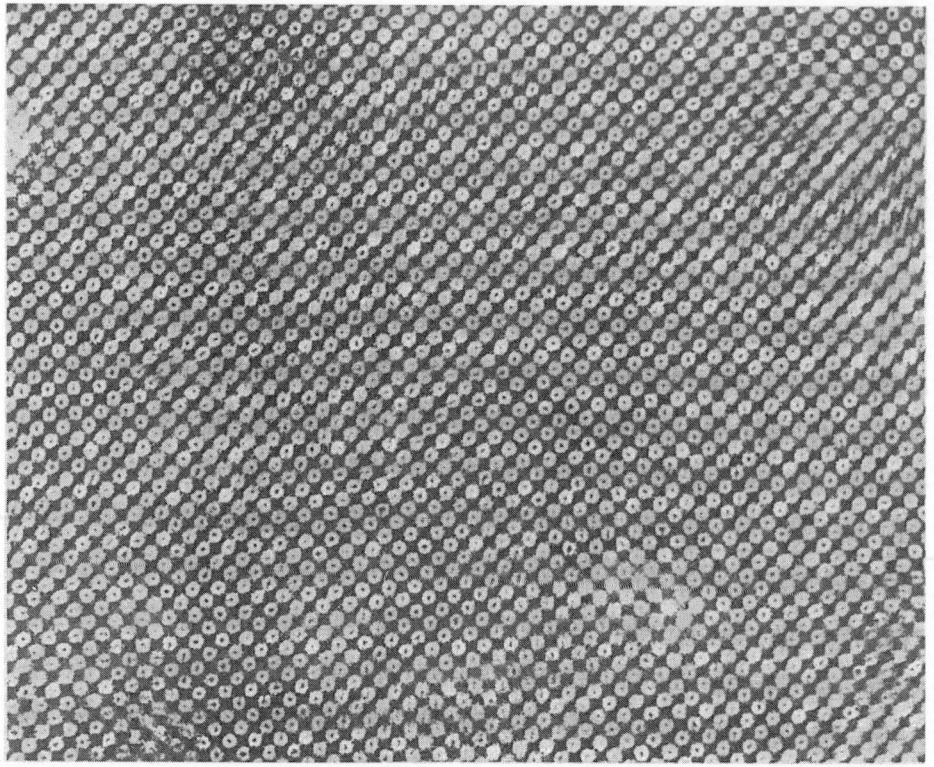

Fig. 8. Transmision electron micrograph of an organic, two-dimensional crystal (influenza neuraminidase) stained with uranyl acetate. The lattice constant is about 10nm; i.e. below the limit for resolution by polymeric resists but a very interesting size range for quantum effects in semiconductor devices. Photograph courtesy of D. C. Wiley [26].

very precise tool at the factory manufacturing the mask blanks. The workpiece is then spin-coated with a layer of positive resist. To pattern the reticle we merely have to expose some portion of the resist above a tile that is to be etched away. After the resist is developed, the chromium is over-etched so that the entire tile is etched away and so the etching stops precisely at the edge of the tile where the resist filling the trench prevents any further encroachement of the etch towards the neighboring tile. In fig. 7 we can see that the 5 X reduction printing of this reticle preserves the pattern and eliminates any image of the trench whose width is below the resolution limit of the stepper. Variations of this technique have been pursued; in one the tiles to be protected from the etchant can be selected by depositing on them an appropriate metal that causes a protective passivating layer to be formed [25].

4.4   NATURAL NANOLITHOGRAPHY

As the title implies, the concept of natural lithography described above can be extended in to the nm regime. One way would be simply to find finer spheres to raft together in a Langmuir-Blodgett trough. A more elegant approach being taken by a few groups is to find a material that can be grown as two-dimensional crystal with a lattice constant of the desired size (e.g. streptavidin)[26]. The crystal can then be stained with a biochemically specific stain that contains a heavy element as in transmission electron microscopy (fig. 8). Thus we will have a lattice of sites of concentration of the heavy element and such a lattice might be transferred to a functional material on the workpiece such as by implanting dopant, or alloying material, through the interstices between the heavy element concentrations. As far as the author us aware this last step has not yet been demonstrated. As with all natural lithography, the choice of patterns will be restricted to those that are periodic. Nonetheless this is the only technique that appears to have promise of combining a patterning resolution below 10nm with throughput that might have some technologically useful application (e.g. in forming arrays of the silicon nano-pillars).

5. Conclusions

Although resolution is the most talked-about parameter in lithography and is the one parameter that will eventually set a limit to optical lithography it is by no means the only, or even the most critical, parameter. Overlay, throughput, feature-size control, reliability, operating and capital cost, and freedom from defects are at least equally relevant. Cost effectiveness is some combination of all the above parameters.

Improvements in resist technology have been critical in allowing optical lithography to be extended.

Direct extension of existing lithographic technology is unlikely to yield 'full service' lithography much below 100 nm; radical new approaches will be needed. Some examples already exist; more new ideas are needed.

Research on (lateral) nanoscale patterning is already yielding dividends in understanding existing manufacturing processes many of which already involve vertical nanmeter-scale patterning.

Feature sizes beow 20 nm are difficult to resolve in polymeric resists. A possible alternative approach for periodic patterns is to use self- or induced- (e.g. Langmuir Blodgett) assembling techniques.

6. References

1. G. Owen *et al.*, (1992), '1/8th Micron Optical Lithography', paper presented at the 1992 International Symposium on Electron Ion and Photon Beam Technology, Orlando FL, USA, June 1992, accepted for publication J. Vac. Sci. Tech. Nov./Dec. 1992.

2. R. P. Feynman, (1960), 'There's Plenty of Room at the Bottom', Engineering and Science, **23**, 22.

3. S. Imai *et al.*, (JEOL), paper presented at 1992 annual mtg. of Bay Area Chrome Users' Group (BACUS), Sunnyvale CA,USA (Sept. 1992).

4. D. R. Allee *et al.*, (1991), 'Limits of Nano-Gate Fabrication', Proc. IEEE, **79**, 1093.

5. S. Y. Chou *et al.*, (1991), 'Lateral Resonant Tunneling Transistors Employing Field-Induced Quantum wells and Barriers', Proc. IEEE, **79**, 1131.

6. G. Sai-Halasz *et al.*, (1987), IEDM Tech. Digest, **1987**, 246.

7. H. I. Smith (1992), Private Communication , 1992.

178

8. F. Cerrina (1992), Paper presented at MicroProcess Symposium, Kanagawa .

9. R. Hill, A. D. Wilson (1988), Paper presented at IEDM 1988.

10.D. L. Spears and H. I. Smith (1972), Electron. Lett., **8**, 102.

12. D. Maydan *et al.* (1975), IEEE Trans Elec. Dev. **ED22**, 429.

12. D. C. Flanders and N. N. Efremow (1983), J. Vac. Sci. Tech. **B1**, 1105

13. J. Deckman and J. Dunsmuir (1983), J. Vac. Sci. Tech., **B1**, 1109.

14. L. T. Canham (1990), Appl. Phys. Lett., **57**, 1046.

15. M. Hybertsen (1991), Paper presented at Fall Mtg. of Materials Research Society .

16. H. Liu (1992), Fabrication of Si Nanostructures for Light Emission Study, paper presented at the 1992 International Symposium on Electron Ion and Photon Beam Technology, Orlando FL, USA, June 1992, accepted for publication J. Vac. Sci. Tech. Nov./Dec. 1992.

17. A. McCarthy (1992), Unpublished report, Stanford University Center for Integrated Systems.

18. H. Bohlen *et al.* (1982), IBM J. Res. Dev. **26**, 568.

19. M. A. McCord and R. F. W. Pease, paper presented at the 1985 International Symposium on Electron Ion and Photon Beam Technology, Portland OR, USA, May 1985; published J. Vac. Sci. Tech. , **B4**, 86.

20. M. A. McCord et al. (1988), J. Vac. Sci. Tech. **B6**, 1877.

21. E. A. Dobisz et al. (1990), J. Vac. Sci. Tech., **B8**, 1754.

22. T. H. P. Chang, D. P. Kern., L. P. Muray (1990), J. Vac. Sci. Tech., **B8**, 1698.

23. L. S. Hordon, R. F. W. Pease (1990) J. Vac. Sci. Tech., **B8**, 1695.

24. N. Maluf, R. F. W. Pease (1991), 'Quantum Lithography', J.Vac Sci. Tech., **B9**, 2986.

25. T. A. Fulton and G. Dolan (1983), 'Brush Fire Lithography', Appl. Phys. Lett., **42**, 752.

26. F. L. Carter (1983), J. Vac. Sci. Tech., **B1**, 959.

27. S. D. Hector *et al.* (1992), paper S4 presented at Int. Symp. on Electron, Ion and Photon Beams, Orlando, FLA. U. S. A. .

GROWTH AND IN-SITU PROCESSING OF LOW DIMENSIONAL QUANTUM STRUCTURES

J. L. MERZ
P. M. PETROFF
Departments of Electrical and Computer Engineering
and Materials
Center for Quantized Electronic Structures (QUEST)
University of California, Santa Barbara, CA , 93106
U.S.A.

ABSTRACT. This paper reviews recent advances in the fabrication and characterization of low-dimensional quantum structures such as quantum wires, boxes, and dots. The emphasis is on ultra-high vacuum compatible techniques, such as direct growth on vicinal substrates using molecular beam epitaxy, and use of in-situ processing: reactive ion etching, radical beam etching, and focused ion beam implantation, all of which can be followed by regrowth to make complicated structures. The current status of research using these methods will be described.

1. Introduction

Much has been said and written about so called quantum wires and dots which could make possible microamp threshold lasers and ultra-high mobility transport devices, and which show a variety of interesting properties, such as highly anisotropic behavior and strong non-linear optical properties. (For example, cf. Merz and Petroff, 1991, and M. Sundaram *et al.*, 1991, and the references contained therein.) However, the realization of such properties requires extreme control of fabrication tolerances. Small deviations of structural dimensions quickly destroy the advantageous properties anticipated. In this paper we describe the fabrication of quantum wires and dots using direct crystal growth by molecular beam epitaxy (MBE) on vicinal substrates, and in-situ processing techniques subsequent to growth which makes possible a variety of etching and regrowth procedures. It is believed that these techniques may lead to the fabrication of uniform quantum structures for which the inherent advantages mentioned above may actually be realized.

In Section 2 of this paper, we present a brief survey of lateral superlattice growth on vicinal (i.e., slightly misoriented) substrates, followed by a description of step-flow MBE growth of "tilted superlattices" (TSL). The difficulties involved in such growth are briefly described, and a novel solution to the most serious of these difficulties (excessive and uncontrollable tilt) is presented: the growth of "serpentine superlattices" (SSL). The optical properties of these SSL are then described. It is found that a high degree of uniformity of these optical properties is observed. The implications of these measurements are then discussed.

In Section 3 the use of in-situ processing techniques is briefly described, including focused ion beam (FIB) implantation, dry etching, and regrowth, all under ultra high vacuum (UHV) conditions. Some preliminary results obtained using these procedures are presented including the fabrication of wires and dot structures using buried stressors, and the fabrication of resonant tunneling diodes (RTDs) with dot emitters having peak-to-valley ratios (PVR) comparable to state-of-the art RTDs made by conventional MBE.

P. Avouris (ed.), Atomic and Nanometer-Scale Modification of Materials: Fundamentals and Applications, 179–190.
© 1993 *Kluwer Academic Publishers.*

2. MBE Growth On Vicinal Substrates

2.1 GROWTH OF TILTED SUPERLATTICES (TSL)

The basic idea of TSL growth by MBE is conceptually simple: one starts with a vicinal substrate misoriented from a high symmetry crystal direction by a small angle, α (cf. Fig. 1).

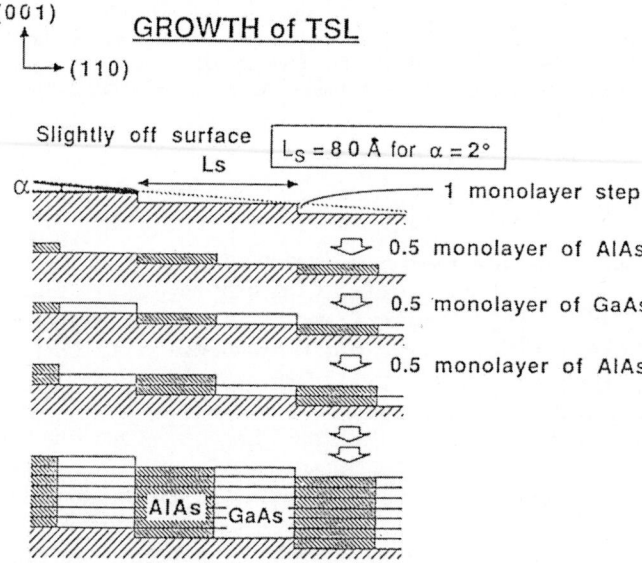

Figure 1. Growth of tilted superlattices (TSL) on vicinal substrates, which form monolayer steps whose spacing is appropriate for the formation of quantum structures. If alternate half-monolayers of different semiconductor compounds are deposited on this substrate, vertical superlattices result with period given by the step spacing.

This misorientation is accommodated by monolayer steps on the sample surface. For an angle of 2°, the average step spacing is only 8 nm, a dimension that is very appropriate (and hard to achieve) for quantum wires and dots. Alternate half-monolayers of GaAs and $Al_xGa_{1-x}As$ are then grown (where x is sometimes unity), leading to a vertical superlattice structure of 4 nm GaAs quantum wells bounded by 4 nm barrier layers of $Al_xGa_{1-x}As$. Quantum wires can be grown using this technique by growing thick barrier layers below and above the TSL region. This scheme, though conceptually simple, requires a degree of MBE control only recently achievable. Slight departures from exact monolayer coverage per cycle can result in large angles of "tilt" of the layer away from the surface normal, as shown in Fig. 2.

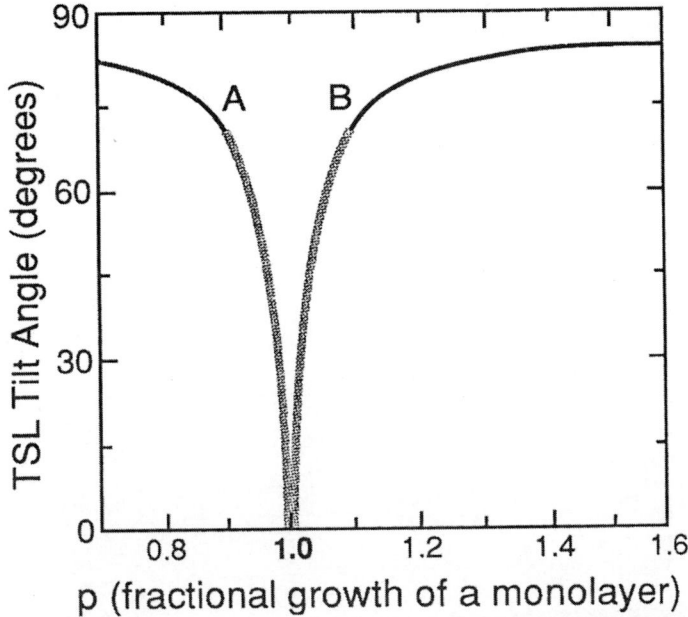

Figure 2. Plot of the TSL tilt angle, measured from the substrate normal, as a function of the fraction of a monolayer (p) grown on the vicinal substrate for each cycle. The strong sensitivity of tilt agle for slight departures from exact monolayer growth ($p = 1.0$) is a fundamental difficulty of this technique.

Nevertheless, excellent results have been achieved using these techniques. The original reports were for growth of GaAs/AlGaAs on [001] substrates by Gaines *et al.* (1988) and Tanaka and Sakaki (1989) using MBE, and Fukui and Saito (1988) using metalorganic chemical vapor deposition (MOCVD). More recently results have been reported for MBE growth on [110] vicinal substrates by Wassermeier *et al.* (1992) and MOCVD growth on [111B] surfaces by Fukui and Saito (1991). The growth of AlGaSb TSL by MBE on [001] wafers has also been reported by Chalmers *et al.* (1990).

2.2. PROBLEMS ASSOCIATED WITH TSL GROWTH

Successful growth of TSL requires near perfect step-flow growth, which is achievable in only a very narrow window of growth conditions. The problem of controlled step-flow growth is exacerbated by the lack of observable reflection high energy electron diffraction oscillation (RHEED oscillations), which are incompatible with pure step-flow growth. Careful growth calibrations must be made outside the "TSL window" and then the growth conditions changed to move inside that window. Other problems inherent in obtaining the required phase separation and superlattice orientation for TSL are (1) the formation of islands on the terraces, (2) exchange reactions of underlying atoms with terrace atoms, or step-edge atoms with terrace atoms, (3) incommensurate terrace/reconstruction, (4) kinks in the direction of the step edge, (5) unequal step separations, and (6) excessive and uncontrollable tilt. It is felt that the first three of these problems can be adequately controlled by appropriate growth conditions. Problem (5) turns out not to be a serious one, because the step edge velocity increases for longer terraces, leading to an equalization of the step-edge separation (Chalmers *et al.*, 1990, and Gossman *et al.*, 1990). The effect of kinks is only now coming under investigation by scanning tunneling microscopy (STM) and other sensitive surface characterization techniques. Fig. 3 shows a very recently-obtained STM image of a step edge on a GaAs vicinal surface; the steps are seen to be smooth over relatively large distances, but many "kinks" can still be observed.

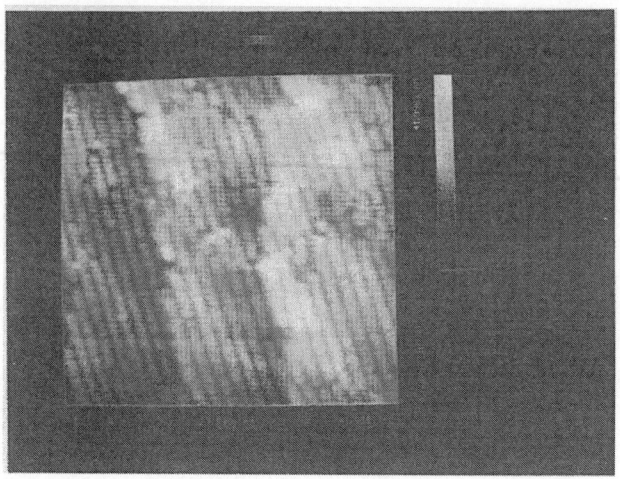

Figure 3. A scanning tunneling microscope (STM) image of a vicinal 1° GaAs (001) surface (after Bressler-Hill, *et al.*, 1992).

It is believed that a close coupling between high resolution characterization techniques such as this, and a comprehensive study of crystal growth parameter space can control the occurrence of kinks to a tolerable level.

The problem of "tilt" of these TSL structures remains a serious one, however. As shown in Fig. 2, a variation of the fraction of a monolayer coverage of only $= 2\%$ (i.e. $0.98 < p < 1.02$, where p is the fractional monolayer coverage) can lead to tilt angle variations outside the range $\pm 45°$. Since $\pm 2\%$ variation represents the range of coverage over a 2" wafer, this sensitivity of tilt angle is simply unacceptable.

2.3   GROWTH OF SERPENTINE SUPERLATTICES (SSL)

A very creative solution to this difficulty was devised by Mark Miller, a graduate student at UCSB, in which the fractional monolayer was periodically varied from below $p = 1.00$ to above that value, and then back (Miller *et al.*, 1992). In the resulting structure, called a serpentine superlattice (SSL), carriers should be confined to the crescent-shaped region which is formed when the monolayer coverage goes through exactly $p = 1.00$. This is shown schematically in Fig. 4.

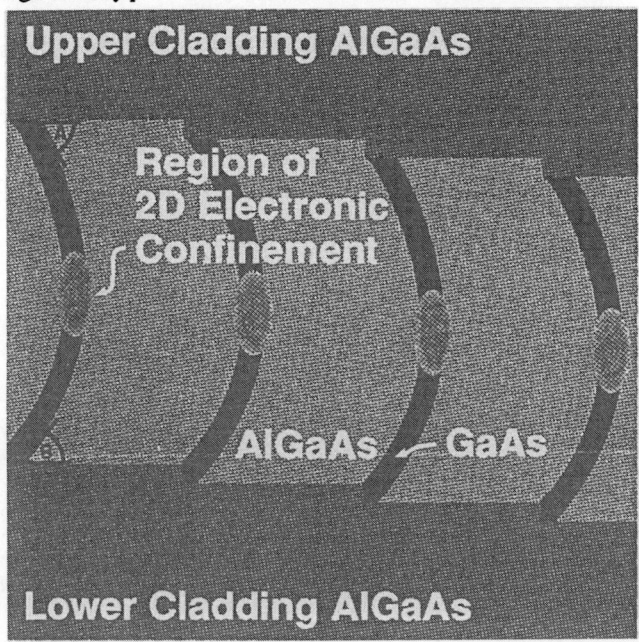

Figure 4. Schematic diagram illustrating the technique of growing serpentine superlattices (SSL) by varying the fractional monolayer growth coverage, p, between $p > 1$ and $p < 1$.

Moreover, the wires formed in this way are expected to be quite uniform. The parabolic areas produced in this fashion can be described as having a radius of curvature, r, at the vertices given by $r = z_o \tan \alpha$, where α is the misorientation angle, and

$$z_0 = \frac{layer\ thickness\ grown}{coverage\ change}$$

A series of samples has been grown with values of r varying from $r = 2.4$ to $r = 12.2$ nm, and comparison was made between the calculated radii of curvature calculated from growth parameters, and radii measured from TEM micrographs. The agreement is found to be excellent (Miller, 1992). These samples were produced by making so-called digital alloy barriers, where p layers of AlAs binary material are alternated with q layers of the GaAs binary, yielding a barrier concentration of Al, $x_{barrier}$, given by

$$x_{barrier} = \frac{p}{p + q}$$

SSLs with values of $x_{barrier}$ between 1/2 and 1/5 have been grown.

One example of the resulting SSL structures grown using these techniques is shown in Fig. 5. The left side of this figure (parts B,C, and D) show a drawing of the SSL, and calculated wave function densities for the $n = 1$ ground state and $n = 2$ first excited state, respectively. The right side of the figure (A) is an actual TEM of the SSL; the agreement with the predicted structure (part B) is remarkable, and led recently to a cover article in *Science* (Sundaram *et al.*, 1991).

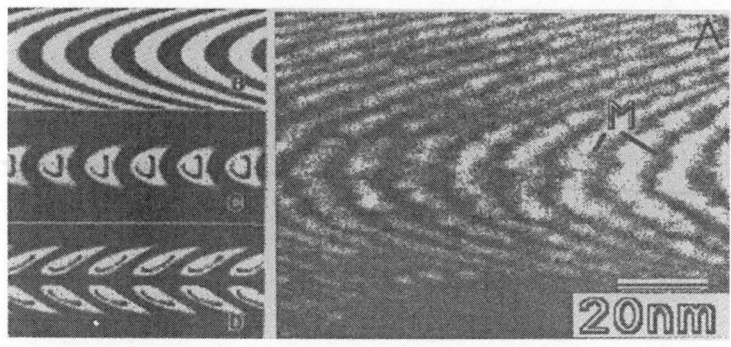

Figure 5. The left side of this figure shows a model of the SSL wires (B), and the calculated ground and excited-state wave functions (C and D). The right side is an actual STM image of the SSL (A).

2.4 CHARACTERIZATION OF SSL: QUANTUM WIRES WITH 2D HOLE CONFINEMENT.

However, these dramatic TEM micrographs of SSLs are clearly not sufficient to demonstrate the 2-D confinement sought for quantum wires, and systematic optical measurements were also carried out, using a photoelastic modulator technique recently described by Wassermeier *et al.* (1992) to measure emission and absorption polarization properties. Comparison was made with theoretical calculations carried out by MacIntyre and Sham (1992) and by Pryor (1991). Polarization was measured in photoluminescence (PL) and photoluminescence excitation (PLE), the latter being a "quasi-absorption" technique, using all three possible orientations of the incident photon wave vector (k) relative to the SSL wires: parallel with the wires, perpendicular to both the wires and the growth direction, and perpendicular to the wires in the growth direction. Results were compared with the calculated values of the polarization expected for the heavy hole and the light hole as a function of $x_{barrier}$. An example is given in Fig. 6 for P_z polarization. All of these experiments yielded very consistent results; for example, in one sample the observed polarization for $k \parallel x, y$, and z (where x is the wire direction, z the growth direction, and y the direction perpendicular to both) corresponded to barrier concentrations $x_{barrier} = 0.225, 0.242$, and 0.205, respectively, for an average Al composition of $x_{barrier} = 0.22 \pm 0.02$.

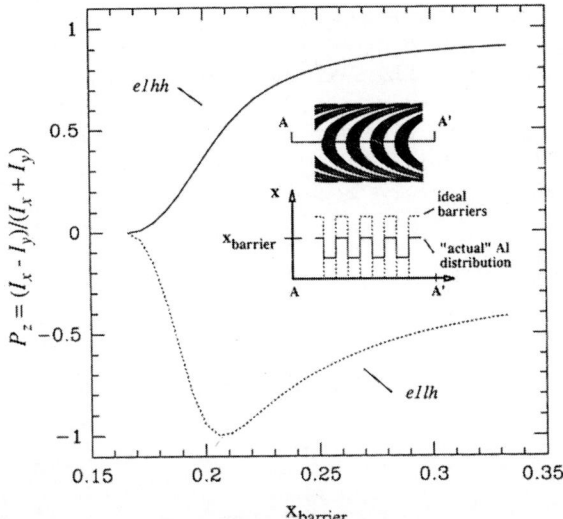

Figure 6. Plot of the calculated luminescence polarization as a function of Al composition of the barrier, for both light and heavy holes. The measured value of P_z corresponds to $x_{barrier} = 0.205$, whereas $x_{barrier}$ (intended) $= 0.33$.

Because of the consistency of these results for three mutually perpendicular directions, as well as the measured lh - hh splitting and the overall energy observed in the PL spectra, it is felt that the observed polarization is a consequence of the 2-D confinement produced by these SSL structures, and that these measurements therefore represent the first unambiguous demonstration of the fabrication of quantum wires. Furthermore, the wires are found to be extremely uniform, as was predicted for this growth technique. For example, the polarization observed for $k \parallel z$ (P_z polarization), measured to be $P_z = 0.205$, has an rms standard deviation less than 10% of this valve ($\Delta P_z \leq 0.02$) over a total sample distance of 1.2 cm. This small variation is believed to result from MBE flux variations, rather than quantum wire dimensional variations, because the PL luminescence peak varies from 1.752 eV to 1.743 eV (e.g., 9 meV) over this same distance.

One serious problem remains, however, for the SSL growth of quantum wires: to date, the segregation of Ga and Al between the quantum wires and barriers is considerably poorer than anticipated. For example, for the sample described above, for which Al composition of the barrier was determined to be $x_{barrier} = 0.22 \pm .02$, the growth compositions were $p_{well} = 0$ (i.e. GaAs quantum well) and $p_{barrier} = 1$, $q_{barrier} = 2$ (i.e., $x_{barrier}$ (intended) $= 0.33$). Hence, one in every three Al atoms intended for the barrier ends up in the quantum well, yielding

$$x_{well}^{actual} = 0.11 \quad \text{and} \quad x_{barrier}^{actual} = 0.22$$

This situation leads to a confining potential of only 80 meV for electrons. This interdifussion of Al and Ga between the wells and barriers of SSLs has been routinely observed for most cases of growth on vicinal [001] substrates that have been investigated to date: approximately 1/3 of the Al atoms intended for the barrier end up in the well. The result appears to be relatively independent of crescent shape, radius of curvature, and barrier composition (Miller *et al.*, 1992). An example of the results obtained for various barrier concentrations is shown in Fig. 7, where the measured vs. intended barrier concentrations are plotted for $p/(p + q) = 1/2, 1/3, 1/4$ and $1/5$.

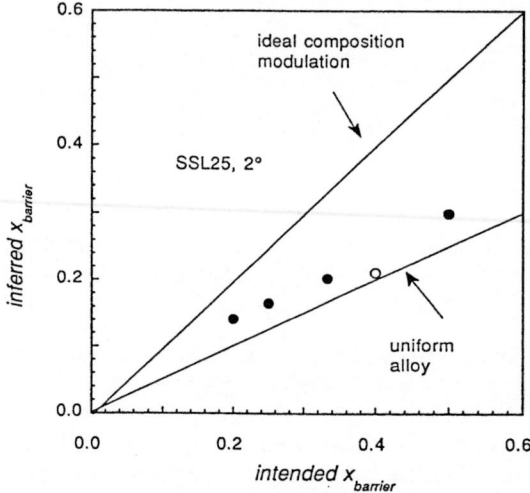

Figure 7. Plot of $x_{barrier}$ (actual) as a function of $x_{barrier}$ (intended) for 5 different samples. Consistent results are obtained for all samples grown at the same temperature (solid circles; the open-circle sample was grown at a lower temperature for which step-flow growth is less successful). The straight lines represent the limits of perfect segregation of well and barrier (upper line), and zero segregation (lower line).

In this figure the upper straight line (unity slope) represents perfect segregation, whereas the lower straight line represents complete interdiffusion and hence the absence of any confining potential. The measured data points suggest that only the hole experiences 2-D confinement in the structures fabricated to date, the electron remains two-dimensional, experiencing only 1-D confinement.

2.5   SUMMARY OF DIRECT GROWTH

The results of these experiments are extremely encouraging, but not totally satisfactory. Uniform quantum wires exhibiting 2-D hole confinement have been fabricated by SSL growth of GaAs/AlGaAs on [001] vicinal substrates. However incomplete segregation results in the cases studied to date, which must be improved before useful quantum-wire devices can be fabricated using the SSL growth technique. Further experiments are in progress to optimize SSL growth, including growth on substrates misoriented from principal axes other than [001]. Preliminary results on [011] vicinal substrates show an improvement in GaAs/AlAs segregation for TSL, but increased difficulty in orienting the structures (Krishnamurthy *et al.*, 1992).

3. In-Situ Processing And Regrowth

The extremely encouraging results reported above, obtained for the direct MBE growth of superlattices on vicinal substrates, suggest that the UHV feature of MBE should be further exploited in the quest for quantum structures by adding in-situ (i.e. UHV) processing techniques and MBE regrowth capabilities. To this end we have assembled at QUEST the system shown in Fig. 8.

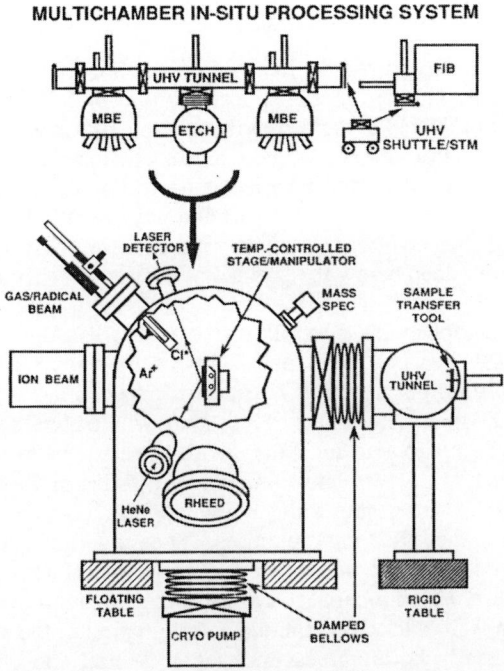

Figure 8. Schematic diagram of the in-situ processing system constructed at QUEST for the fabrication of quantum structures.

In addition to the two Varian Gen II MBE machines illustrated, a major UHV processing system has been constructed with load locks to the MBE chambers. This chamber includes an *Ar*+ ion etching gun, a source of Cl* radicals (excited neutral atoms) for damage-free chemical etching, temperature controlled sample stage and manipulator, and provision for in-situ optical reflection measurements to monitor etching. Samples can be moved from this system to a FIB implanter by means of a UHV shuttle which includes a vacuum-type STM for surface characterization and imaging with atomic resolution.

It is felt that this system offers an unusual degree of flexibility for innovative in-situ growth/processing/regrowth. One can envision many different scenarios involving surface preparation and/or characterization, MBE growth, etching to produce various patterns of features, FIB implantation or sputtering with resolution less than 100 nm, and regrowth of additional epitaxial layers by MBE. Additional facilities are available to complement these capabilities, including a large array of surface-chemistry/physics characterization tools (also accessible via the UHV shuttle), and an MOCVD reactor for the regrowth of phosphides and other compounds.

At present a variety of fabrication experiments are underway at this facility to fabricate structures which could eventually benefit from increased quantum confinement. Only two of these projects will be described (briefly) in this paper: (1) production of "buried stressors" to produce lateral localization of electronic carriers, and (2) fabrication of resonant tunneling diodes (RTDs) having quantum-dot emitters.

3.1   BURIED STRESSORS

Stress fields have been effectively used by Kash *et al.* (1988) to further localize electrons in quantum wells in one or more lateral directions, producing 2- or 3-D confinement (i.e., quantum wires or dots). One of the problems associated with this technique, however, is the fact that the quantum well in which the electrons are to be confined must be located near the stressor. Since the stressor is usually on the surface, this produces unwanted surface effects on carriers in the quantum well. A solution to this problem provided by Petroff and his co-workers and made possible by the in-situ processing system described above, is to bury the stressor below the quantum well, which can then be placed arbitrarily deep below the surface of the sample (Xu and Petroff, 1991, Xu *et al.*, 1992, Petroff *et al.*, 1992).

The technique used to accomplish this is as follows. A 6 nm InGaAs stressor layer is grown on a suitable buffer (e.g., GaAs/AlGaAs superlattice followed by thick GaAs layer), to reduce the propagation of defects up from the substrates. A 2 nm GaAs cap layer is grown on the stressor. The wafer is then transfered under UHV to the FIB system, where the GaAs and InGaAs layers are partially removed by 50 keV Ga ions to form the desired stressor pattern by ion sputtering. The wafer is then transfered back to the MBE machine where a series of GaAs quantum wells can be grown with different well thicknesses, and with suitable AlGaAs barriers (e.g., 200 nm).

The resulting structure has the following properties. The remaining InGaAs stressor segments, under compressive strain because the lattice constant of the InAs is larger than that of GaAs, create a complicated strain pattern which emanates upwards to the quantum wells. The different thicknesses of the quantum wells allow for a unique determination of the effect of stress on each of them by measurement of their photoluminescence (PL) spectra, and because each well is at a different distance from the stressor layer, these wells serve as a "depth gauge," providing information about the stress field as a function of separation from the stressor.

Interesting bandgap modulation effects have been observed for this system. Significant bandgap decreases have been measured directly above the stressors, resulting in a 6 mev redshift of the PL spectra. On the other hand, a blue shift is observed in the regions away from the stressor which results from the partial disordering caused by the diffusion of defects produced by the FIB sputtering. Although it has not yet been demonstrated that the redshift produced by this technique is sufficient for 2- and 3-D localization of electrons, observed shifts are in good agreement with theory using finite-element elasticity calculations, and larger shifts are predicted under appropriate conditions. The blue shift described above can be added to the bandgap reduction under the stressor, further increasing the confining potential barrier. For these reasons this technique appears quite promising for quantum structure fabrication.

3.2 DOT EMITTER RESONANT TUNNELING DIODE

In this experiment dot-like regions of n^+ material are created by FIB implantation of Si^+ ions into an n-type GaAs layer grown by MBE on an n^+substrate (Petroff *et al.*, 1992). The RTD structure, consisting of GaAs wells and AlGaAs barriers, is then grown on the implanted wafer by MBE, after UHV transfer using the shuttle. The entire structure is then treated by rapid thermal annealing, mesas are etched and contacts applied to make the tunneling diode.

Surprisingly good characteristics have already been measured in these structures, particularly for so preliminary a stage of experiment. Peak-to-Valley ratios as high as 10 at 77 K and 3.33 at 300 K have been observed, which are comparable in performance to RTDs made by conventional MBE techniques.

4. Summary

We have reviewed a novel crystal growth technique using MBE on vicinal substrates to produce lateral confinement of carriers. The original approach of making tilted superlattices (TSL) has been improved by periodically varying the beam flux during growth to make crescent-shaped or serpentine superlattices (SSL). The major problems with this technique are described, as well as current efforts to solve them. It is found that surprisingly uniform structures are obtained which show 2-D confinement for holes, but only 1-D confinement for electrons, but that complete segregation of the species used to make the quantum well and barrier has not yet been achieved.

Preliminary results of UHV in-situ processing experiments are also described. Significant red and blue shifts have been obtained in quantum wells by burying patterned stressors below the wells, and a resonant tunneling device with a dot emitter has been made by focused ion beam implantation whose performance in some respect rivals the state-of-the-art achieved with MBE. The cluster of MBE, etching, and FIB machines is believed to provide great flexibility for a variety of fabrication and processing procedures yet to be explored.

REFERENCES

Bressler-Hill V., Lorke A., Wassermeier M., Pond K., Maboudian R., Petroff P.M., Weinberg H. (unpublished).

Chalmers S. A. , Kroemer H. and Gossard A. C. (1990) Appl. Phys. Lett. **57**, 1751.

Chalmers S. A., Gossard A. C., Petroff P. M., Gaines J. M. and Kroemer H. (1990) J. Vac. Sci. Technol. **B7**, 1357.

Fukui T. and Saito H. (1988) J. Vac. Sci. Technol. **B6**, 1373.

Fukui T. and Saito H. (1991) J. Cryst. Growth **107**, 231.

Gaines J.M., Petroff P. M., Kroemer H., Simes R.J., Geels R. S., and English J.H. (1988) J. Vac. Sci. Techol. **B6**, 1378.

Gossman H. G., Siden S. W. and Feldman L. C. (1990) J. Appl. Phys. **67**, 745.

Kash K., Worlock J. M., Sturge M. D., Grabbe P., Harbison J. P., Scherer A. and Lin P. S. D. (1988) Appl. Phys. Lett. **53**, 782.

Krishnamurthy M., Wassermeier M., Weman H., and Merz J. L. (1992) in K. S. Liang, M. Anderson and R. Bruinsma (eds.) Interface Growth Dynamics, MRS Symp. **237,** 473.

Krishnamurthy M., Wassermeier M., Williams D., and Petroff P. M. (1992) Appl. Phys. Lett. (to be published)

McIntyer C. and Sham L. (1992) Phys. Rev **B45,** 9443.

Merz J. L. and Petroff P. M. (1991) Materials Science and Engineering **B9,** 275.

Miller M. S., (1992) Ph.D. dissertation (unpublished) Universtity of California Santa Barbara.

Miller M. S., Weman H., Pryor C. E., Krishnamurthy M., Petroff P. M., Kroemer H. and Merz J. L. (1992) Phys. Rev. Lett. **68,** 3464.

Petroff P. M., Xu Z., Li Y. J. and Wassermeier M. (1992) Proceedings, International Workshop on Quantum-Effect Physics, Eelctronics and Applications, Luxor, Egypt, 6-10 Jan. 1992, Institute of Physics, Adam Hilger Publishers, London (to be published).

Pryor C. (1991) Phys. Rev. **B44,** 12912.

Sundaram M., Chalmers S. A., Hopkins P. F. and Gossard A. C. (1991) Science **254,** 1326.

Tanaka M. and Sakaki H. (1989) Appl. Phys. Lett. **54,** 1326.

Wassermeier M., Weman H., Miller M. S., Petroff P. M. and Merz J. L. (1992) J. Appl. Phys. **71** 2397.

Xu Z. and Petroff P. M. (1991) J. Appl. Phys. **69,** 6564.

Xu, Z., Wassermeier, M., Li, Y. J. and Petroff, P. M. (1992) Appl. Phys. Lett. **60,** 586.

QUANTUM DOT FABRICATION BY OPTICAL LITHOGRAPHY AND SELECTIVE ETCHING

J.A. ADAMS, J.N. WANG, P.H. BETON AND M. HENINI
Department of Physics
University of Nottingham
Nottingham NG7 2RD
U.K.

ABSTRACT. We have developed a process for the fabrication of 100 nm scale dots using conventional optical lithography together with selective wet etching. Resonant tunnelling diode (RTD) and quantum well based dots have been fabricated from GaAs/AlAs heterostructures on (100) and (111)B GaAs substrates. The GaAs is selectively etched by NH_4OH/H_2O_2 to fabricate deep sub-micron structures. The selective nature of the etch facilitates control of the vertical dimension of the dots, and the preferential nature uncouples the etch front geometry from the photoresist mask geometry. The suitability of these dots for probing with a scanning tunnelling microscope will be discussed.

1. Introduction

A semiconductor structure in which the active region is sufficiently small ($\ll 1\mu m$) to quantise the motion of electrons and/or holes in all three dimensions is known as a quantum dot. The energy levels of a quantum dot are discrete and characterised by three quantum numbers. This quantisation of carrier motion is expected to result in a variety of novel electronic and optical properties. For example enhanced luminescence and low threshold current for quantum dot lasers has been predicted [1]. In addition, due to their small size the capacitive energy, $e^2/2C$, of quantum dots can be comparable with the thermal energy, kT, leading to single electron charging effects [2,3].

The most common fabrication process for quantum dots is a combination of electron beam lithography (EBL) and reactive ion etching. This technique has been used to fabricate quantum dots for photoluminescence studies [4], and also to investigate resonant tunnelling in structures with sub-micron lateral dimension [5]. Equivalent structures have also been fabricated using focussed ion beams [6], and exploiting strain caused by surface patterning [7]. More recently a device incorporating field-effect gating has been used to create a submicron resonant tunnelling diode with variable area [8]. In addition, quantum dots have been fabricated on modulation doped heterostructures by using surface Schottky depletion gates, either covering arrays of resist dots defined by holographic lithography [9], or defined as split gates by EBL [10,11]. Quantum dots have also been produced by wet chemical etching [12] using a metal etch mask defined by EBL and recently by epitaxial growth on patterned substrates [13].

In this paper we describe a process for fabricating dot structures on GaAs/AlAs heterostructures by preferential, selective, wet chemical etching. One of the applications we

191

P. Avouris (ed.), Atomic and Nanometer-Scale Modification of Materials: Fundamentals and Applications, 191–197.
© 1993 *Kluwer Academic Publishers.*

envisage for these dots is investigation and modification by a scanning tunnelling microscope (STM), both at room temperature and low temperature. Particular advantages of this method for fabricating structures to be probed by the STM are the overcut sidewalls and absence of the damage which normally results from reactive ion etching.

2.   Fabrication Process

Photoresist of thickness 1.2 μm was spun coated and hexagonal arrays of circular holes opened up by exposure using a dark field mask and development. Three arrays, each of total area 200 μm x 200 μm, were used. The nominal diameter of the holes were 6, 3.5, and 2 μm with centre to centre spacing of 16, 9.5, and 6 μm respectively. One of the three principal axes of the array of holes was aligned to a major crystal orientation (identified by cleave plane or wafer flat). The resist mask and dots are shown schematically in Figure 1. The etch pits had shapes determined by the etch preference for different crystal planes. The etch selectively stopped on the AlAs layers, and the etch front then proceeded laterally. Free-standing dots remain when the etch fronts from adjacent resist holes converge (see Figure 1).

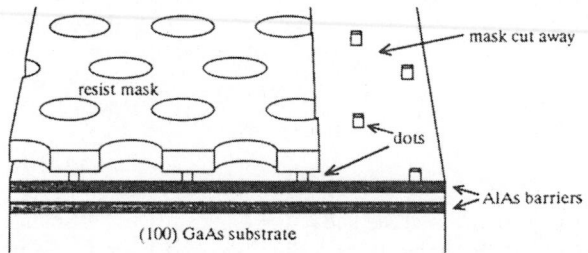

Figure 1. Schematic of quantum dot fabrication using a photoresist mask on a (100) wafer.

GaAs/AlAs heterostructures were grown by molecular beam epitaxy (MBE) on (100), (111)A, ($\bar{1}\bar{1}\bar{1}$)B planes. Etching of GaAs/AlAs heterostructures on (110) and (311) wafers was also investigated, but the planes revealed by preferential etching did not yield symmetrical dots, and so will not be further considered here.

In this paper we will present results for a resonant tunnelling diode (RTD) structure on a (100) substrate and a single quantum well (SQW) structure on a ($\bar{1}\bar{1}\bar{1}$)B substrate. The RTD structure consisted of 7.5 nm AlAs barriers, on either side of a GaAs well of thickness 9 nm. Each barrier was separated by an undoped GaAs spacer 20.4 nm thick from n-type GaAs contact regions where the Si doping varied from 2×10^{16} cm^{-3} close to the barriers to 2×10^{18} cm^{-3} over a thickness of 330nm. A 1.8μm n-type GaAs buffer doped to 2×10^{18} cm^{-3} separated the structure from the n+ GaAs (100) substrate. Thus the total thickness of the GaAs above the AlAs barriers was 350 nm. The SQW structure consisted of 19.9 nm AlAs barriers, on either side of a GaAs well of thickness 5.1 nm. The barriers were separated from the surface by an undoped GaAs cap 200 nm thick and from the semi-insulating GaAs ($\bar{1}\bar{1}\bar{1}$)B substrate by an undoped GaAs buffer 1.5 μm thick. In both cases the AlAs top barriers acted as the etch stop.

In addition a 1μm thick undoped GaAs layer with no etch stop was grown on a (111)A substrate. This layer required a high As$_4$ flux during growth in order to yield a good surface morphology.

H_2O_2 (30 weight percent):NH_4OH was used in the volume ratio 95:5. Samples were etched for up to 2 minutes and then rinsed in de-ionised water. The etch rate could be reduced by dilution of the etch with water, however this gave rise to a roughening of the etch front. The etch rate for (100) GaAs was 4.5 μm per minute, while that for AlAs was typically 10^3 times lower. This allowed the etch to stop on the 7.5nm AlAs top barrier of the resonant tunnelling diode after removing the GaAs cap layer. The etch rate of AlAs appeared to depend on the area of exposed surface, being slightly faster in the centre of large (100μm scale) etch windows. The GaAs etch rate generally varied gradually across each hexagonal array of dots, resulting in a range of dot sizes. This may be a consequence of the bubble formation which was noted during the etch. In this work, the etch conditions were optimised for a study of the shape and sizes of the dots, rather than etch rate uniformity.

Due to the lack of inversion symmetry of GaAs there are two types of {111} surfaces (faces) which have either all Ga or all As atoms exposed. These faces are referred to as {111}A and {111}B respectively. {100} faces have an equal proportion of Ga and As atoms. There are two free electrons available for chemical reactions on an As face, but none on a Ga face. This leads to a difference in the reactivity of the two surfaces with 95:5 H_2O_2:NH_4OH. The magnitude of the etch rates are ordered {111}B > {100} > {111}A [14]. Exposure to this etch is expected to reveal the slowest etching {111}A faces. This simple description is adequate for large structures, although we found that a {100} face was revealed at the external intersection of {111}A faces. However, for small structures, the situation is different. The etch faces which form the small dots cannot be predicted on the basis of the simple model above. Figures 2 and 3 show schematic diagrams of the faces revealed by etching large and small dots on the (100) and ($\bar{1}\bar{1}\bar{1}$)B wafers. The angle between the {111}A and {111}B planes is $2(90-\tan^{-1}\sqrt{2}) = 70.5°$ and the angle between the {111} and {100} planes is $\tan^{-1}\sqrt{2} = 54.7°$.

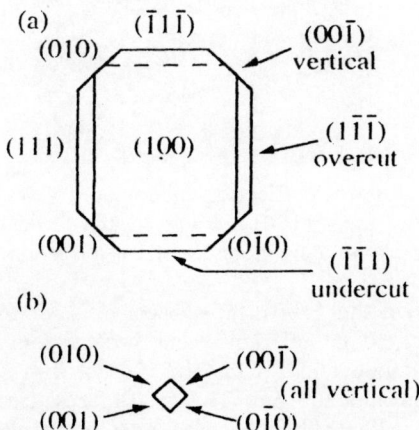

Figure 2. Crystal faces revealed by etching of (a) a large and (b) a small dot on a (100) wafer viewed along the <$\bar{1}$00> direction.

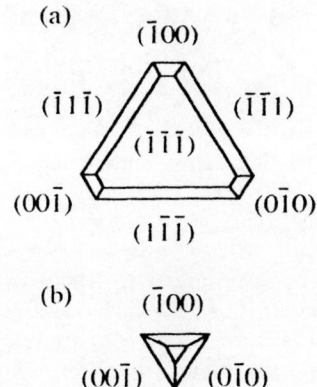

Figure 3. Crystal faces revealed by etching of (a) a large and (b) a small dot on a ($\bar{1}\bar{1}\bar{1}$)B wafer viewed along the <111> direction.

Note that although the resist mask had circular holes, the preferential nature of the etch governed the overall shape of both the etch pits and the quantum dots. The etch pits were rectangular for the (100) wafers and triangular for the {111} wafers. In addition, we found that on the (100) wafers, the etch rate of the (111)A and (1$\bar{1}\bar{1}$)A faces was greater than the rate of the ($\bar{1}\bar{1}$1)A and ($\bar{1}$1$\bar{1}$)A faces. This resulted in the formation of approximately square dots at the points between four holes in the hexagonal array of the mask.

3.   Experimental Results

Figure 4 is an electron micrograph of a large dot on the (100) RTD wafer with lateral dimensions of 1.3 and 1.5 μm. Comparing with Figure 2(a), the {111}A faces are the dominant remaining structure, with {100} vertical faces at the intersection of the {111}A faces. Figure 5 is a micrograph of a smaller dot with a 250 nm lateral dimension chosen to demonstrate the vertical {100} faces at the left and right of the dot.

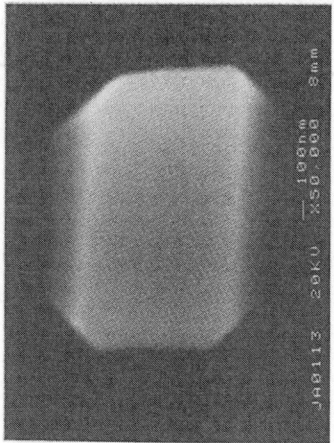
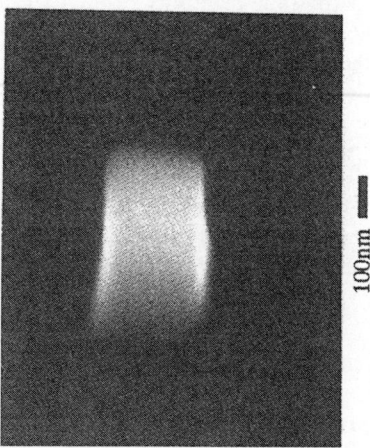

Figure 4. Electron micrograph of a dot on the (100) RTD wafer viewed along the <$\bar{1}$00> direction.

Figure 5. Electron micrograph of a dot on the (100) RTD wafer viewed approximately along the <$\bar{1}$01> direction.

As a dot is progressively etched towards zero size the {100} faces become the dominant feature. For (100) dots this typically occurs when the lateral dimensions become less than 500 nm. Figures 6(a) and 6(b) are respectively a side view and a plan view of a dot with a 120 nm tip and maximum lateral dimensions of 150 nm and 200 nm. The {100} faces are now dominant as in Figure 2(b), giving almost vertical walls, although some trace of the sloping (111)A and (1$\bar{1}\bar{1}$)A faces is evident as fluting at the bottom of the column in Figure 6(a). Figure 7 is a side view of one of the smallest dots on the (100) RTD wafer. This dot has an 80 nm tip.

Figure 8 shows the corner of an array of partially etched dots with a 4 μm lateral dimension on the (1$\bar{1}\bar{1}$)B SQW wafer (tilted to about 45° from a plan view). The dominant faces can be identified by reference to Figure 3(a) as {111}A. As with the (100) wafer, the {100} faces can be observed at the external intersections of the {111}A faces, in this case at the corners of each triangular dot. Figures 9(a) and 9(b) show micrographs (also taken with 45° tilt) of smaller etched dots from the same wafer. The dot in figure 9(a) should be compared with Figure 3(b), where the {111}A faces are absent. The dot in Figure 9(b) has been further etched to a lateral dimension of 200 nm. Its sloping sidewalls can be expected to facilitate probing by a STM.

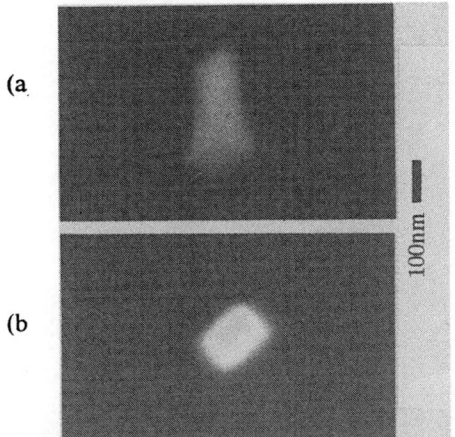

(a

(b

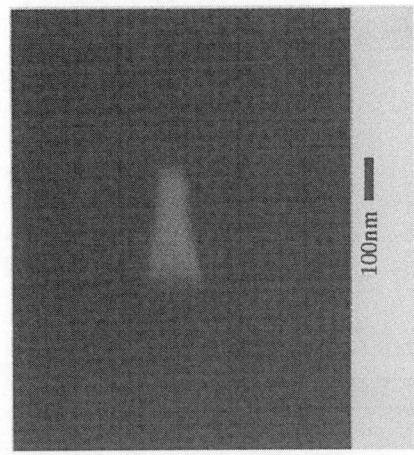

Figure 6. Electron micrographs of a dot on the (100) RTD wafer viewed along (a) approximately the $<\bar{1}1\bar{1}>$ direction and (b) the $<\bar{1}00>$ direction.

Figure 7. Electron micrograph of a dot on the (100) RTD wafer viewed approximately along the $<\bar{1}1\bar{1}>$ direction.

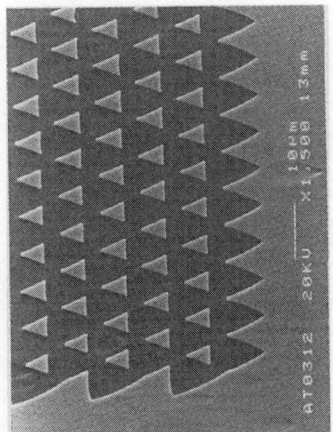

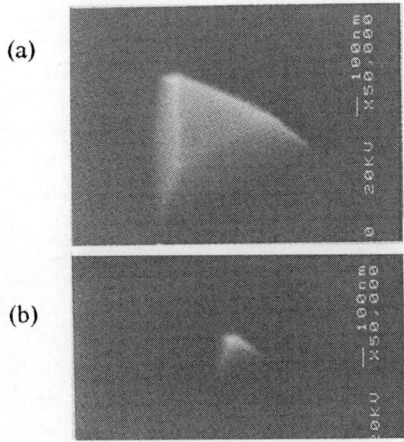

(a)

(b)

Figure 8. Electron micrograph of the corner of an array of partially etched dots on the $(\bar{1}\bar{1}\bar{1})$B SQW wafer viewed along a direction approximately between $<111>$ and $<0\bar{1}1>$.

Figure 9. Electron micrographs of two dots on the $(\bar{1}\bar{1}\bar{1})$B SQW wafer viewed approximately along the $<\bar{1}01>$ direction.

Figure 10 shows an array of inverted tetrahedral dots, which result from the application of the etching process to a (111)A GaAs wafer. Without an AlAs stop layer, the etch was deep. The lower apex of each dot has a lateral dimension of approximately 300 nm. We are currently evaluating GaAs/AlAs heterostructures grown on (111)A wafers in order to gain control of, and reduce the vertical dimension of the dots.

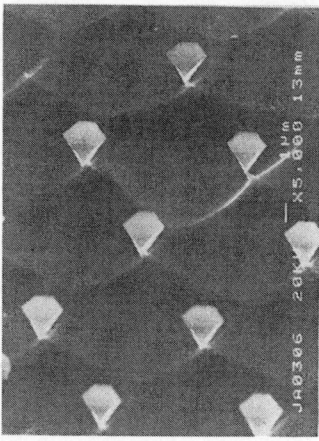

Figure 10. Electron micrograph of an array of dots on the (111)A undoped wafer.

4.   Discussion

We have shown that dots with lateral dimensions as small as 80 nm can be produced using conventional optical lithography. These dots have several advantages for fabrication of quantum dot devices including a wet etching process which avoids damage associated with reactive ion etching. Larger dots have a different shape to smaller dots. On a (100) wafer, large dots have a combination of overcut and undercut, whereas for less than approximately 0.5 μm in lateral dimension, the vertical {100} faces dominate the shape. On a ($\bar{1}\bar{1}\bar{1}$)B wafer, large dots have overcut {111} and {100} faces, but the overcut {100} faces were seen to dominate for lateral dimensions less than approximately 1μm. This lack of undercut could reduce the possibility of STM tip crashes while probing ($\bar{1}\bar{1}\bar{1}$)B dots. Furthermore, these dots could offer a substrate suitable for the technique of epitaxial overgrowth described in Ref. 13. The preliminary results for a (111)A wafer are encouraging in that large dots show symmetrical undercut of the {111} planes. This interesting structure might be optimised to give natural point contacts to heterostructures with a size only limited by mechanical stability. Furthermore, we can envisage devices based on self-aligned deposition of contacts using the undercut profile of the (111)A dots.

5. Acknowledgements

We are grateful for the technical assistance of P. O'Byrne, and the mask design of P. Steenson. This work is supported by United Kingdom Science and Engineering Research Council (SERC). One of us (P.H.B.) thanks the Royal Society (London) for financial support.

1 Arakawa, Y., and Sakaki, H. (1982) "Multidimensional quantum well laser and temperature-dependence of its threshold current", Appl. Phys. Lett. 40, 939-941.

2 Kouwenhoven, L. P., Johnson, A. T., Vaart, N. C. van der, Harmans, C. J. M. P., and Foxon, C. T., (1991) "Quantized current in a quantum-dot turnstile using oscillating tunnel barriers", Phys. Rev. Lett. 67, 1626-1629.

3 Heitmann, D., Kern, K., Demel, T., Grambow, P., Ploog, K., and Zhang, Y. H. (1992) "Spectroscopy of quantum dots and antidots", Surface Sci. 267, 245-252.

4 Clausen, E. M. Jr., Craighead, H. G., Worlock, J. M., Harbison, J. P., Schiavone, L. M., Florez, L., and Gaag, B. van der (1989) "Determination of nonradiative surface layer thickness in quantum dots etched from single quantum well GaAs/AlGaAs", Appl. Phys. Lett. 55, 1427-1429.

5 Reed, M. A., Randall, J. N., Aggarwal, R. J., Matyi, R. J., Moore, T. M., and Wetsel, A. E. (1988) "Observation of discreet electronic states in a zero-dimensional semiconductor nanostructure", Phys. Rev. Lett. 60, 535-537.

6 Tarucha, S., Hirayama, Y., Saku, T., and Kimura, T. (1990) "Resonant tunneling through one- and zero-dimensional states constricted by $Al_xGa_{1-x}As$ /GaAs/$Al_xGa_{1-x}As$ heterojunctions and high-resistance regions induced by focused Ga ion-beam implantation", Phys. Rev. B 41, 5459-5461.

7 Kash, K., Bhat, R., Mahoney, D. D., Lin, P. S. D., Scherer, A., Worlock, J. M., Gaag, B. P. van der, Koza, M., Grabbe, P. (1989) "Strain-induced confinement of carriers to quantum wires and dots within an InGaAs-InP quantum well", Appl. Phys. Lett. 55, 681-683.

8 Dellow, M. W., Beton, P. H., Langerak, C. J. G. M., Foster, T. J., Main, P. C., Eaves, L., Henini, M., Beaumont, S. P., and Wilkinson, C. D. W. (1992) "Resonant tunneling through the bound states of a single donor atom in a quantum well", Phys. Rev. Lett. 68, 1754-1757.

9 Sikorski, Ch. and Merkt, U. (1989) "Spectroscopy of electronic states in InSb quantum dots", Phys. Rev. Lett. 62, 2164-2167.

10 Wees, B. J. van, Kouwenhoven, L. P., Harmans, C. J. P. M., Williamson, J. G., Timmering, C. E., Broekaart, M. E. I., Foxon, C. T., and Harris, J. J. "Observation of zero-dimensional states in a one-dimensional electron interferometer" Phys. Rev. Lett. 62, 2523-2526.

11 Taylor, R. P., Sachrajda, A. S., Adams, J. A., Zawadzki, P., Coleridge, P. T., and Davies, M. (1992) "Classical and Quantum Mechanical Transmission Effects in Submicron-Size Dots.", Surface Science 263, 247-252.

12 Su, B., Goldman, V. J., Santos, M., and Shayegan, M. (1991) "Resonant tunneling in submicron double-barrier heterostructures", Appl. Phys. Lett. 58, 747-749.

13 Fukui, T., Ando, S., Tokura, Y., and Toriyama, T. (1991) "GaAs tetrahederal quantum dot structures fabricated using selective area metalorganic chemical vapor deposition", Appl. Phys. Lett. 58, 2018-2020.

14 Gannon, J. J. and Nuese, C. J. (1974) "A chemical etchant for the selective removal of GaAs through SiO_2 masks", J. Electrochem. Soc. 121, 1215-1219.

High Frequency (MHz) Nanoactuators for Tips and Tip-Arrays

N. C. MacDonald
School of Electrical Engineering and
The National Nanofabrication Facility,
Cornell University, Ithaca NY 14853.

ABSTRACT. Integrated nanostructure mechanisms should add a new dimension to mesoscopic physics. Here we discuss the fabrication and operation of movable, integrated nanostructure mechanisms that produce precision motion. Self supporting, movable single crystal silicon springs and electric-field drive structures with integrated, movable field emission or tunneling tips are described. These devices include resonant mass sensors; precision x-y stages for instrument and sensor applications; and scanning tunneling and atomic force tip-structures for sensors. Movable mechanical structures with cross-sectional dimensions of 150 nm x 1000 nm have been fabricated using single crystal silicon. The nanofabrication process includes high resolution e-beam and optical lithographies, anisotropic reactive ion etching, and selective oxidation of silicon. For these very stiff, low mass nanostructures, the mechanical resonant frequency, $f_r = (k/m)^{1/2} = (\text{stiffness/mass})^{1/2}$ is 5 MHz. We demonstrate nanometer-scale motion of these nanostructures and scanning probe devices. This technology is a significant array-technology that supports on-chip array addressing, control and amplification. Large, dense arrays of 10 - 20 nm diameter tips for field emission or scanned-probe applications have been fabricated using these processes. Such nanoelectromechanical array-architectures are important for the fabrication of tunnel sensors, scanned-probe actuators for information storage systems, and microrobotic mechanisms for moving small objects and atoms.

1. Introduction

The imaging [1-3] and transport of individual atoms or small groups of atoms using macroscopic scanned-probe instruments is leading the way to new and revolutionary instrument concepts [4-10]. However, the present macroscopic actuators are slow and the one-tip serial nature of manipulating atoms limits the throughput (atoms positioned/second) of this serial process. By scaling mechanical devices to nanometer dimensions [11,12], we can build very low mass, high frequency actuators and large arrays of these electromechanical actuators that can be used in turn to build high throughput atom-transport systems. Moreover, the smaller actuators are more compatible with the atomic or molecular scale. Since the mechanical devices are constructed of only a few hundred layers of atoms in the vertical and the lateral dimensions, the addition of

P. Avouris (ed.), Atomic and Nanometer-Scale Modification of Materials: Fundamentals and Applications, 199–209.
© 1993 Kluwer Academic Publishers.

one atomic layer to such structures can change the mechanical properties and the mechanical resonant frequency of the device. Such low mass, nanometer-scale structures can be used to measure atomic and molecular forces and to position atoms or molecules.

Nanometer-scale tips have recently received worldwide attention via the scanning tunneling microscope (STM) [1,3] and the atomic force microscope (AFM) [2]. The tips and tip-like structures are used for electron tunneling devices; for atom probes; for atom "tweezers"; for field emission electron and ion sources; and for ultra small capacitors. Tip-like nanostructures show great promise for nanoelectronics, scanned probe instrumentation [13-15], electron and ion beam instruments, flat panel displays and terabit information storage. To integrate actuators, tips and electronics, we have developed single crystal silicon (SCS) processes that allow the fabrication of movable nanostructures and allow the integration of transistors for control of the tip position and amplification of the tip signals. Processes to produce vertical "tip-above-a-tip" and vertical "triple-tip" structures, and dense arrays of stationary and movable tips are now available.

Array architectures for scanned-probed devices offer many exciting possibilities. Arrays of nanoelectromechanical devices may be used to improve sensitivity or dynamic range; to increase the maximum displacement amplitude or the maximum force amplitude; or to produce massively parallel sensing instruments and information storage devices. Reliable, high frequency (MHz) mechanical mechanisms that produce nanometer-scale motion are of prime importance in realizing scanned-probe array architectures; thus the choice of material is important. The potential of array architectures drives our processing research toward processes compatible with SCS.

Materials and processes that scale to nanometer dimensions and survive high frequency fatigue cycles are required to make reliable high frequency nanomechanisms. Two single crystal silicon (SCS) processes called COMBAT [12,16] and SCREAM [17] have been developed that can be scaled to submicrometer dimensions. We present examples of devices made using the COMBAT process including capacitor actuators, resonant mode devices, and a "micro scanned-probe device" or STM "on-a-chip"-- including actuators and integrated tunneling tips.

2. Nanoactuators

Motion and mechanical force are easily obtained from the energy stored in an electrical field. Thus, capacitor actuators offer substantial motion and force for a very small volume consumed by the actuator. Two types of capacitor actuators are easily integrated with springs and other movable, structures. Figure 1 shows the two nanoactuators: the parallel plate capacitor actuator, Fig. 1(a); and the comb drive capacitor actuator [18] Fig. 1(b).

The force produced on the spring (plate) by each capacitor is the derivative of the energy stored in the capacitor. For the parallel plate capacitor, one plate of the capacitor is assumed fixed and the second, movable plate is attached to a spring (Fig. 1(a)). Thus, the force on the spring is inversely proportional to the capacitor-gap squared. Consequently the motion of the actuator is linear or controllable only over a small range of displacement. As the plates come ever closer together the plates collide and latch until

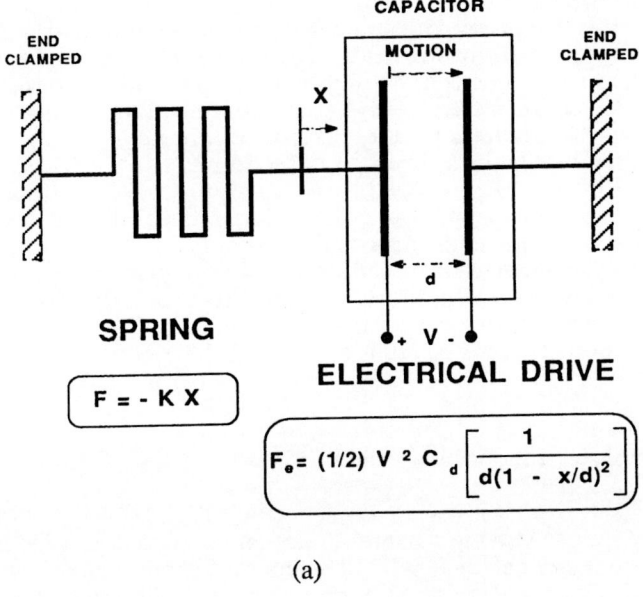

(a)

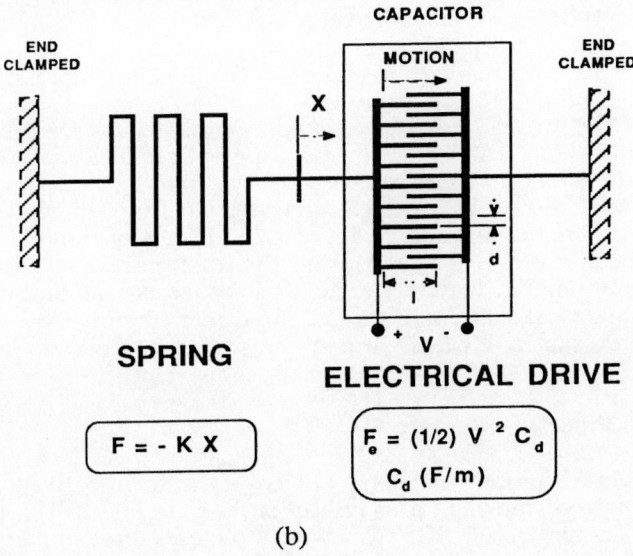

(b)

Figure 1. Two nanoactuators that use the energy stored in a capacitor to produce motion: a parallel plate capacitor-actuator (a) and a linear comb drive capacitor-actuator (b)

the voltage is set well below the value necessary to bring them together. To prevent a short circuit, we coat the plates with a thin layer of silicon nitride or silicon dioxide. The parallel plate capacitor allows for simultaneous displacements in three dimensions. However, the large displacements are highly nonlinear, and the maximum displacement is the gap-distance between the plates at zero applied voltage. An example of an STM sensor device that incorporates parallel plate capacitor actuators for three-dimensional displacement of a movable tip is presented in section 2.1.

The comb drive capacitor actuator can produce large, linear displacements at low applied voltage. The maximum travel distance is limited only by the length of each comb tooth (Fig. 1(b)). Only one-dimensional displacement in the plane of the actuator is possible, but a small vertical displacement in the vertical direction is possible by incorporating a capacitor plate above or below the comb drive structure. For two-dimensional motion a compound stage design would be required; that is, the x actuator would "ride on" the y actuator.

3. Single Crystal Silicon Nanomechanisms

Single Crystal Silicon (SCS) is a natural material candidate to build nanomechanisms. SCS should exhibit low defect density, low internal friction, and high fatigue strength. However, it is difficult to develop SCS processes that allow the formation of complex geometry, freely suspended, submicron structures. The etch rate of wet chemical etching is usually highly dependent on crystal orientation, consequently larger SCS structures are more easily fabricated if the minimum feature size and feature spacing are compatible with the tolerances of the etch-dependent, irregular surfaces. Such large (>10 μm) chemically etched structures have been utilized to make accelerometers and pressure sensors.

We have developed two SCS processes named: Cantilevers by Oxidation for Mechanical Beams and Tips (COMBAT) [12,16] and Single Crystal Reactive Etch And Metal (SCREAM) [17]; these processes can be scaled to submicron dimensions. In this paper only the COMBAT process is discussed. Since our goal is to integrate active devices on moving structures and to make large addressable arrays of moving structures, special attention has been given to the integration of isolation and contacts consistent with submicrometer minimum features. The SCS processes include the integration of submicron contacts and thermal silicon dioxide isolation, and the processes are compatible with standard silicon processes for integration of MOS and bipolar transistors.

3.1. COMBAT Process

The SCS, COMBAT process sequence [12,16] used to form fixed and movable SCS beams with integrated, moving tips is briefly outlined in Fig. 2. This process allows the fabrication of nanometer scale SCS suspended structures through the lateral oxidation of silicon islands etched from SCS [12]. The oxidation process can be extended to form multiple vertical tips. Figure 3 shows the process sequence to produce a vertical triple-tip process for an atomic force microscope [19]. The lower, tip-above-a tip structure is used to sense the vertical displacement of the top tip due to a (atomic) force. Using this

ONE TIP "COMBAT" PROCESS:

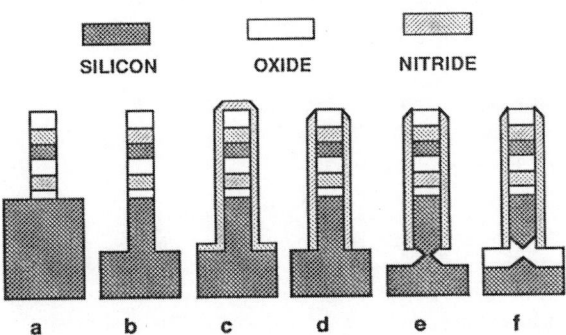

Figure 2. COMBAT Process: Silicon sidewall-masked oxidation process sequence illustrating the fabrication process steps to make released silicon cantilever beams with an integrated, tip-above-a-tip tunneling structure.

MULTIPLE- TIP "COMBAT" PROCESS:

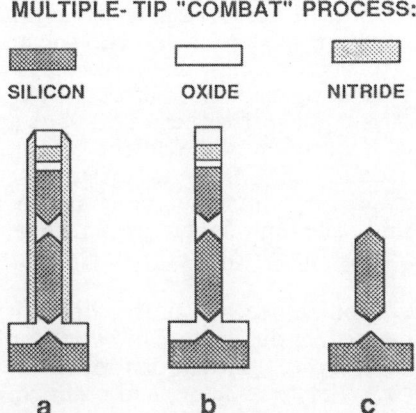

Figure 3. Extended COMBAT process to form a vertical triple-tip structure. The lower tip-above-a-tip tunneling structure is used to sense the force exerted on the top tip or the vertical displacement of the top tip.

process, we can form suspended SCS structures with integrated tunneling tips (Fig. 2(a)) and capacitor actuators. The process also includes methods to electrically isolate segments of the structures, and to form electrical contacts to the 100 nm SCS suspended structures [12].

Examples of devices fabricated using the SCS oxidation process called COMBAT include the Scanning Tunneling Microscope (STM) sensor device shown in Figs. 4-5, and a high frequency resonant structure shown in Fig. 6.

3.2. Scanned-Probe Sensor Or STM "On-A-Chip"

Using COMBAT, we have fabricated a single crystal silicon (SCS) scanned-probe sensor device that occupies an area of 40 μm x 40 μm [12]. One version of the device is shown in Fig. 4 which is a voltage contrast, SEM micrograph. The low mass (2E-13 kg) device, includes a novel self-aligned tip-above-a-tip tunneling structure (Fig. 5) and capacitive x-y translators (Figs. 1(a) and 4). The tunneling structure consists of two vertically self-aligned tunneling tips which are nominally 20 nm in diameter. The top tunneling tip is positioned relative to the fixed or "benchmark" tip by the activation of integrated, x-y capacitive translators. The x-y translators produce a maximum x-y displacement of ± 200 nm. The movable plates are attached to released, movable beams with the vertical beams intersecting the horizontal beams at the center of the device. The suspended, tip-above-a-tip tunneling mechanism is located underneath the intersection of the beams (Fig. 5).

The x-y scanning mechanism consists of eight sets of capacitor plates, four sets for the x axis drive and four sets for the y axis drive. The z axis drive (not visible) is achieved by biasing the suspended structure with respect to the silicon substrate. Newer versions of the devices include electrically isolated capacitor plates mounted on the movable beam to achieve z axis motion. Using a scanning electron microscope (SEM) we measured the x-y displacement of the beams. A potential difference of 50 volts applied across one set of capacitor plates displaces the center section of the device 200 nm away from the origin; therefore, a total x or y displacement of 400 nm is possible. An 8-volt peak-to-peak sinusoidal signal produced ±200 nm x-displacement at the fundamental resonant frequency of the device of 5 MHz. The high fundamental resonant frequency is afforded by the high stiffness of the structures (55 N/m in x, y; 30 N/m in z).

The integrated SCS, scanned-probe device can be considered to be a generic sensor device which can be modified to accept different excitations including sound, photons, acceleration, mass changes or any excitation that can move the 2E-13 kg mass of the devices at least 0.1Å in x, y or z. A plate structure and other complex structures can be easily integrated with the basic scanned-probe mechanism to tailor the device to respond to a specific excitation.

3.3. Nanoelectromechanical Resonators

Resonant structures with low mass, high resonant frequency and high quality factor can be used to detect monolayer changes in mass. We have fabricated laterally driven silicon resonant sensors [20] that move in a plane parallel to the silicon substrate. The minimum feature size (MFS) of the composite SCS beams is nominally 100 nm; the beams are conformally coated with either 50 nm or 100 nm silicon nitride. The devices move as a

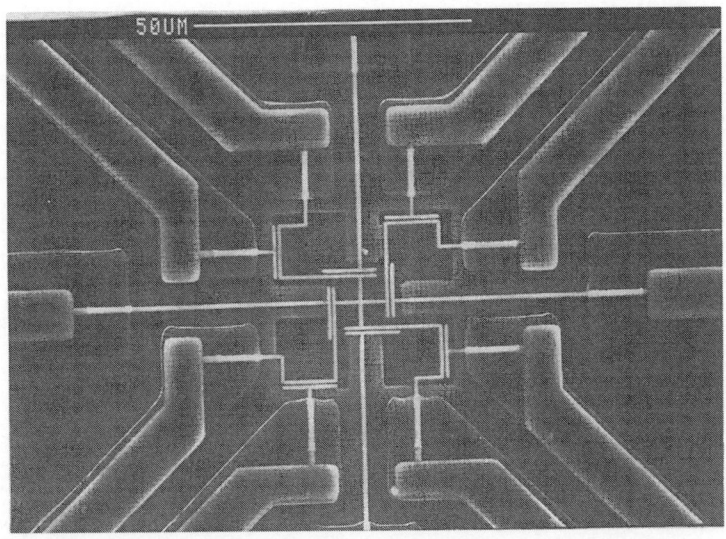

Figure 4. An SEM micrograph of the scanned-probe sensor device with eight sets of x-y parallel plate capacitor actuators; four sets for x displacement and four sets for y displacement. The cross in the center of the structure supports a tip that moves above a stationary tip, i.e., the tip-above-a-tip structure (Figure 5). The single crystal silicon beams (springs) are suspended (and movable) above the silicon substrate. One plate of each capacitor is fixed and connected with an aluminum conductor.

'IELD EMISSION: $I = a\ V^2 exp\ (-b/V)$

'UNNELING: $J = (e^2/h)(k\ /\ 2\pi\ z)\ V\ exp\ (-2k\ z)$

:APACITANCE: $I = V/(\ 2\pi\ f\ C_{tip})$

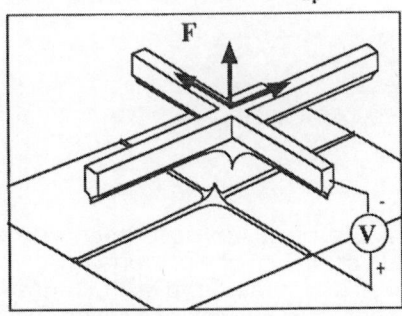

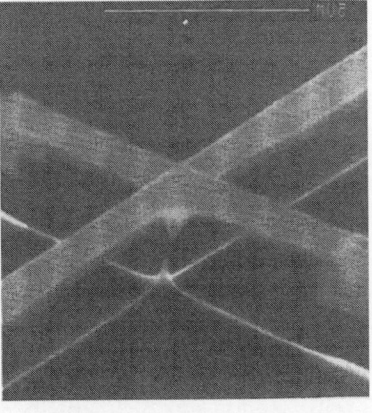

Figure 5. Schematic showing the tip-above-a-tip geometry beneath the movable-cross beam structure (Figure 4). (a) Tip-above-a-tip structure illustration with equations for possible sensing signals. (b) SEM micrograph of a tip-above-tip structure.

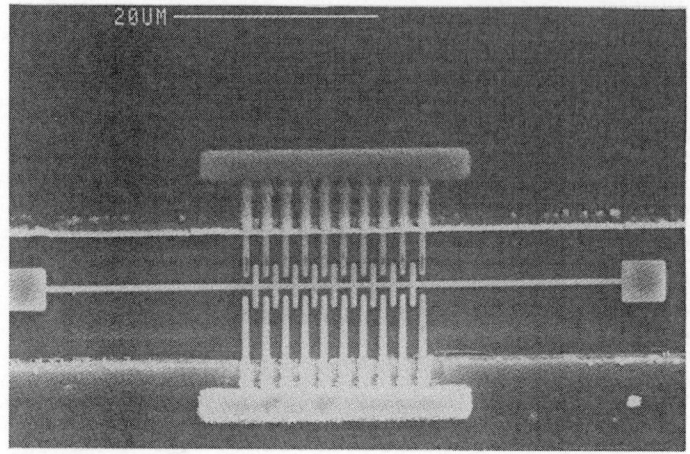

(a)

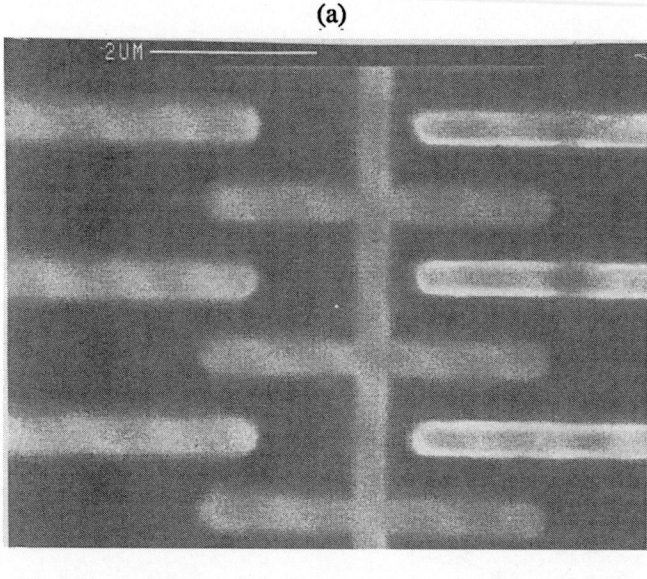

(b)

Figure 6. (a) SEM micrograph of a resonant device with a comb drive capacitor actuator. The beams in this structure are made of 100 nm single crystal silicon conformally coated with 50 nm silicon nitride. (b) SEM micrograph of the comb drive with a constant potential difference of 80 Volts applied between the capacitor plates. The center fingers of the comb are displaced a nominal 0.6 μm to the right. The resonant frequency of this device is 0.6MHz.

result of the force derived from the change in the energy stored in the capacitance fingers of the comb drive (Fig. 1 (b)), or a parallel plate drive structure (Fig. 1 (a)). The spacing between the fingers, or the plates, is less than 500 nm. The COMBAT process is used to fabricate the resonators and sensors. The structures are defined in the silicon substrate and then isolated and released from the substrate using the silicon-sidewall oxidation process and a release etch (Fig. 2 (f)).

The sensitivity of the resonant device to a change in mass is proportional to the (resonant frequency)/(mass). Thus, low mass, mechanically stiff structures are required for maximum sensitivity. An array of such devices with array elements of varying mass and stiffness can be made and used to increase the dynamic range of such resonant sensors.

3.4 Tip-Arrays With Control Electrodes

Arrays of tips with integrated apertures to control the electric field at the tip can be fabricated using an extended version of the COMBAT process [21] as shown in Fig. 7. Such tip-arrays or a single tip with an integrated control electrode are used as field emission electron sources for electron beam instruments or can be used to achieve external control of the electric field on a tunneling tip. Our tip fabrication sequence produces an integrated, self aligned control electrode which is unique. Low voltage operation is achieved by using cathodes with very sharp tips and with very small tip-to-extraction-grid distances. Tip height, radius of curvature, and profile must be uniform over the array if all cathodes are to contribute to the total current.

We have fabricated 2500-element arrays of defect free, single crystal silicon tips with a control electrode. The formation of these tips is achieved using the COMBAT process. Tips formed using this process are inherently uniform in height, profile, and radius of curvature. This tip-uniformity is due to three primary mechanisms. First, the process of thermal oxidation is uniform. Second, the flux of reactants to the silicon interface decreases as the thickness of the SiO_2 increases; this reduction in flux causes the oxidation rate to decrease. Third, the oxidation rate is slower at regions of high curvature. Consequently, as the formation of a tip nears completion, the rate of formation slows. This rate limiting property is unique to this process. Instead of having to stop the process at or near the endpoint, processing can be continued beyond the endpoint without loss of uniformity or tip blunting. Arrays of tips, therefore, can be designed with different endpoints and, consequently, different tip heights. A significant advantage of this process is that the tips remain "capped" once their formation is complete. The cap prevents tip-damage that may result from tip exposure to subsequent processing.

4.   Conclusions

Nanoactuators with integrated tips have been fabricated using a new single crystal silicon process called COMBAT for Cantilevers by Oxidation for Mechanical Beams and Tips. We have fabricated a three-dimensional actuator with integrated tunneling tips that can scan at MHz rates; high frequency low mass mechanical resonators; and both single tips and tip-arrays with integrated, self aligned control electrodes. The COMBAT process is compatible with modern silicon integrated circuit technology, thus fabrication of unique sensors and sensor-arrays, and instruments "on-a-chip" should be possible using the COMBAT processes.

208

Tips With Control Electrodes

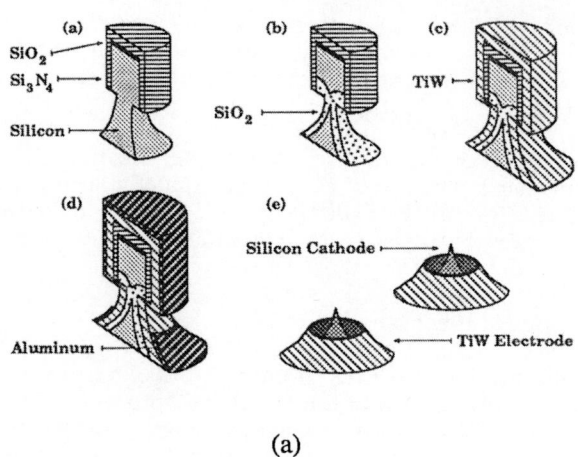

(a)

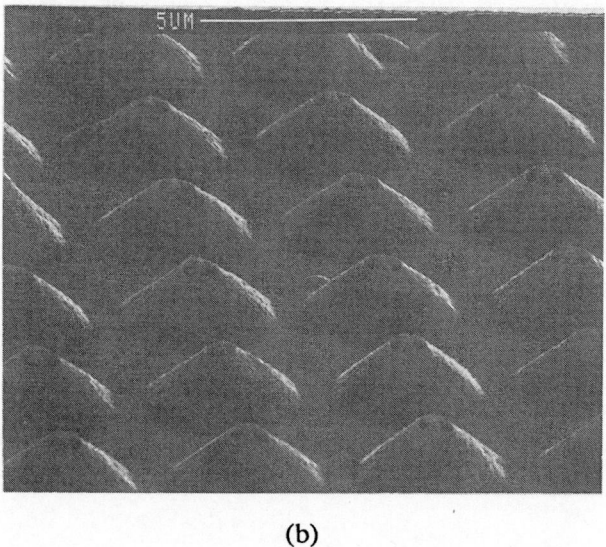

(b)

Figure 7. Tip-array process including integrated electrode to control electric field at the tip. (a) The process flow using a modified version of the COMBAT process which is used to form the tips by the thermal oxidation of silicon. (b) SEM micrograph of the completed tip array with tips protruding through the 0.7 μm diameter apertures.

REFERENCES

[1]G. Binnig and H. Rohrer, "Scanning Tunneling Microscopy from Birth to Adolescence," Reviews of Modern Physics, vol. 59, pp. 615-625, 1987.

[2]G. Binnig, C. F. Quate, and Ch. Gerber, "Atomic Force Microscope," Phys. Rev. Lett., vol. 56, pp. 930-933, 1986.

[3]C. F. Quate, "Vacuum Tunneling: A New Technique for Microscopy," Phys. Today, vol. 39, pp. 26-33, August 1986.

[4]D. M. Eigler and E. K. Schweizer, "Positioning Single Atoms with a Scanning Tunneling Microscope," Nature, vol. 344, pp. 524-526, 1990.

[5]J. A. Stroscio and D. M. Eigler, "Atomic and Molecular Manipulation with the Scanning Tunneling Microscope," Science, vol. 254, pp. 1319-1326, 1991.

[6]H. J. Mamin, S. Chiang, H. Birk, P. H. Guethner, and D. Rugar, "Gold Deposition from a Scanning Tunneling Microscope Tip," JVST., vol. B9, pp. 1398-1402, 1991.

[7]H. J. Mamin, P. H. Guethner, and D. Rugar, "Atomic Emission From a Gold Scanning-Tunneling-Microscope Tip," Phys. Rev. Lett., vol. 65, pp. 2418-2421, 1990.

[8]I. Lyo and Ph. Avouris, "Atomic Scale Desorption Processes Induced by the Scanning Tunneling Microscope," J. Chem. Phys., vol. 93, pp. 4479-4480, 1990.

[9]I. Lyo and P. Avouris, "Field-Induced Nanometer- to Atomic-Scale Manipulation of Silicon Surfaces With the STM," Science, vol . 253, pp. 173-176, 1991.

[10]R. Gomer, "Possible Mechanisms of Atom Transfer in Scanning Tunneling Microscopy," IBM J. Res. Develop. ,vol. 30, pp. 28-30, 1986.

[11]T. W. Kenny, S. B. Waltman, J. K. Reynolds, and W. J. Kaiser, "Micromachined Silicon Tunnel Sensor for Motion Detection," Appl. Phys. Lett., vol. 58, p. 100, 1991.

[12]J. J. Yao, S. C. Arney, and N. C. MacDonald, "Fabrication of High Frequency Two-Dimensional Nanoactuators for Scanned Probe Devices," J. of Micro-electromechanical Systems, vol. 1, no. 1, pp. 14-22, March 1992.

[13]S. Akamine, T. R. Albrecht, M. J. Zdeblick, and C. F. Quate, "Microfabricated Scanning Tunneling Microscope," IEEE Elect. Device Lett., vol. 10, pp. 490-492, 1989.

[14]Y. Kuk and P. J. Silverman, "Scanning Tunneling Microscope Instrumentation," Rev. Sci. Instrum. vol. 60, no. 2, pp. 165-180, 1989.

[15]T. R. Albrecht, S. Akamine, T. E. Carver, and C. F. Quate, "Microfabrication of Cantilever Stylii for the Atomic Force Microscope," J. Vac. Sci. Technol. vol. A8, pp. 3386-3396, 1990.

[16]S. C. Arney and N. C. MacDonald, "Formation of Submicron Silicon on Insulator Structures by Lateral Oxidation of Substrate-Silicon Islands", J. Vac. Sci. Technol., vol. B6, no. 1, pp. 341-345, 1988.

[17]Z. L. Zhang and N. C. MacDonald, "An RIE Process for Submicron, Silicon Electromechanical Structures," J. Micromech. Microeng., vol. 2, no. 1, Mar. 1992.

[18]W. C. Tang, T.-C. H. Nguyen, and R. T. Howe, Sensors and Actuators, vol. 20, pp. 25-32, 1989.

[19]S. C. Arney, J. J. Yao and N. C. MacDonald, "Fully-Integrated Vertical Triple-Tip Structure for Atomic Force Microscopy," 38th National Symp and Topical Conf. of the American Vacuum Society, in Seattle 11-15 Nov. 1991.

[20]J. A. McMillan, J. J. Yao, S. C. Arney and N. C. MacDonald, "High Frequency Single Crystal Silicon Resonant Devices," 38th National Symp. and Topical Conf. of the American Vacuum Society, in Seattle 11-15 Nov. 1991.

[21]J. P. Spallas, J. H. Das and N. C. MacDonald, "Emission Properties of Self-Aligned Silicon Field Emission Cathode Arrays Formed by Lateral High Temperature Thermal Oxidation", Fifth International Vacuum Microelectronics Conference, Vienna, Austria, July 13-17, 1992. Paper submitted to J. Vac. Sci. Technol.

LIGHT PRESSURE LITHOGRAPHY

N. P. BIGELOW
Department of Physics and Astronomy
and Laboratory for Laser Energetics
The University of Rochester
Rochester, NY 14627-0011
USA

ABSTRACT. I describe a new technique for the controlled deposition of atoms onto a substrate. Light pressure forces are used to manipulate the density profile of an atomic beam as it impinges onto a surface. This technique uses nearly resonant light fields and is therefore specific to a particular atomic species, requiring no direct contact with the target substrate. A particularly important feature of this technique is that the patterning process is massively parallel offering angstrom level positioning accuracy over macroscopic areas. Experimental results are described for the deposition of metallic sodium films onto quartz and silicon substrates and considerations for the application and generalization of this technique are discussed.

1. Introduction

In recent years there has been significant interest in the manipulation of the positions and velocities of a collection of gas phase atoms using nearly resonant light fields [1]. Experiments have demonstrated the ability to slow and cool atoms and to confine them in space (to trap them). In 1986 several laboratories spectroscopically showed [2] that, in a one dimensional optical standing wave, atoms could be localized along the intensity nodes (or anti-nodes). Several years later, in our laboratory at A. T. & T. Bell Labs [3] and at NIST [4], experiments showed that under appropriate experimental conditions the atoms could be similarly localized in 3-d optical interference patterns. It was this result that originally lead me to suggest [5] that these wavelength scale localization effects might be used in the microfabrication of novel materials.

The principle of the technique is to use a chosen optical interference pattern as a stencil for atoms over large areas yet with submicron resolution. The spatial structure of the light field placed between an atomic source and a target surface is transferred through the atom-field interaction to the density distribution of the atomic vapor. The distribution of atoms which impinge on the target reflect this distribution. Preliminary experiments have

P. Avouris (ed.), Atomic and Nanometer-Scale Modification of Materials: Fundamentals and Applications, 211–226.
© 1993 *Kluwer Academic Publishers.*

been carried out on the controlled deposition of metallic sodium films which demonstrate the efficacy of this technique on a micron length scale [6]. Additional experiments have recently demonstrated submicron resolution[7]. In this paper I will review this work with a clear bias towards my own original vision of this technique.

2. Light Pressure Forces

In general the force experienced by a two level atom interacting with a monochromatic light field can be expressed as $F=F_{sp}+F_{st}=P_i\alpha+P_q\beta$ where the first and second terms reflect the spontaneous and stimulated components of the force respectively. α and β are related to the gradient of the dot product between the electric field E with the induced dipole moment P [8]. Specifically if $g=ih/2\pi(P \cdot E)$ then $\nabla g=(\alpha+i\beta)$, $P_i=\Delta p/2(1+p)$ and $P_q=\Gamma p/2(1+p)$ where $p=|g|^2/\gamma^2$, $\gamma^2=\Gamma^2+\Delta^2$ and Γ is the excited state lifetime and Δ is the detuning of the light field from resonance. To successfully apply light pressure forces to lithography both stimulated and spontaneous forces must be considered in order to correctly predict the final atomic distribution on the surface.

2.1 SPONTANEOUS FORCES

The spontaneous force is the workhorse of many laser cooling experiments and so-called optical molasses [9]. F_{sp} can understood in terms of the momentum transferred to an atom as photons are absorbed and reemitted spontaneously. When an atom absorbs a photon it receives a momentum kick and recoils along the wavevector k of the absorbed photon. If the subsequent reemission is via spontaneous emission, then the emission direction is randomly oriented in space (spatially isotropic). Averaged over many events, the momentum transfer to the atom on emission averages to zero,whereas the momentum transfer on absorption is cumulative, producing a finite average force along k. The size of this force is of the order of the recoil momentum per absorption ($hk/2\pi$) multiplied by the transfer rate Γ (the inverse lifetime). For sodium, this corresponds to about 10^{-20} N, or an acceleration of $\sim 10^6$ m/s! Note that the length scale over which F_{sp} varies is determined by the variations in k (particularly its direction) and by the spatial variations in the field intensity. An important feature of the spontaneous force lies in the inherently the random nature of the spontaneous emission process. This randomness gives rise to a random walk in atomic momentum space much like Brownian motion. As a result, although the atomic momentum may have a zero average value it has a non-zero mean square value. Over time one effect of spontaneous emission is to cause an increase in the average kinetic energy of the atom referred to as diffusive heating. This effect can be described by an effective diffusion coefficient [8].

If the light field is an optical standing wave, and the field frequency is tuned below (to the red of) the two level transition frequency then spontaneous processes can give rise to a damping of higher atomic velocities. The velocity damping (cooling) arises from the fact that an atom with sufficient velocity will be Doppler shifted into resonance and will absorb and reemit more photons than a slower atom with a smaller Doppler shift. Clearly then, the faster atom will generally experience the larger force. The result is a velocity dependent photon scattering rate. This configuration is referred to as optical molasses [9]. The 'terminal' velocity for an atom in this case is determined by the balance between the Doppler cooling and the spontaneous (diffusive) heating. For light pressure lithography, optical molasses can be a useful tool in preparing a narrow velocity distribution atomic source, but is not necessarily well suited to nanofabrication as will be discussed below.

2.2 STIMULATED FORCES

Stimulated forces arise from the interaction between the quadrature component of the dipole moment induced in the atom by the field and the gradient of the field. Unlike F_{sp}, F_{st} is a conservative force. As pointed out by Ashkin in 1978 [10] the motion of a two level atom due to F_{st} can be described in terms of an effective or pseudo potential $U(x)=h\Delta/2\pi$ $\ln\{1+p(x)\}$ where $p(x)=I(x)/I_{sat}\Gamma^2/\{\Gamma^2+\Delta^2\}\equiv p_mI(x)/I_{max}$. Here I_{sat} is the saturation intensity for the driven transition. There are several important points which make the stimulated force component the best choice for submicron light pressure lithography: (a)The sign of the potential depends on Δ, so that the atom will either be attracted to ($\Delta<0$) or repelled from ($\Delta>0$) the regions of high intensity (b)F_{st} (and hence the depth or height of U) is not bounded since the rate of photon momentum transfer is not limited by the spontaneous lifetime, but by the available laser power (c)F_{st} varies on the same length scale as $I(x)$. This (c) is perhaps the most important feature of F_{st}. Consider a standing wave optical field; here $F_{st}(x)$ has half optical wavelength periodicity ($\lambda/2$) corresponding to, at worst, a 300 nm resolution for an alkali metal such as sodium!

3. Micron Scale Lithography

The ability to use light pressure forces to control the deposition of atoms onto a substrate has been successfully demonstrated [6]. In this work, a standing wave light field propagating parallel to a quartz substrate was used to pattern the growth of a metallic sodium film. The effect of the light was to produce either a depletion line in the film (a groove) or an accumulation line (a "wire") in the film with a characteristic dimension of 200 μm (see figure 1). This result has been described in detail elsewhere[6]. The essential features of this experiment are best understood in terms of the pseudo-potential associated with the intensity variations across the focused light beam (see figure 1). For a light field detuned above resonance U is repulsive and the atoms are deflected away from the center of the light field. In this case there is depleted region in the film below the light field. The resulting pattern can be described as a shadow of the light field; the light field has acted as a defocussing (negative focal length) lens for the incident atomic beam.

The feature size depends not only on the shape of U(x) but on the duration of the interaction with the light field (t_{int}) and on the free flight trajectory after the atoms leave the light field and before they stick to the substrate. In the experimental geometry of figure 1 the dependence of the film morphology on t_{int} is particularly pronounced for $\Delta<0$, hence when U(x) is attractive. If an atom enters the potential with low enough energy it can become trapped in the potential well. The localization of atoms in such optical potentials is referred to as channeling. More importantly if the atom is channeled it may oscillate in transverse position as it moves along the trough of the potential. When the motion along the well (the z direction in the figures) is oscillatory, both the position and transverse velocity $v_\perp$ will be a function of t_{int}. Here $v_\perp$ is the velocity component parallel to the substrate and perpendicular to the incident atomic beam flux (along the x direction in all of the figures).

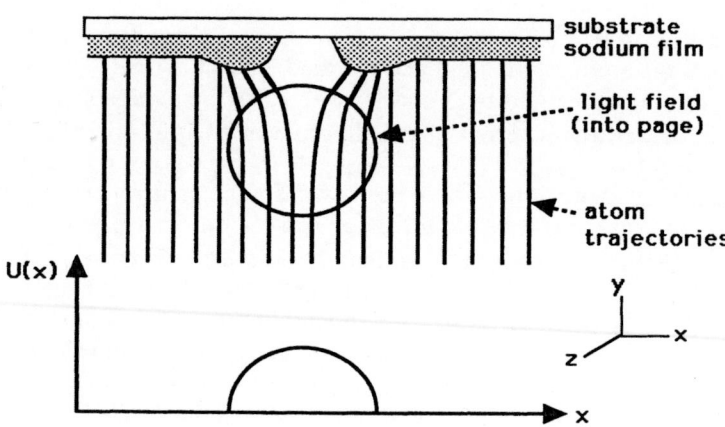

Figure 1. Configuration for micron scale light pressure lithography. Upper portion of figure shows arrangement of light field, incident atomic beam, target substrate and shows typical atomic trajectories. Lower portion depicts corresponding effective potential $U(x)$ used to describe atomic motion in light field with $\Delta>0$.

If the optical field is a standing wave, then when the number of spontaneous emissions N_s during the interaction is greater than unity ($N_s>1$) additional velocity dependent dissipative processes can become important, particularly at high field intensities. For high intensities ($I \gg I_{sat}$) one might expect that F_{st} could be made to dominate F_{sp}. However, if the atom has (or acquires) a non-zero velocity along the optical field, there can be additional processes in which both stimulated and spontaneous emission conspire to give a velocity dependent component to the light pressure force [11]. This force contribution is important because it can be *opposite* in sign to the force associated with $U(x)$. This effect has been experimentally observed spectroscopically [12] and lithographically [6]. These observation are important because such effects place constraints on $I(x)$ and on the transverse velocity distribution of the atomic beam particularly if one plans to control the atoms solely using $U(x)$. Note that in this discussion we have implicitly averaged over the variations in $U(x)$ which occur on the scale of the optical wavelength. Such variations are present if the optical field is a standing wave and are crucial in the application of light pressure lithography to submicron structure fabrication (see below).

4. Submicron Direct Write Optical Lithography

4.1 QUANTUM WIRES

The importance of the use of light pressure forces and particularly F_{st} to control surface deposition lies in the the ability to position atoms on the scale of an optical wavelength. To achieve this high accuracy it is most attractive work under conditions where the conservative stimulated force component dominates. In this case, the motion of the atoms can be entirely described in terms of $U(x)$. For moderate field intensities, the spatial dependence of of $U(x)$ is determined by that of $I(x)$. For a two level atom in 1-d standing wave with $I_{max} \sim I_s$, $U(x)$ is sinusoidal with period $\lambda_{opt}/2$. For the sodium D_2 line this length is 294.5 nm.

Consider figure 2. A uniform 1-d SW is oriented parallel to the substrate and perpendicular to the atomic beam. Under the correct conditions the light will cause the atoms to become localized in a series of parallel stripes so that the film which grows on the substrate will be a series of parallel wires. This experiment is currently under way and results [7] on light scattering from a sodium film deposited in this geometry indicate the presence of a 294.5±.3 nm grating over an area of several square millimeters.

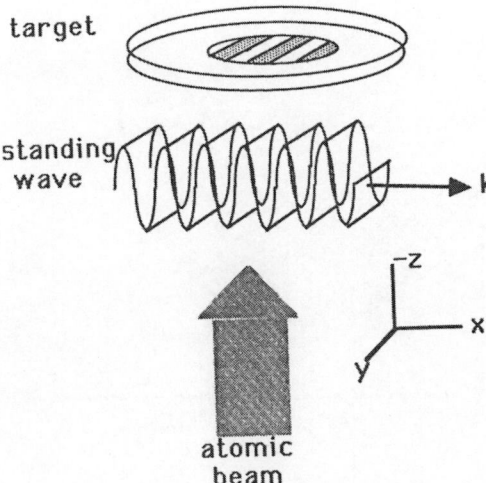

Figure 2. Experimental configuration for submicron light pressure lithography. Here an optical standing wave is used to create a parallel array of lines on target substrate. Periodicity of final metallic structure is one-half the optical wavelength.

4.2 FEATURE SIZES

An important question is: what are the expected widths of these lines? Consider one well of the 1-d SW. If the atoms fill the well as a static equilibrium gas, then the width of the distribution is determined by the comparison of the average kinetic energy to the well depth (see figure 3). However, as mentioned above, the atomic motion can be important [13]. If an atom enters the well at a point away from the well center (the bottom) as is travels along the well (z direction) it will oscillate around the well center. In the harmonic approximation the oscillations will be periodic at some frequency $f=1/2\pi\sqrt{\kappa/m}$ determined by the atomic mass m and an effective spring constant $\kappa=hk^2\Delta p_m/2\pi$. Consider a transverse slice across the atomic beam. Each atom will enter the potential at a different position and begin to 'roll' toward the well center. A time $t_c=1/4f$ (one-quarter oscillation period) later, all of the atoms will *simultaneously* cross the well bottom (see figure 4) [14]. For atoms with velocity $v_\parallel$ (along the well and perpendicular to the substrate) if the substrate is placed a distance $t_c v_\parallel$ after the start of the potential, the resulting surface distribution will be a delta function, located at the position of the optical anti-node (node for $\Delta>0$).

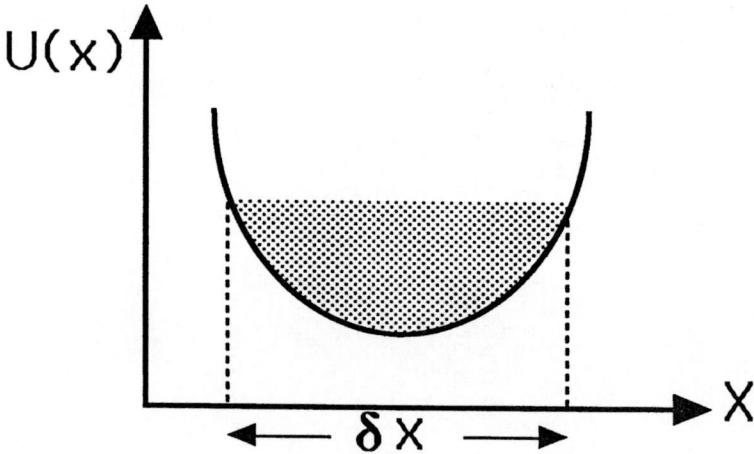

Figure 3. Equilibrium distribution of atoms in U(x) neglecting atomic motion in well. Atoms are treated as simple kinetic gas.

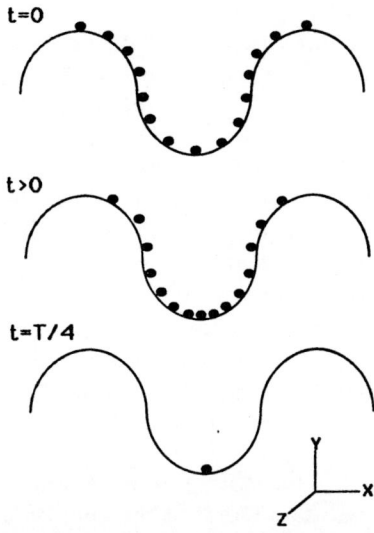

Figure 4. Time evolution of atomic positions transverse to U(x) showing narrowing of distribution after one quarter oscillation period. Incident atomic beam oriented into page (see orientation shown in figure 1).

In practice however, there are additional effects which together with the oscillatory motion will determine the true width of the atomic distribution.

4.2.1 Finite Transverse Velocity ($v_\perp$). In the above discussion it was assumed that the atomic beam entered U(x) with $v_\perp=0$, however, in a real atomic beam $v_\perp \neq 0$. The atom which enters U(x) with $v_\perp \neq 0$ has different initial condition for the oscillatory motion as compared to an atom which entered with $v_\perp=0$ and hence it oscillates with a different relative phase. The result is that the two atoms will not converge at well bottom at $t=t_c$. The importance of the transverse velocity will depend on where in the standing wave the atom enters the potential. The effect will be most important for an atom which enters at the well minimum. An atom which enters the well at the minimum yet with a finite $v_\perp$ will be *furthest* from the well bottom one-quarter oscillation period later. For small $v_\perp$ ($v_\perp \ll \sqrt{2U/m}$; $v_\perp \ll$ the maximum velocity during an oscillation cycle) the broadening is estimated as $\delta x \sim v_\perp/4f$.

4.2.2 Finite Distribution of Longitudinal Velocities ($v_\parallel$). The position along the channel at which the atoms converge at t_c is determined by $v_\parallel$. In a thermal atomic beam there will be a spread of $v_\parallel$ values described by a Maxwell-Boltzmann distribution. As a result the critical quarter oscillation convergence point will occur at different interaction lengths for different velocity groups (see figure 5). For a finite interaction length (i.e. a fixed target surface position), there will be a resulting broadening of the linewidth. In a lens analogy this occurs because the different velocity classes focus at different focal lengths due to chromatic (velocity dependent) abberations. If the incident atomic beam has a thermal (Maxwell-Boltzmann) distribution of longitudinal velocities the resulting distribution can remain surprisingly well peaked. To verify this a numerical simulation [13] have been be made (see figure 6). This persistence of a relatively sharp peak is a direct consequence of the fact that the thermal distribution is also peaked. For sodium atoms, the harmonic oscillator frequencies are typically on the order of 2.5 MHz (for optical field with $I/I_{sat}=100$ and $\Delta=-10\gamma$). Hence the interaction length at the peak in the Maxwell-Boltzman distribution is approximately $l_{int} \approx 50$ μm. To estimate the spot size consider an atom with $v_{\parallel}=v_o +\delta v$ where v_o is the peak velocity. For a sodium source at 200°C, $v_o \approx 550$ m/s. The width of the distribution is estimated as $\delta x \sim [(\lambda/4) / l_{int}] \delta v \, t_c \sim 3 \times 10^{-10} \delta v$ (in meters). Hence a 50 m/s deviation in velocity gives rise to a 150 angstrom half width, worst case.

4.2.3 Anharmonicities. In making a harmonic approximation to the well shape the true sinusoidal shape of $U(x)$ is neglected. The effect of the sinusoidal shape is to add a time dependent phase to the oscillatory motion such that the atoms do not converge at the bottom simultaneously. In fact, after an arbitrarily long interaction times (i.e. many oscillations) the relative phases of the atoms becomes sufficiently random that the distribution resembles that of a equilibrium gas (see discussion at opening of section 4.2). Note that the effects of anharmonicities are included in the simulations in figure 6.

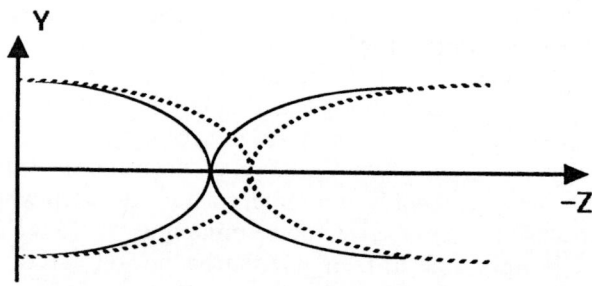

Figure 5. Effect of different longitudinal velocity of interaction length corresponding to clustering of atoms after one quarter oscillation period. The faster the longitudinal velocity, the longer the interaction length for 1/4 period (dashed lines; atoms move left to right). For fixed substrate position this causes broadening of the final surface distribution.

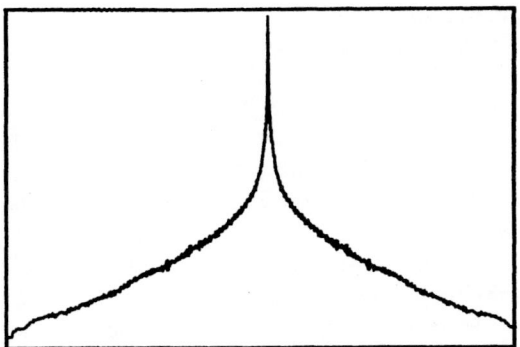

Figure 6. Numerical simulation of expected atomic surface distribution in sinusoidal potential and including thermal spread in longitudinal velocities (see text). Here $\Delta=100$ MHz, $l_{int}=50$ μm and $I/I_{sat}=800$ [13]. X-axis ranges from 0 to $\lambda/2$, Y-axis arbitrary units.

4.2.4 Spontaneous Heating. For short interaction times ($N_s \ll 1$) the effects of spontaneous emission can be neglected. However, at longer interaction times the effects of spontaneous emission must be considered. One effect of spontaneous emission is the heating effects described earlier. Another important consequence of spontaneous emission is the dephasing of the oscillation of the atoms. For the sodium system, if $\Delta=-10\gamma$, $I/I_{sat}=500$ and $t_{int}=0.1$ μs then $N_s \equiv I/I_{sat}(\gamma/2)^3 t_{int}/\{\Delta^2+(\gamma/2)^2\} \sim 0.5$. For $N_s \ll 1$ during the first quarter oscillation cycle, the experimental conditions must satisfy $\{\Delta^2+(\Gamma/2)^2\} \gg (\pi/32)[\{I/I_{sat}\}m/hk^2]$.

4.2.5 Collisions and High Flux Sources. A fundamental limit to the linewidth can be due to collisions among atoms as they channel. and collisions between channeling atoms and atoms not participating in the channeling process. In particular these can be other atoms intentionally introduced into the atomic beam for co-deposition. The point is that although the optical field may selectively interact with only one species in a growth chamber, collisional feature broadening is sensitive to total atomic density in the gas phase. Note also that as the total number of atoms interacting with the optical field increases, the effects of absorption must be considered. From the point of view of feature width, attenuation of the light field by absorption causes variations in U(x) and hence in harmonic oscillator frequency which is not uniquely determined by the light field parameters. Consequently, what may be the correct interaction length for a quarter period oscillation at one initial position in the field will not be at another different position.

4.2.6 Vibrations and Surface Mobility. Relative motion of the substrate and the optical field which positions the atoms can destroy the fidelity of the resulting pattern. To minimize these effects the optical interference pattern and the substrate must be kept in register to within the expected feature width, typically several tens of angstroms. A novel solution to this problem is to mount a Fabry-Pérot interferometer such that one end mirror is rigidly attached to the substrate and the other end mirror is mounted on the same mount as the mirror used to form the standing wave which acts as to pattern the atoms as they impinge on the substrate (see figure 7). The length of the reference cavity is held fixed by a servo loop which monitors the cavity transmissivity and uses the error signal to drive a piezo electric transducer on the cavity/standing wave mirror mount. In this design, the mirror mount moves to track the motion of the substrate, thus keeping the standing wave fixed relative to the substrate and its support. This technique has the advantage that, if the laser used to form the light pressure field is simultaneously used in the Fabry-Pérot, then parasitic problems such as the frequency jitter of the laser does not contribute. Moreover, an offset added to the Fabry-Pérot operating point can be used to scan the position of the optical field relative to the substrate, and hence can be used to interlace different growth distributions on the substrate surface. This stabilization technique was developed by us at Bell and has been described in elsewhere [6].

Surface mobility of deposited atoms must also be considered. For sodium experiments the atoms are highly mobile on many substrates at room temperature. A simple solution to this problem is to cool the substrate and hence "freeze" the deposited atoms into place. Because the surface state wavefunctions of the atoms is sufficiently different from the gas phase, the atom decouples from the light field as it sticks to the surface making light pressure induced surface drift negligible.

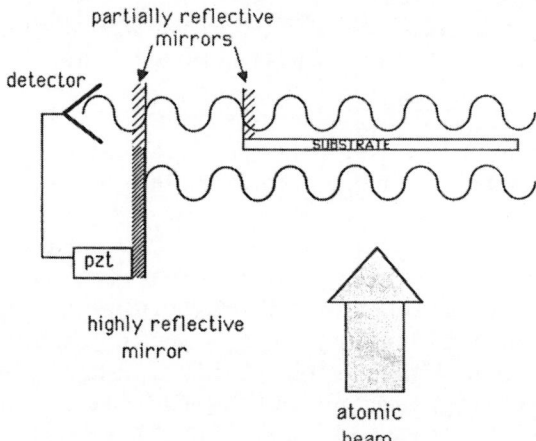

Figure 7. Experimental configuration depicting use of piezo electric positioner (pzt) on mirror mount to stabilize the position of the optical standing wave relative to the substrate. The upper two mirrors (wide hatching) are partially transmissive and form a Fabry-Pérot resonant cavity. The left Fabry-Pérot mirror and the standing wave mirror (high reflector, narrow hatching) are rigidly mounted on the same support, the position of which is controlled by the pzt.

4.2.7 Energy Quantization in Wells and Hopping. In an optical standing wave the confining potential can be sufficiently narrow as compared to the particle mass that the spacing between the vibrational quantum levels in the wells can become significant. In recent experiments on laser cooled cesium vapors the quantization of atomic energies in the wells of U(x) has been observed [15]. In these elegant experiments workers showed that the atoms were localized to approximately $\lambda/15$. For the representative sodium parameters considered above there are in excess of 20 bound states in the channeling wells. In the harmonic approximation, classical particle like wavepackets can typically be formed from 5 bound state wavefunctions, so that relatively classical motion of channeling atoms can be expected under most experimental conditions. More important, however, can be the loss of oscillation phase coherence that can arise due to transitions amongst these levels and between adjacent wells. For a representative transition in the cesium system, Castin has estimated [16] the hopping rate τ and compared it to the harmonic oscillation frequency f. and found $f\tau = 6\hbar k |\Delta| / [2\pi\Gamma\sqrt{\{2Um\}}]$. If this quantity is significantly greater than unity then the atom is more likely to hop to an adjacent well than to remain channeled long enough to oscillate one quarter period. Choice of parameters which minimize this rate will improve pattern feature sizes. A simple measure of the importance of other quantum effects is a comparison of the thermal deBroglie wavelength ($\lambda_{th} = h/2\pi\sqrt{[k_B Tm]}$) to the extent of the potential well ($\lambda/2$). The smaller the relative size of the thermal wavelength the more classical the behavior of the system.

4.2.8 Field Saturation ($I \gg I_{sat}$) An important feature of U(x) is that it depends logarithmically on the intensity of the channeling field. In the limit of high intensities, the two level transition becomes saturated, and the potential flattens, except in the regions near the field nodes (see figure 8). This effect may be useful in increasing the localization of the atoms over increasingly small length scales. As mentioned earlier, such high saturation levels can also give rise to undesirable velocity dependent forces which are associated with combined spontaneous and stimulated processes. These processes can drastically alter the force on the atom and are undesirable. The importance of such effects will be minimized if the interaction time is chosen such that $N_s \ll 1$.

5. Atoms With More Than Two Levels

Up to this point we have described the atoms interacting with the light as simple two-level objects, having only one excited and one ground state. This, however, is rarely the case. In many atomic systems it is possible, by choice of laser frequency and polarization and by judicious use of a static magnetic fields to selectively isolate a pair of levels such that the atom is well approximated in the two level picture. However, there are situations where this is not possible. Moreover, there are situations in which it may in fact be desirable to interact with more than one atomic level. The physics of multi-level atoms interacting with photons is a rich and complex subject. We consider here only one particular aspect of this situation.

222

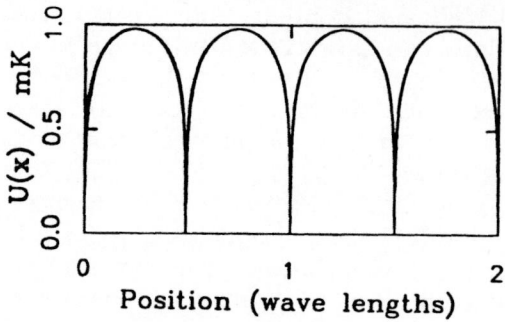

Figure 8. Saturation of pseudo-potential U(x) at high intensities. Field parameters here are Δ=10 MHz and I=400 mW in a 400 μm diameter standing wave.

Consider first the interaction of an atom with several Zeeman sublevels. In this case it is necessary to include the polarization of the optical standing wave if we introduce a pseudopotential U(x) to describe the atomic motion. The most natural description is to decompose the optical field into left and right hand circularly polarized standing wave light fields. Notice that if the counter propagating light fields which together form the standing wave have opposite circular polarization, then there is *no standing wave formed*. If however, the traveling wave light fields are linearly polarized, then the resulting standing wave field can be decomposed into two standing waves of opposite circular polarization. The key point is that or the atoms, these two standing waves can form two different effective potentials, each with periodicity $\lambda/2$ but whose minima are spaced by $\lambda/4$! This situation has been treated in detail theoretically [16] and played a key role in the interpretation of recent experiments on channeling [17]. The consequences of this point for light pressure lithography are crucial. In the recent work at Bell, effort was made to force the atom to appear as a two level atom by creating a standing wave of only one sense of circular polarization. As a result, a $\lambda/2$ periodicity was introduced into the thickness of the sodium film deposited on the target. By simple modification of the light field polarization, this periodicity should become $\lambda/4$, providing conclusive evidence of the control of the experimental situation. In general the degree to which the atoms will behave as a two level systems depends critically of the presence of a uniform magnetic field and on having precisely controlled optical polarizations. This ability to achieve this level of control must be considered in any practical applications of this technique.

6. A View to the Future: Generalizations etc.

6.1 TWO DIMENSIONS

Perhaps the most attractive feature of light pressure lithography is that it is massively parallel. In particular, a large surface area of the substrate can be patterned in one step without any physical contact with the target substrate. The inherent precision of the optical field over millions of wavelengths gives rise to similar precision in the resulting surface distribution. A logical experimental continuation of this work will be to two dimensional interference patterns, and the fabrication of quantum dots. An substantially more difficult problem will be to extend the technique to the formation of the more complex patterns needed for semiconductor circuit fabrication. Possible solutions to this problem may include the use of holographically formed interference patterns as the light pressure "stencil".

6.2 ATOMIC SCALE ACCURACY?

In considering application of this technique to microstructure fabrication an important point is that light pressure lithography is inherently unsuitable for fabrication of structures with atomic scale precision. The fundamental statistical nature of the atom-field interaction is not compatible with such precision. In the recent experiments at Bell [7] which demonstrated the fabrication of submicron scale gratings in metallic sodium, diffraction of an optical probe beam by the grating was used to evaluate grating characteristics. A striking feature of the diffraction pattern was the appearance of sharp diffraction spots. One clear implication of this sharpness is that the gratings are regular over large areas. Another implication is that the gratings are highly sinusoidal as would arise from rather broad grating features. Without a doubt as the experiments continue the exact distributions of atoms on the substrate will be determined and optimized.

Even if the features cannot be made atomically sharp, there are many applications which do not require atomic scale accuracy. For example, light pressure lithography could be used to control dopant placement during crystal growth. As device sizes become increasingly small, device to device performance on a given chip become sensitive to the positions of individual dopants. Uniformity of devices could be significantly improved if the dopant placement could be controlled. In addition, as smaller devices are produced, and smaller interconnects are fabricated, questions of Fermi level pinning at sharp interfaces may become important. These problems can be circumvented if diffuse junctions boundaries are used.

6.3 TECHNOLOGICALLY RELEVANT ATOMS

Sodium and other alkalis are not currently of major technological interest. Indeed, most MBE owners would be horrified at the prospect of introducing sodium into their systems. An important step for the future of this technology will be the development of light pressure technology for other, more technologically interesting materials. To make an atom suitable for the methods discussed in this paper it must meet several criteria. First the atom must have an optical transition which is accessible with existing laser technology. Unfortunately, unlike the alkalis, most metals have outer electronic transitions in the blue or UV spectrum. The visible transitions of the alkalis are associated with their weakly bound outer electrons. It is also desirable to have a relatively strong oscillator strength so that it does not require excessive laser power to saturate the transition. Of current materials, aluminum seems to offer the most promise for future development. Aluminum has suitable transitions around 400 nm with oscillator strengths about one order of magnitude less than the sodium transition used in the preliminary experiments.

6.4 ATOMIC PHYSICS

In addition to the application of light pressure forces to nanofabrication, the detailed understanding of the surface distribution of the atoms will yield a new and exciting probe of the atom field interaction [17,18]. To date, the study of atomic motion in optical fields has been through spectroscopic measurements. These probes average over may atoms, diluting the information about individual atomic position. By studying the morphology of the patterned film with an atomic scale probe such as an atomic force microscope or a scanning tunneling microscope one may hope to gain new insight into the fundamental atomic physics of light pressure forces.

7. Acknowledgements

I would like to thank Prof. Mara Prentiss of Harvard University for the many important and insightful discussions which originally led to the ideas presented in this paper. I also acknowledge Greg Timp and R. E. Behringer at A. T. & T. whose intense technical effort made the experiments described here possible. I am also grateful to Dr. P. Avouris and the Engineering Foundation for the invitation to participate in the workshop.

8. References

1. See for example the November special issues of J. Opt. Soc. Am. B (1985 and 1986).
2. Prentiss, M. G. and Ezeikel, S. (1986), 'Observation of Intensity Dependent Lineshape Asummetry for Two Level Atoms in a Standing Wave Field', Phys. Rev. Lett., **56**, 46 and also Salomon, C., Dalibard, J., Aspect, A., Metcalf, H., and Cohen-Tannoudji, C., (1987), 'Channeling Atoms in a Laser Standing Wave', Phys. Rev. Lett., **59**, 1659. Finally we also note the substantial research of V. Balykin; see for example: Balykin, V., Lethokov, V. S., Ovchinnikov, Yu. B., Sidorov, A. I., and Shul'ga, S. V., (1988), 'Channeling of Atoms in a Standing Spherical Light Wave', Opt. Lett, **13**, 958.

3. Bigelow, N. P. and Prentiss, M. G., (1990) 'Observation of Channeling of Atoms in the Three Dimensional Interference Pattern of Optical Standing Waves', Phys. Rev. Lett., **65**, 29.
4. Westbrook, C. I., Watts, R. N., Tanner, C. E., Rolston, S. L., Phillips, W. D. and Lett, P. D. (1991), 'Localization of Atoms in Three Dimensional Standing Waves', Phys. Rev. Lett., **65**, 32.
5. Bigelow, N. P., Prentiss, M. G., Timp, G. L., Behringer, R. E. and Cunningham, J. E., (1991),'Light Pressure Lithography ILS 1991, Monterey CA.
6. Prentiss, M. G., Timp, G. L., Bigelow, N. P., Behringer, R. E. and Cunningham, J. (1992), 'Light as a Stencil', Appl. Physics Lett., **60**, 1027.
7. Timp, G. L., Behringer, R. E., Tennent, D. M., Cunningham, J. E., Prentiss, M. G., and Berggren, K. K. (1992), 'Using Light as a Lens for Submicron, Neutral Atom Lithography', Phys. Rev. Lett., **69**, 1636.
8. Gordon, J. P. and Ashkin, A. (1980), 'Motion of Atoms in a Radiation Trap', Phys. Rev. A, **21**, 1606.
9. see for example, Chu, Steven, Hollberg, L., Bjorkholm, J. E., Cable, Alex and Ashkin, A. (1985), 'Three Dimensional Viscous Confinement and Cooling of Atoms by Resonant Radiation Pressure', Phys. Rev. Lett., **55**, 48.
10. Ashkin, A. (1978), 'Trapping of Atoms by Resonant Radiation Pressure', Phys. Rev. Lett., **40**, 729.
11. Dalibard, J. and Cohen-Tannoudji, C. (1985), 'Dressed Atom Approach to Atomic Motion in Laser Light: the Dipole Force Revisited', JOSA **B**, **2**, 1707.
12. Aspect, A., Dalibard, J., Heidmann, A., Salomon, C. and Cohen-Tannoudji, C.(1986), 'Cooling Atoms with Stimulated Emission', Phys. Rev. Lett., **57**, 1688.
13. Berggren, K., Prentiss, M., Timp, G. L., Behringer, R. E. and Bigelow, N. P., 'Semi-Classical Simulations of the Atomic Density in an Optical Standing Wave: Theoretical Limits on Narrowness of Submicron Direct Write Optical Lithography' (1992), Int'l. Conf on Quant. Elect., Tech. Digest Ser. **9**, 264.
14. Recent spectroscopic studies performed by Bigelow and Prentiss (unpublished) of channeling atoms has shown that this essentially classical model of atomic motion is realistic in typical experimental conditions.
15. Verkerk, P., Lounis, B., Salomon, C., Cohen-Tannoudji, C., Courtois, J.-Y., and Grynberg, G. (1992), 'Dynamics and Spatial Order of Cold Cesium Atoms in a Periodic Optical Potential', Phys. Rev. Lett., **68**, 3861 and also Jessen, P. S., Gerz, C., Lett, P. D., Phillips, W. D., Rolston, S. L., Spreeuw, R. J. C. and Westbrook, C. I. (1992), 'Observation of Quantized Motion of Rb Atoms in an Optical Field', Phys. Rev. Lett., **69**, 49.
16. Castin, Y., Dalibard, J. and Cohen-Tannoudji, C. (to be published), 'The Limits of Sisyphus Cooling', LIKE proceedings, Elba, May 1990.

17. I would like to note that a version of this technique was proposed by the group at the Ecole Normale Supérieure for study of channeling effects, but not pursued.
18. Bigelow, N. P., Prentiss, M. G., (1991), 'Neutral Atom Photography', Quant. Elect. Laser Sci. 1991, Tech. Digest Ser., OSA **11** 22.

SEMICONDUCTOR QUANTUM DOT RESONANT TUNNELING SPECTROSCOPY

Mark A. Reed
Department of Electrical Engineering
Yale University
New Haven, CT 06520, USA

John N. Randall, James H. Luscombe,Yung-Chung Kao, and Tom M. Moore
Central Research Laboratories
Texas Instruments Incorporated
Dallas, Texas 75265, USA

William R. Frensley
Erik Jonsson School of Engineering
University of Texas at Dallas
Richardson, TX 75083, USA

Richard J. Matyi
Department of Metalurgical and Mineral Engineering
University of Wisconsin - Madison
Madison, WI 53706, USA

ABSTRACT. Recently, 3-dimensionally laterally-confined semiconductor quantum wells ("quantum dots") have been realized. These structures are analogous to semiconductor atoms, with energy level separation of order 25 meV, and tunable by the confining potentials. A systematic study reveals a $(radius)^{-1}$ dependence on the energy separation. The spectra corresponds to resonant tunneling from laterally-confined emitter contact subbands through the discrete 3-dimensionally confined quantum dot states. Momentum non-conservation is observed in these structures.

227

P. Avouris (ed.), Atomic and Nanometer-Scale Modification of Materials: Fundamentals and Applications, 227–233.
© 1993 Kluwer Academic Publishers.

1. Introduction

Carrier confinement to reduced dimensions has led to numerous important developments in basic semiconductor physics and device technology. Advances in microfabrication technology now allow one to impose quantum confinement in additional dimensions, typically done by constricting or confining in lateral dimensions an existing 2D
carrier system. Though the technology to create lateral confining potentials is considerably less advanced than the degree of control that exists in the vertical (epitaxial, or interface) dimension, remarkable progress has been made in elucidating the relevant
physics of these systems.

Recently, three-dimensionally confined "quantum dots" have been realized.[1]
These structures are analogous to semiconductor atoms, with energy levels tunable by
the confining potentials. These structures pose an experimental paradox distinct from
2D and 1D structures; to allow transport through the single electronic states the states
cannot be totally isolated; i.e., the confining potential must be slightly "leaky", and
thus the states are "quasi-bound". Additionally, contact to carrier resevoirs are non-
trivial from an experimental and analysis viewpoint. We have adopted a configuration
where the quasi-bound momentum component (and thus the resultant transport direction) is epitaxially defined in the form of a resonant tunneling structure, and additional
confinement is fabrication-imposed. This configuration is distinct from all the above
referenced quantum wire cases where the (unbound) current flow is along the interface;
here it is through the interface. Because of this constraint, the system can be operated
far from equilibrium.

2. Quantum Dot Fabrication and Tunneling

Our approach used to produce quantum dot nanostructures suitable for electronic transport studies is to laterally confine resonant tunneling heterostructures. This approach
embeds a quasi-bound quantum dot between two quantum wire contacts. An ensemble
of AuGe/Ni/Au ohmic metallization dots (single or multiple dot regions) are defined by
electron-beam lithography on the surface of the grown resonant tunneling structure.
Creation of dots less than 500Å is possible, though we will show that the appropriate
range for the typical epitaxial structure and process used is in the range 1000 - 2500Å in
diameter. A bi-layer polymethelmethacrylate (PMMA) resist and lift-off method is
used. The metal dot ohmic contact serves as a self-alligned etch mask for highly anisotropic reactive ion etching (RIE) using BCl_3 as an etch gas. The resonant tunneling
structure is etched through to the n+ GaAs bottom contact, defining columns in the
epitaxial structure . A SEM of a collection of these etched structures is seen in Figure 1.
Contact is made to just a single structure, so that the resulting transport is tunneling
spectroscopy of a single dot.

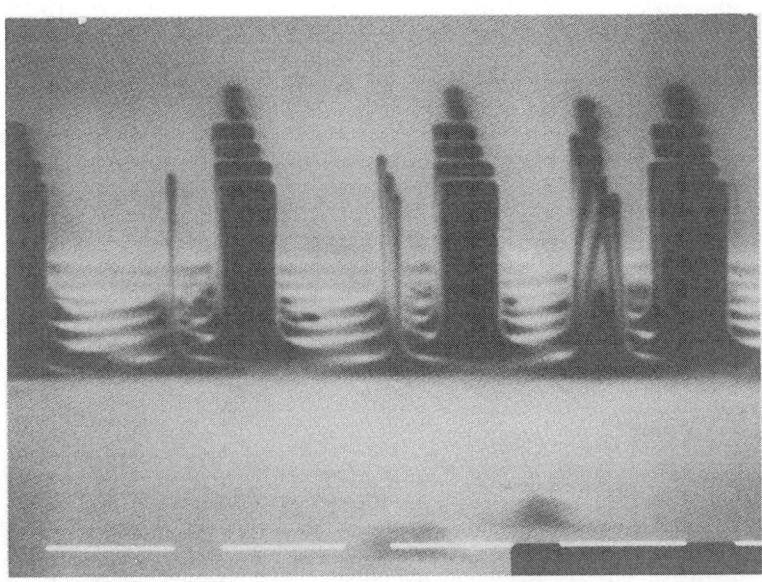

Figure 1. Scanning electron micrograph of an array of anisotropically etched columns containing a quantum dot. The horizontal white marker is 0.5 micrometer.

A spectrum of discrete states would be expected to give rise to a series of resonances in transmitted current as each state is biased through the source contact. To observe lateral quantization of quantum well state(s), the physical size of the structure must be sufficiently small that quantization of the lateral momenta produces energy splittings > 3kBT. Concurrently, the lateral dimensions of the structure must be large enough such that pinch-off of the column by the depletion layers formed on the side walls of the GaAs column does not occur. Due to the Fermi level pinning of the exposed GaAs surface, the conduction band bends upward (with respect to the contacts Fermi level), and where it intersects the Fermi level determines in real space the edge of the central conduction pathcore. When the lateral dimension is reduced to twice the depletion depth or less, the lateral potential becomes parabolic though conduction through the central conduction path core is pinched off. Observation of tunneling through the discrete states of a quantum dot necessitates fabrication within a narrow design criterion.

Figure 2 shows the current-voltage-temperature characteristics of a quantum dot resonant tunneling structure successfully fabricated within these constraints. The structure lithographically is 1000Å in diameter and epitaxially is a n$^+$ GaAs contact/AlGaAs barrier/InGaAs quantum well structure.[1] At high temperature, the characteristic negative

230

differential resistance of a double barrier resonant tunneling structure is observed. As the temperature is lowered, two effects occur. First, the overall impedence increases presumably due to the elimination of a thermally activated excess leakage current. Second, a series of peaks appears above and below the main negative differential resistance (NDR) peak. In the range of device bias 0.75V-0.9V, the peaks appear equally spaced with a splitting of approximately 50 mV. Another peak, presumably the ground state of the harmonic oscillator potential, occurs 80 mV below the equally spaced series.

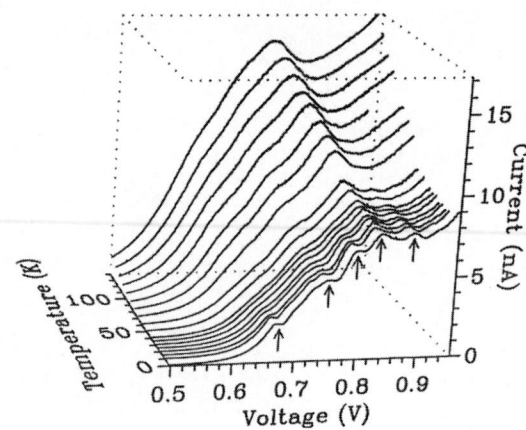

Figure 2. Current-voltage characteristics of the 1000A single quantum dot nanostructure as a function of temperature, indicating resonant tunneling through the discrete states of the quantum dot. The arrows indicate peak positions for the T = 1.0K curve.

The existence of the fine structure in the tunneling characteristics of this, and other, laterally confined resonant tunneling structures indicates the formation of laterally confined electronic states. The current-voltage-temperature characteristics of these structures typically exhibit a clear series of additional NDR peaks at low temperatures, as shown in Figure 2. Variation of the lateral diameter by more than 1 order of magnitude verifies an approximate R^{-1} energy splitting dependence, indicating energy quantization by the lateral potential instead of single electron charging, which would have a R^{-2} dependence.[2] However, a full indexing of the spectrum is needed to verify that the structure in the electrical characteristics is the discrete levels. To do this, a full 3D screening model of the quantum dot system is necessary.

We have used a full 3D zero-current, finite-temperature Thomas-Fermi screening model to obtain the energies of the 1D emitter (and collector) and 0D quantum dot states under arbitrary applied bias.[3,4] Figure 3 shows the crossings of the emitter subband level (n') with the quantum dot levels (n) as a function of applied bias, transposed onto the 1.0K experimental spectra of Reference 1. The relevant dot states all have the same n_z (=1) quantum number (n_2=2 dot states are virtual), which is hereafter suppressed. There is general agreement between the experimental and predicted peak voltage positions. The 3' - 3 transition is not seen since the dot states become virtual at ~0.95V. The 1D DOS in the emitter becomes thermally smeared when $3k_BT > \Delta E$ (the emitter subband spacing), and thus at high temperatures the only surviving transition is the 1' - 1. This indeed occurs when $3k_BT \sim 21$ meV, in good agreement with the subband spacing of 25 meV.

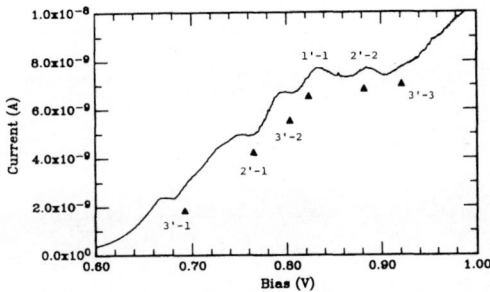

Figure 3. The crossings of the emitter subband level (n') with the quantum dot levels (n) as a function of applied bias, transposed onto the 1.0K experimental spectra of Figure 2.

The observation of the momentum-nonconserving transitions (n' not equal to n) shows that k-perpendicular is not a good quantum number in this system, due to the hourglass topography of the electron energy surface. This absence of a n,n' (radial) selection rule is natural from the radially changing, cylindrically symmetric geometry that breaks translational symmetry.

3. 0D - 0D Tunneling

An intriguing situation would be to examine the case of 0D-0D tunneling, compared to the 1D-0D situation. The fabrication of such structures is relatively straightforward, using a a double well - triple barrier initial epitaxial structure. The structure was designed so that state overlap small (with respect to the lateral splitting energy), so that the 1D-

0D and 0D-0D transitions can be compared. Lateral fabrication is similar to before.

Figure 4 shows the 4.2K current-voltage characteristics of this structure. A similar modeling of the state crossings as described above was performed, and the spectra of the transitions are superimposed on the Figure. The notation used here is 1(emitter subband) - (final state), for either the first (n') or second (n'') (i.e., further downstream from the emitter) dot.

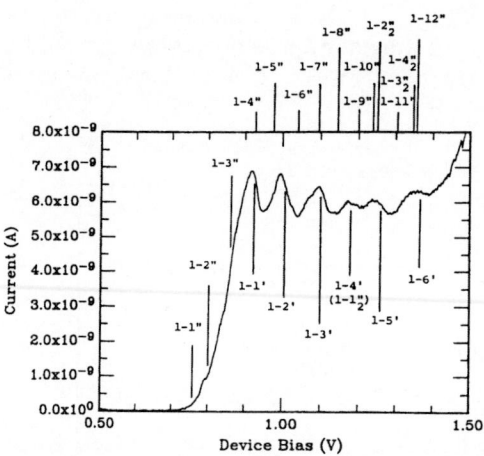

Figure 4. Current-voltage characteristic at T=4.2K of the "0D-0D" sample, with predicted resonant peak positions superimposed. The notation is 1(emitter subband) - (final state), for either the first (n') or second (n'') dot. The subscript (i.e., n_2'') denotes the $n_z=2$ excited state.

The spectra is well fit by resonances occuring for crossings of the 0D states with the lowest 1D subband in the contact. The agreement is good for the major (1-n') resonances, significantly stronger than the n'' states due to the thinner tunnel barrier. The position of the minor second set, 1-n'', predicts the structure below the 1-1' peak. The peaks are not apparent at higher voltages, probably due to overlap with the 1-n' resonances.

The predicted 0D-0D crossing spectra gives bad agreement both qualitatively (spacings and number of peaks) and quantitatively (peak positions); the predicted spacing of 0D-0D crossings are approximately 5mV throughout the region examined. However, since the lateral potential of the adjacent dots is nearly identical, k-perpendicular is a good quantum number in the 0D-0D transitions, compared to the dramatic lateral potential variation in the 1D-0D transitions. Thus, the 0D-0D transitions would only occur for n'=n'', which is not experimentally accessible in this structure. Thus, we only observe the 1D-0D transitions (each seperately), and not the 0D-0D transitions (the selection rule holds for n'=n'', and not for n=n',n'') since the lateral potential changes from the contact-to-dot region, but not for the dot-to-dot region. To this date, there have not been any successful 0D-0D dot transition experiments reported.

4. Summary

Quantum dot structures provide a unique laboratory for the exploration of quantum transport through nanostructured semiconductors. The "atomic states" can be varied with structural variables, allowing for the exploration of the fundamentals of quantum-confined electronic states and of transitions involving those states. Coupling of "molecule" states has been investigated, and these structures should provide an intriguing tool for the investigation of quantum-localized electronic states.

5. Acknowledgement

We are indebted to our collaborators R.J. Aggarwal (MIT) and A.E. Wetsel (Harvard) during the initial phases of this work while on summer internships. We thank R.T. Bate for constant encouragement and support, and R.K. Aldert, E.D. Pijan, D.A. Schultz, D.L. Smith, P.F. Stickney, and J.R. Thomason for technical assistance. This work was sponsored in part by ONR, ARO, and WRDC.

6. References

[1] M. A. Reed J. N. Randall, R. J. Aggarwal, R. J. Matyi, T. M. Moore, and A. E. Wetsel, "Observation of discrete electronic states in a zero-dimensional semiconductor nanostructure", *Phys. Rev. Lett.* **60**, 535-537 (1988).

[2] A. Groshev, "Single electron trapping in ultrasmall double barrier semiconductor heterostructures", *Proc. 20th Int. Conf. Physics. Semic.*, ed. E.M. Anastassakis and J.D. Joannopoulos (World Scientific, 1990), p. 1238-1241.

[3] M. A. Reed, W. R. Frensley, W. M. Duncan, R. J. Matyi, A. C. Seabaugh, and H-L. Tsai, "Quantitative Resonant Tunneling Spectroscopy : Current-Voltage Characteristics of Precisely Characterized RTDs", *Appl. Phys. Lett.* **54**, 1256-1259 (1989).

[4] M. A. Reed, J. N. Randall, J. H. Luscombe, W. R. Frensley, R. J. Aggarwal, R. J. Matyi, T. M. Moore, and A. E. Wetsel, "Quantum Dot Resonant Tunneling Spectroscopy" (Proceedings of Arbeitskreis Festkörperphysik bei der Deutschen Physikalischen Gesellschaft), *Festkorperprobleme* **29**, ed. U.Rößler (Vieweg, Braunschweig/Wiesbaden 1989), pp. 267-283.

SINGLE ELECTRON EFFECTS IN SMALL METALLIC TUNNEL JUNCTIONS

L.J. GEERLIGS
Delft Institute for Microelectronics and Submicron technology DIMES/S
Lorentzweg 1, 2628 CJ Delft, The Netherlands

ABSTRACT. This paper will review some charging effect experiments on metal tunnel junction circuits. The Single Electron Transistor will be discussed, controlled transfer of single charge carriers in normal and superconducting junctions, and co-tunneling in multi-junction circuits.

1. Introduction

With nanofabrication it is possible to create metallic or semiconductor islands, separated from leads by tunneling barriers, with capacitances C of the order of 10^{-16} F or even smaller. At temperatures below 1 K the charging energy $E_C \equiv e^2/2C$ is larger than the energy $k_B T$ of thermal fluctuations. The charge on the island is a well-defined quantity and adding or removing a single electron is possible only under certain bias conditions. At zero temperature tunneling is forbidden if the total electrostatic energy of the circuit (including voltage sources, which are viewed as large capacitors) would increase. This phenomenon, now known as Coulomb blockade of (electron) tunneling, was appreciated and known experimentally as early as 1951.[1, 2, 3, 4].

I will give a short discussion of experiments on metal tunnel junction systems, mainly at Delft University (and partly in collaboration with the CEA Saclay). Recent reviews of charging effects have been written by Averin and Likharev [5], and Schön and Zaikin [6]. Up to date information on charging effects is also collected in the proceedings of the NATO ASI on single charge tunneling in Les Houches.[7]

A planar metal-oxide-metal tunnel junction is a leaky parallel plate capacitor with a capacitance of about 10^{-15} F per $(100 \text{ nm})^2$ junction area. The smallest planar junctions that have been produced so far (about $(30 \text{ nm})^2$) were all fabricated from aluminum.[8, 9] The conventional fabrication by the method of shadow evaporation is shown in Figure 1. The tunnel resistances can be varied between a few kΩ and infinity by changing the barrier oxidation pressure. A photograph of a double metal tunnel junction fabricated by this shadow evaporation technique is given in Figure 2.

Other types of devices widely used for charging effect experiments include STM-grain combinations, and vertical and planar semiconductor double barrier structures. In metal tunnel junction circuits, the charging energies are dominated by the junction capacitances, *i.e.*, the capacitances between leads and island (or the nearest neighbour capacitance between islands for an array of more than two junctions). For decreasing lead capacitance, the junction capacitance can eventually be far exceeded by the self-capacitance of the island.

P. Avouris (ed.), Atomic and Nanometer-Scale Modification of Materials: Fundamentals and Applications, 235–246.

236

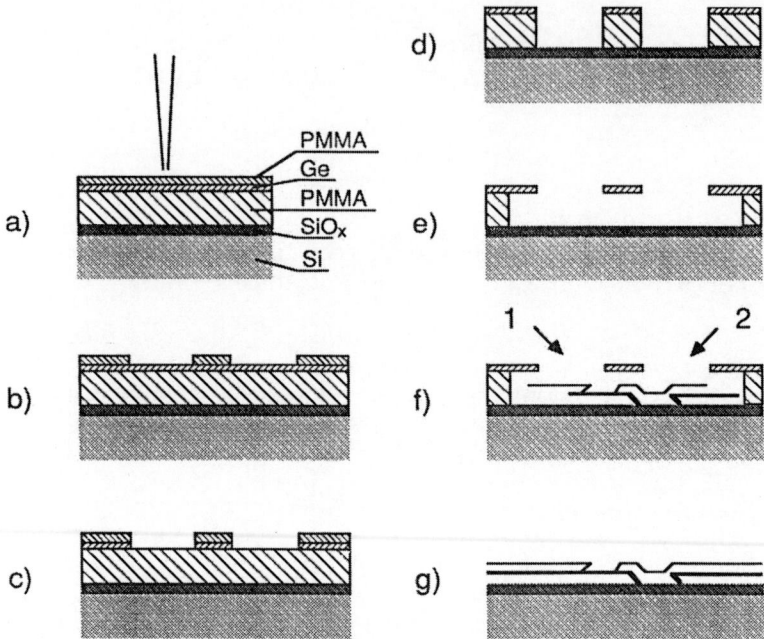

Figure 1: Sideview of the fabrication of a small metal tunnel junction. The pattern is written in the top PMMA layer of a 3-layer resist with an e-beam (a), developed (b), etched into the middle Ge layer (c), and the bottom resist layer (d). Isotropic etching creates a free-hanging bridge (e). Oblique angle evaporation of two metal layers (f) results after liftoff in a planar junction (g).

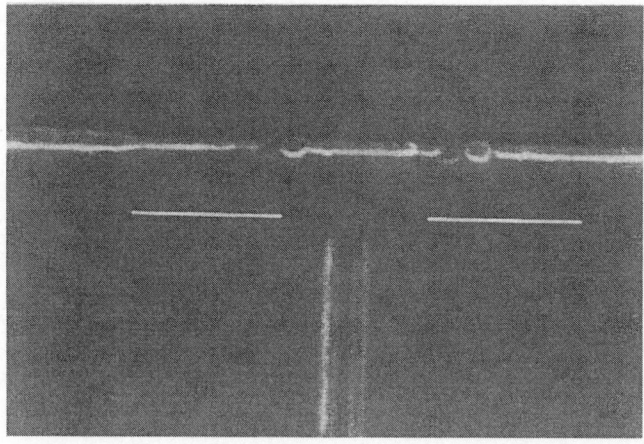

Figure 2: SEM photograph of a double metal tunnel junction, with a gate electrode at the bottom. The white bar is 1 μm.

This is generally the case for the STM-grain tunneling configuration and split-gate defined islands in semiconductor heterostructures.

Tunnel junctions with high tunnel resistance, $R_t \gg h/e^2$, are the easiest case to consider. Under the additional constraint that the impedance of the leads is very low, the golden rule and first order perturbation theory yield for the tunneling rate (with a Fermi distribution over a constant density of states, as is appropriate for rather large metal electrodes):

$$\Gamma = \frac{\Delta E}{e^2 R_t} \left[\exp\left(\frac{\Delta E}{k_B T}\right) - 1 \right]^{-1} \tag{1}$$

$-\Delta E$ is the energy change of the tunneling electron. R_t is the tunnel resistance of the junction (the resistance in the absence of charging effects). In the limit $k_B T \ll |\Delta E|$ the tunneling rate reduces to

$$\Gamma = \begin{cases} \dfrac{-\Delta E}{e^2 R_t} & \text{for} \quad \Delta E < 0 \\ \\ 0 & \text{for} \quad \Delta E > 0 \,. \end{cases} \tag{2}$$

Charging effects come into play when we realize that the kinetic energy transferred to an electron, when it crosses a tunnel barrier over which a voltage drop V exists, can be less than eV. The difference is $e^2/2C_\Sigma$ where C_Σ is some relevant capacitance in the system (we will specify this below). Usually this difference can be neglected, because C_Σ is very large and/or the temperature is relatively high. However, at low temperature and for small capacitance an excess voltage drop is necessary for tunneling to occur.

For example, for two tunnel junctions in series the electrostatic energy, composed of capacitive energy and work performed by voltage sources, is:

$$E = \frac{Q_0^2}{2C_\Sigma} - q_\ell V_\ell + q_r V_r + \frac{Q_0}{C_\Sigma}(C_g V_g + C_\ell V_\ell + C_r V_r) \tag{3}$$

V_ℓ is the potential of the left lead, C_ℓ is the capacitance of the left junction and q_ℓ is the charge passed through this junction. The index r denotes similar parameters for the right junction and g for a gate electrode which can change the potential of the central island, but which has no tunneling conductance to the island. $C_\Sigma = C_\ell + C_r + C_g$ is the total island capacitance.

If we write the energy change of all four possible tunneling events in the form $\Delta E = e \Delta V$, with $\Delta V = \pm(V_{\ell,r} - \varphi)$, the effective island potential is either $\varphi = Q_0'/C_\Sigma + (C_\ell V_\ell + C_r V_r)/C_\Sigma - e/2C_\Sigma$ for tunneling of an electron into the island, or $\varphi = Q_0'/C_\Sigma + (C_\ell V_\ell + C_r V_r)/C_\Sigma + e/2C_\Sigma$ for tunneling out of the island. Since the last term $e/2C_\Sigma$ changes sign for tunneling into versus out of the island, it represents the Coulomb barrier of electron tunneling (the excess voltage mentioned above). We have written Q_0' for $Q_0 + C_g V_g$, to show that, although the net island charge clearly can only change by e due to tunneling, it is as if a gate voltage provides a means of changing the island charge over a continous range.

For $Q_0' = +e/2$ ($Q_0' = -e/2$) the Coulomb barrier vanishes for tunneling into (out of) the island. If $Q_0' = +e/2$, tunneling will change this to $Q_0' = -e/2$, hence, tunneling into and out of the island can alternate, yielding charge transport at any finite voltage: The Coulomb gap has disappeared. These conductance peaks at halfinteger Q_0'/e are called Coulomb oscillations.

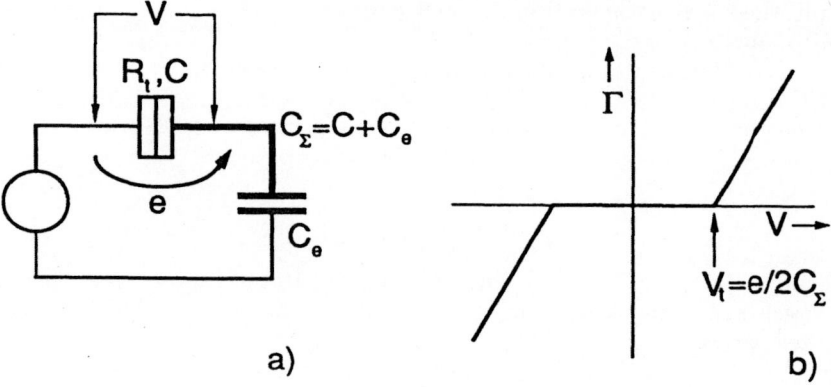

Figure 3: (a) Equivalent circuit to describe tunneling through any junction. (b) The resulting tunneling rate versus voltage drop over the junction.

For systems with a quasi-continuous energy spectrum the periodicity of Coulomb oscillations is e/C_g. For a discrete spectrum the position of the conductance peaks versus gate voltage provides information about the energy spectrum. The lineshape of the Coulomb oscillations should show a concomitant change from Fermi function derivative to a resonant tunneling (Lorentzian) type. McEuen *et al.* [10, 11] and others (see, *e.g.*, [7]) have studied these effects in detail.

The Coulomb barrier $e/2C_\Sigma$ can be generalized for any tunnel junction whose environment can be replaced by an equivalent voltage source and capacitor (Figure 3). The capacitance C_Σ gives the threshold voltage $V_t = e/2C_\Sigma$ or junction charge CV_t which has to be exceeded for tunneling through that particular junction to be possible.

2. The single electron transistor

We now turn to a few experiments on (metallic) double tunnel junctions, or Single Electron Tunneling (SET)-transistors.[8]

Figure 4(a) gives the measured $I(V)$ characteristic for a double junction, for the two gate voltages where the Coulomb gap is maximum and minimum. Figure 4(b) and (c) give the corresponding potential landscapes, following the remarks in the previous section. In the case of the maximum Coulomb gap (solid curve) the conduction below the threshold voltage e/C_Σ is very low, although not completely zero. By applying a gate voltage, the Coulomb gap can be completely suppressed (dashed curve, Q_0'/e half-integer), but at high voltages the same voltage offset e/C_Σ is recovered. As a function of gate voltage the $I(V)$ curve evolves continuously between the two extremes shown with the period corresponding to a change of the island charge of e. With the average current through the device fixed at a low level, the voltage versus gate voltage can be recorded (Figure 5(a)). The Coulomb oscillations for small excitation are given in Figure 5(b).

The $I(V)$ curve shown in Figure 4 is typical for symmetrical metal double junctions. It does not show the Coulomb staircase which is well known from *e.g.* STM-grain tunneling experiments.[12, 13, 14, 15] On increasing the voltage bias of a double junction, additional

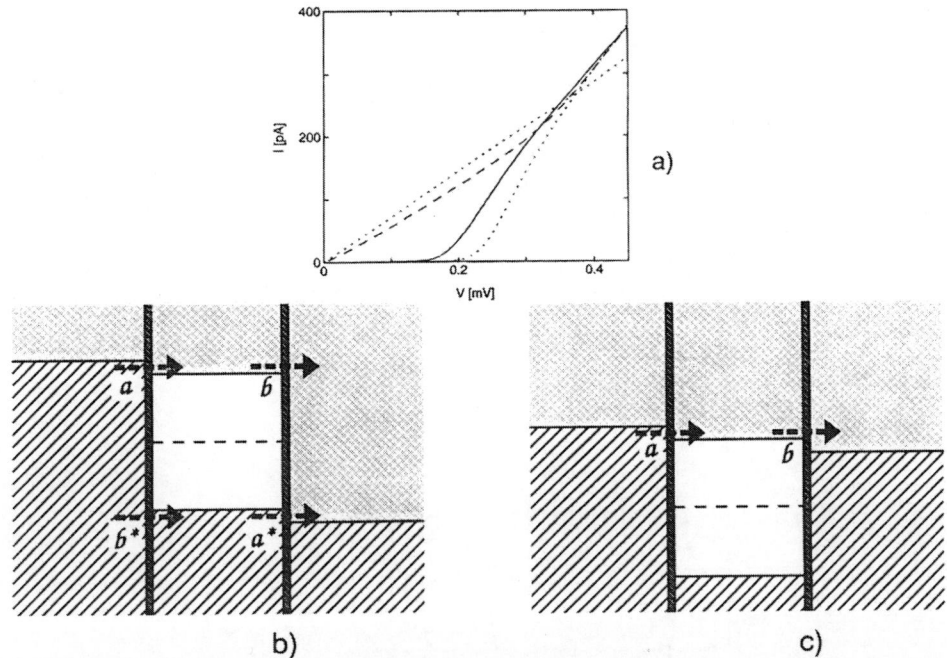

Figure 4: (a) Current voltage characteristic of a double junction at 20 mK, for gate voltages 0 (solid line) and $e/2C_g$ (dashed line). The dotted lines give the theoretical curves for a symmetrical system of the same total capacitance and resistance. (b) and (c) are potential landscapes in the conducting state, for integer and half-integer Q_0'/e, respectively. The dashed regions represent occupied electronic states, the grey regions unoccupied states. In the island these are effectively separated by e/C_Σ. Due to tunneling the potential landscape of the island will shift over e/C_Σ.

charged states of the island (e, $2e$, $3e$, etc.) start to contribute to the current, resulting in a steplike increase of the $I(V)$. The slope of a step edge is determined by the junction with the highest tunneling rate. On further increase of the voltage, the current only increases slowly since it is limited by the junction with the slowest tunneling rate. This effect is most prominent when the RC-times of the two junctions are very different.

Figure 6 shows an analysis of the sub-gap leakage. The current below the Coulomb gap is proportional to V^3 and R_t^{-2}. The process responsible for the leakage is conventionally called co-tunneling. In metal tunnel junctions this is an inelastic event where an electron-hole pair is created on the central island, and two different electrons tunnel through the two junctions in a single event.[16] For longer arrays the rate depends on the number of junctions N involved:

$$\Gamma \propto \frac{V^{2N-1}}{h} \prod \frac{\alpha_i}{4\pi^2} , \qquad (4)$$

$\alpha_i = R_K/R_t$ for junction i, and the product is over all junctions. Several experiments have confirmed the occurrence of this process. [17, 18, 19]

The large slope of the $V(V_g)$ curve (Figure 5(a)) indicates the potential of the SET-

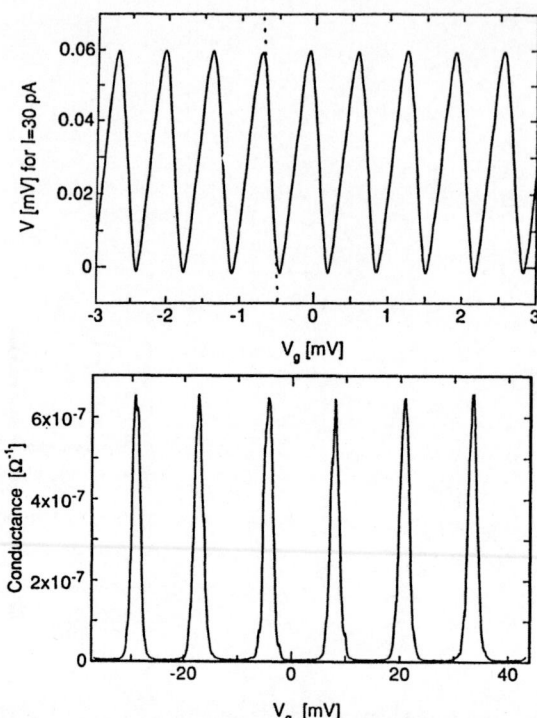

Figure 5: (a) Voltage across a double junction, versus gate voltage, at a fixed current level. (b) Conductance versus gate voltage, with small excitation voltage.

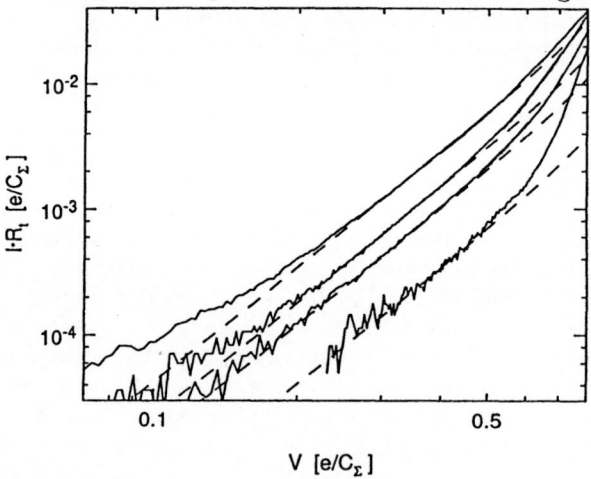

Figure 6: $\log(I)$–$\log(V)$ measurements for four double junctions with R_t varying between $42k\Omega$ (top) and $340k\Omega$ (bottom). The measurements have been scaled to the dimensionless voltage VC_Σ/e and current IR_tC_Σ/e to allow for easy comparison (the curves should be the same to first order in R_t^{-1}). The inset shows schematically the co-tunneling process.

transistor as a voltage/charge detector. A double junction configured as a 3-terminal device is a very sensitive charge detector and a high impedance voltmeter. It can be used to count electrons on the gate electrode, like a DC SQUID is used to count flux quanta. Like the SQUID the sensitivity is better than the electronic charge quantum ($\leq 10^{-4}e/\sqrt{\text{Hz}}$ [20]) The SET-transistor has recently been applied in several experiments. One experiment [21] connected the island between a tunnel junction and a capacitor to the gate of a SET-transistor and observed directly the incremental charging of this island by single electrons. Another experiment, by NIST in collaboration with the same group in Saclay, uses the device in a metrological experiment where an electron-pump device is the charge source for a calibrated capacitor.[22] The SET-transistor is the sensor in a feedback loop to keep the voltage at one electrode of the capacitor at zero. The fact that the electrometer can be placed very close to the capacitor (on the same chip) and operates at mK temperatures is essential in this experiment.

3.   Controlled electron transfer

The controlled transfer of electrons one by one, clocked by a radio frequency (r.f.) gate voltage is an important possibility of small tunnel junction circuits. It results for a clock frequency f in a current $I = ef$ or a multiple.

When a clocked device needs an applied bias voltage to transfer a current it is usually named a turnstile. When it operates with zero bias voltage, or even transfers electrons against the direction of the driving bias voltage, it is called a pump. We will consider the operation of a turnstile device and a pump device constructed of metal junctions. Recently, a Coulomb island in a two-dimensional electron gas has been operated in both turnstile and pump mode.[23]

A double junction is for given bias and gate voltage either insulating or conducting. This is due to the fact that either precisely one charge configuration is stable, or two "complementary" charge configurations Q'_0 and $Q'_0 + e$ are unstable. In the turnstile device [24, 25] (Figure 7) this problem is solved by adding one junction on each side of the gate capacitor (each arm must have at least 2 junctions). The charging energy of the side islands increases the thresholds for electron tunneling through the arms. Compared to the double junction, the threshold for entering of an electron is shifted to more positive gate voltage, whereas the threshold for leaving of an electron is shifted to more negative gate voltage. An electron entering because of high V_g will remain trapped on the central island until V_g is decreased to such a low value that an electron is forced to leave the island (through the other arm). The new charge state of the central island is also stable, until V_g is increased again. Cyclically changing the bias conditions by applying an alternating voltage to the gate capacitor moves one electron per cycle through the array. A finite bias voltage V is necessary to determine the current direction by introducing asymmetry in the voltage drops over the turnstile arms. The same charging energy that traps charge on the central island of course also prohibits more charge to tunnel to the central island. Only by making the gate voltage more positive (in this example, by another $2e/C$) will a second electron be trapped on the central island. Increasing the amplitude will thus move more electrons per cycle.

Figure 8 shows the $I(V)$ of a device with tunnel junction capacitance about 0.38 fF and tunnel resistance 340 kΩ (($R_t C$)$^{-1} \approx 8$ GHz). The gate capacitance C_g is 0.3 fF. The values

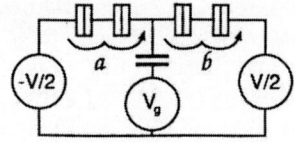

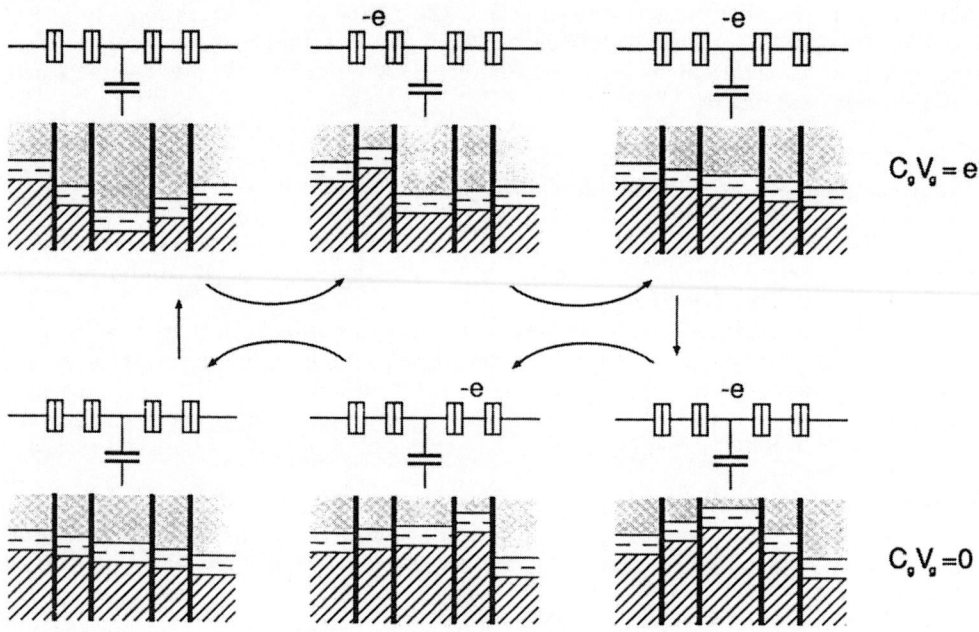

Figure 7: Schematic of a turnstile device and potential landscape for block wave modulation. Effective gaps in the available electron states due to Coulomb blockade yield the necessary hysteresis for turnstile operation.

of R_t and C were determined from the large scale $I(V)$ curve, and C_g was determined from the period ΔV_g of the current modulation by the gate voltage as $C_g = e/\Delta V_g$. Without ac gate voltage, a large zero-current Coulomb gap is present. With ac gate voltage of frequency f, current plateaus develop at a current level $I = ef$ for bias voltage $V > 0$, and at $I = -ef$ for $V < 0$. The plateaus even extend to voltages outside the gap.

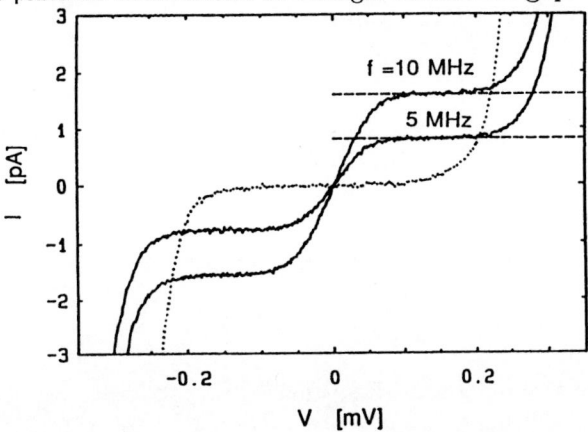

Figure 8: Current-voltage characteristics of a four-junction turnstile without rf (dotted) and with gate voltage modulation at two different frequencies.

An alternative current source based on charging effects is the single charge pump developed by the Saclay group.[25, 26] It is a linear array of three tunnel junctions (Figure 9). The two islands between the three junctions are capacitively coupled to gate electrodes. With bias voltage V, and the two gate voltages U_x and U_y, one can induce an arbitrary number of excess charge carriers on either of the islands (this is in contrast to the turnstile, where only the central island charge is controlled). By adjusting the gate voltages one can thus cause a single electron to tunnel through any of the three junctions. It results in current plateaus like those of the turnstile, which however in this case cross the $V = 0$ axis. A further advantage is that by time-reversing the gate voltage modulation the sign of the current can be reversed.

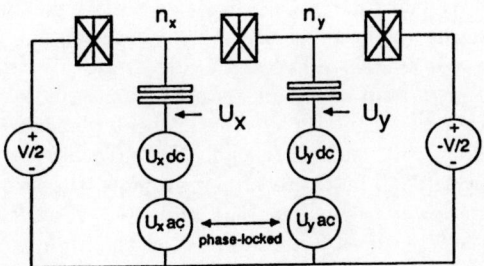

Figure 9: Circuit of a pump of single charges.

Typical $I(V)$ characteristics in the normal state are presented in Figure 10 for clocking frequencies of 10 and 15 MHz. Plateaus at positive and negative current are found for the appropriate phase difference between the two gate voltags. The dashed lines are at the

244

expected current values $\pm ef$.

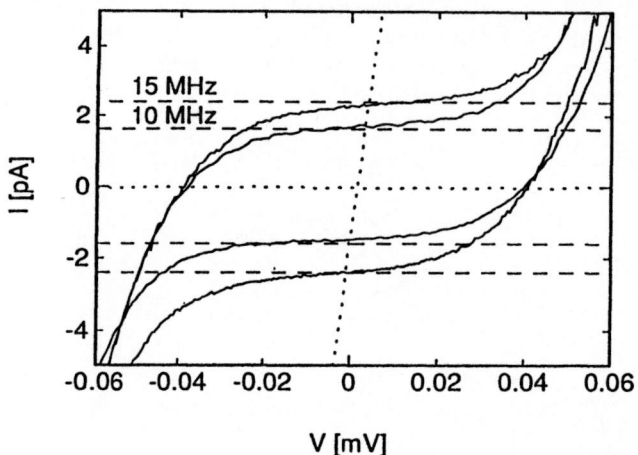

Figure 10: $I(V)$ curves at two different rf frequencies of a single-electron pump. Plateaus at ef or $-ef$ are obtained depending on the phase difference between the two gate voltages.

A main reason to perform experiments on the pump is also that it is expected to operate with superconducting charge carriers. For a superconducting tunnel junction, a tunneling matrix element of magnitude $E_J/2$ couples charge states differing by one Cooper pair.[27] The Josephson coupling energy E_J is for a 10 kΩ aluminum junction of order 1 K, and is proportional to R_t^{-1}. The Cooper pairs all occupy the same quantum state and unlike electrons can not occupy states over a wide range in kinetic energy. Therefore a Cooper pair can only tunnel if the charge states before and after tunneling have the same electrostatic energy. In the pump the tunnel junctions are swept through such a resonance one after the other, in a controlled way. This results in $2e$-charge pumping

Aluminum is superconducting below 1.3 K, unless a magnetic field is applied to drive it to the normal state. $I(V)$ characteristics of the same device as Figure 10, but now in the superconducting state, are given in Figure 11 for clocking frequencies between 2 and 20 MHz. Just as in the normal state, positive and negative current plateaus are obtained by appropriate choice of the phase difference between the two rf gate voltages. In contrast to the normal state in which the plateau goes smoothly through the I-axis, in the superconducting case there is a step in the plateau at $V \approx 0$. The rounding of the current plateaus in the normal state and the step in the current plateau in the superconducting state are due to co-tunneling, and exemplify the difference between superconducting and normal state charging effects [28]. The dotted lines indicate the expected values $I = \pm 2ef$ (for 2 to 10 MHz). For the lower frequencies the higher of the two levels in a plateau (in absolute value) is close to $2ef$. At high frequencies even the high level is significantly lower, showing that the system cannot follow the modulation adiabatically.

4.   Acknowledgements

I especially thank V. Anderegg, D. Averin, U Geigenmüller, L Kouwenhoven, J Mooij, M

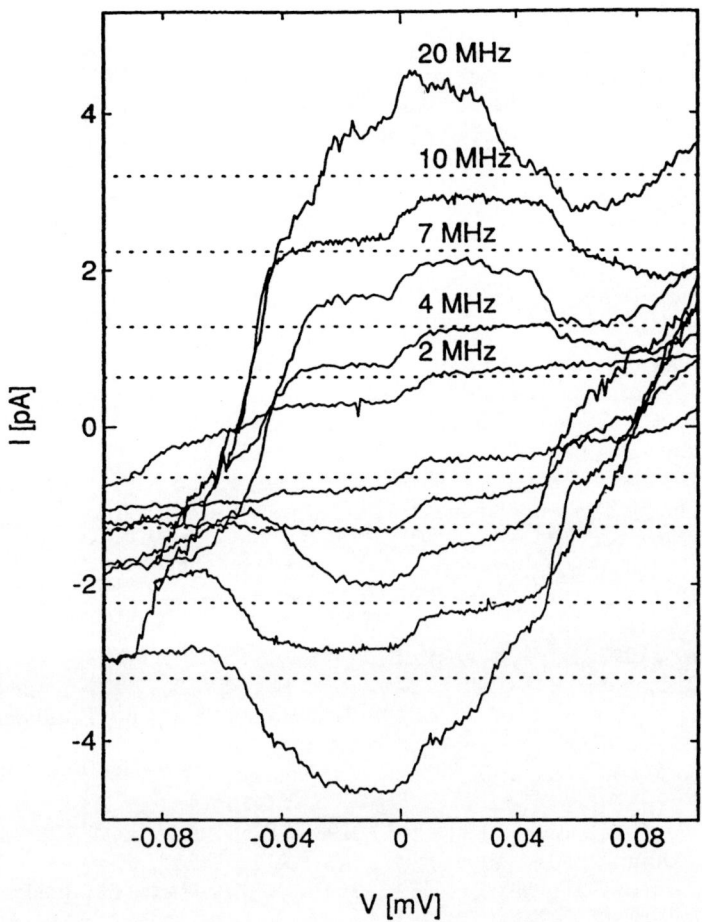

Figure 11: $I(V)$ curves at several different rf frequencies of a single Cooper-pair pump. Plateaus at about $2ef$ or $-2ef$ are obtained depending on the phase difference between the two gate voltages.

Peters, G Schön, and N van der Vaart in Delft, and M Devoret, D Esteve, H. Pothier and C Urbina in Saclay for very valuable collaboration, advice, and discussions in the course of these experiments. Facilities and research results of the Delft Institute for Microelectronics and Submicron Technology (DIMES) were used for fabrication of the devices. This work was supported by the Dutch Foundation for Fundamental Research on Matter.

5.   References

1. Gorter, C J (1951) Physica **17**, 777
2. Giaever I, and Zeller, H R, (1968) Phys. Rev. Lett. **20**, 1504.
3. Zeller H R, and Giaever I (1969) Phys. Rev. **181**, 789.
4. Lambe, J, and Jaklevic, R C (1969) Phys. Rev. Lett. **22**, 1371.
5. Averin, D V and Likharev, K K (1991), in: Mesoscopic Phenomena in Solids, eds. B L Altshuler, P A Lee, and R A Webb (Elsevier, Amsterdam) chapter 6.
6. Schön G and Zaikin A D (1990) Physics Reports **198**, 237.
7. Grabert, H, and Devoret, M H (eds) (1991) proceedings of the NATO Advanced Study Institute on Single Charge Tunneling (Les Houches, France), published as Z. Phys. B, vol. **85**, and by Plenum, New York.
8. Fulton, T.A, and Dolan, G.J. (1987) Phys. Rev. Lett. **59**, 109.
9. Dolan, G J, and Dunsmuir, J H (1988) Physica B **152** 7.
10. McEuen, P L, et al. 1991. Phys. Rev. Lett. **66**, 1926.
11. McEuen, P L, et al. 1992. Phys. Rev. B **45**, 11419.
12. Barner, J B, and Ruggiero, S T (1987) Phys. Rev. Lett **59**, 807.
13. Van Bentum P J M, van Kempen H, van de Leemput L E C and Teunissen P A A, 1988, Phys. Rev. Lett. **60**, 369.
14. Van Bentum, P J M, Smokers, R T M, and van Kempen, H (1988) Phys. Rev. Lett. **60**, 2543.
15. Wilkins, R, Ben-Jacob, E, and Jaklevic, R C (1989) Phys. Rev. Lett. **63**, 801.
16. Averin D V, and Odintsov A A (1989) Phys. Lett. A **140**, 251.
17. Geerligs L J, Averin D V and Mooij J E (1990) Phys. Rev. Lett. **65**, 3037.
18. Glattli, D C, et al. (1991) Z. Phys. B **85**, 375.
19. Eiles, T M, et al. (1992) Phys. Rev. Lett. **69**, 148.
20. Korotkov, A N, Averin, D V, Likharev, K K, Vasenko, S A (1992) in: proceedings of SQUID'91 (eds. Koch H, and Lübbig H) Springer, Berlin, 45.
21. Lafarge, P, Pothier H, Williams E R, Esteve D, Urbina C and Devoret M (1991) Z. Phys. B **85**, 327.
22. Williams, E, Martinis, J, Devoret, M, Esteve, D, Urbina, C, Pothier, H, Larfarge, P, Orfila, P (1991) to be published.
23. Kouwenhoven L P, Johnson, A T, Van der Vaart N C, Harmans C J P M, Foxon, C T (1991) Phys. Rev. Lett..**67**, 1626
24. Geerligs, L J, Anderegg, V F, Holweg, P A M, Mooij, J E, Pothier, H, Esteve, D, Urbina, C, and Devoret, M H (1990) Phys. Rev. Lett. **64**, 2691.
25. Urbina C, Pothier H, Lafarge P, Orfila P, Esteve D, Devoret M H, Geerligs L J, Anderegg V F, Holweg P A M, Mooij J E (1991) IEEE Trans. Magn. **27**, 2578.
26. Pothier H, et al. (1991) Europhys. Lett. **17**, 249.
27. Josephson B D (1962) Phys. Lett. **1** 251.
28. Geerligs L J, et al. (1991) Z. Phys. B **85**, 349.

FABRICATION AND ELECTRIC CONDUCTANCE OF A FINITE ATOMIC GOLD WIRE: A THEORETICAL STUDY.

C. JOACHIM[1], X. BOUJU[2] and C. GIRARD[2]

[1]CEMES
29, rue J. Marvig
P.O. Box 4347
31055 Toulouse Cedex
France

[2]Lab. de Physique Moléculaire
Université de Franche-Comté
La Bouloie
25030 Besançon Cedex
France

ABSTRACT. The possibility of making an atomic wire by pushing with an AFM tip some gold atoms adsorbed onto a NaCl(100) surface is discussed. Constant force images of the NaCl(100) surface, of a single gold atom and of a chain of gold atoms adsorbed onto the NaCl(100) surface are presented. A method of pushing a gold atom with the apex of an AFM tip is proposed. The conductance of a gold atomic wire is calculated using the Elastic ScatteringQuantum Chemistry (ESQC) technique.

1.   Introduction

There are several possible ways of fabricating an atomic wire, that is, a finite chain of equally spaced metal atoms, of one atom in width, connected via metallic contacts (a few microns in width) to a current source. As proposed by D. Eigler [1], this wire can be assembled at low temperature by pushing (or sliding) some metal atoms previously deposited onto the surface of an insulator. The atomic structure of this wire is stabilized by the surface. At room temperature and by pulsing the bias voltage, an STM tip is capable of desorbing well defined atoms on a surface. This is the case of sulfur atoms on MoS_2 or selenium atoms on WSe_2 [2]. Patterns drawn using this technique can serve as a mask to stabilize the metal atoms of an atomic wire. In this case, the material used must be an insulator at low temperature and, at least, a semiconductor at room temperature so that its surface can be imaged by STM. The atomic wire can also be assembled and stabilized by chemical synthesis using for example organometallic [3] or organic oligomers [4]. In this case, the fabrication problem to be solved is no longer the stabilization of the wire structure but its connection to the micron metallic contacts [5].

Since neither of these fabrication procedures has been tested experimentally, it is worthwhile evaluating the feasability of one of them. Because gold (Au) deposited on NaCl has been well discussed in the past in relation to electron microscopy, we have chosen such a system to discuss the fabrication of a Au atomic wire on a NaCl(100) surface. This paper follows the steps we have to perform in order to make such a wire. In section 2, AFM constant force images of an NaCl(100) surface, obtained using a diamond tip, are calculated. The equilibrium positions of an Au atom on the NaCl(100) surface are discussed. This section ends with the calculation of the constant force AFM image of a single Au atom on NaCl(100). In section 3, the interaction between the AFM tip apex and the adsorbed Au

247

P. Avouris (ed.), Atomic and Nanometer-Scale Modification of Materials: Fundamentals and Applications, 247–261.

atom is discussed. The conditions required to push an Au atom on an NaCl(100) surface are then deduced. A calculated constant force AFM image of an Au atomic wire is proposed. In section 4, the elastic conductance of an Au atomic wire of 7 atoms in length, connected to larger gold contacts, is calculated using the Elastic Scattering Quantum Chemistry (ESQC) technique.

2.   The AFM image of a gold atom adsorbed onto a NaCl(100) surface

2.1.   THE CALCULATION OF THE AFM IMAGES OF NaCl(100).

The atomic structure of an NaCl(100) surface is well known [6]. On the surface, there is a small departure from the bulk fcc structure of the NaCl crystal due to the difference in the polarizabilities of the Na^+ and Cl^- ions [7]. The relative displacement of Cl^- with respect to Na^+ is a few tenths of an angström. This leads to a rumpling structure on the NaCl(100) surface [8]. However, this relaxed geometry is not essential for our discussion of the displacement of an Au atom with an AFM tip, since it causes only a small increase of the Au binding energy [9]. Therefore, only a non relaxed NaCl(100) surface will be considered in the following. A more complete calculation will be provided in a forthcoming publication.

For the NaCl(100) surface, He-scattering experiments indicate a 0.72 Å surface corrugation [10] and theoretical calculations predict a corrugation smaller than 0.85 Å [9]. This corrugation is attributed to the Cl^- ions which have a larger ionic radius than the Na^+ ions [10]. Constant force AFM images of the NaCl(001) surface have been obtained experimentally by C. Meyer and N. Amer [11]. The corrugation found in a 10 nN force image is 0.5 Å. This corrugation was attributed to the Cl^- ions. There was no Na^+ corrugation in the images [11].

Constant height AFM images of a NaCl cluster (obtained using a metallic tip) were calculated by one of us [12]. However, experimental tips are made of materials such as silicon nitride and diamond [11]. Therefore, we have calculated constant force AFM images of the NaCl(100) surface using a diamond tip with a C3v apex composed of 4 layers of carbon atoms (figure 1). This tip was already used to reproduce, with a good agreement with the experiments, constant force AFM images of graphite by using a full relaxed molecular mechanic approach [13].

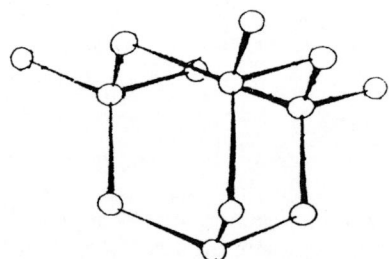

Figure 1: Representation of the C3v diamond tip used to scan on an NaCl(100) surface

Both the apex and the surface relaxation are frozen so as to focus more on describing the displacement of Au atom than on analysis in details the AFM corrugations of the NaCl(100) surface. The van der Waals interactions between the apex carbon atoms and the NaCl(100) surface atoms are described by pairwise potentials of the Lennard-Jones type with $C_6 = 6913$ meV Å^6 and $C_{12} = 2.586 \times 10^7$ meV.Å^{12} for the Na - C interactions, $C_6 = 59292$ meV Å^6

and $C_{12} = 9.037 \times 10^7 meV.Å^{12}$ for the Cl - C interactions. No electrostatic correction has been taken into account since, by hypothesis, our model of the NaCl(100) surface does not consider the relaxation of the surface induced by the local surface electric field. We have also neglected the inductive energy induced by this field on the tip apex.

The calculated corrugation depends on the force used during the scans (table I). A complete image is presented in figure 2 for a 0.01 nN force. In calculating this image, only the z-component F_z of the force between the apex and the surface has been optimized. The F_x and F_y frictional components of this force have not been optimized. This is justified for an atomically flat surface because, in this case, F_x and F_y are usually one order of magnitude smaller than F_z. The square lattice of the Cl^- ions is clearly seen in figure 2. Its corrugation is 0.353 Å. The square lattice of the Na^+ ions is also imaged but its corrugation is of 0.172 Å. This second lattice cannot be imaged experimentally even at larger forces. The difference between the values obtained by our calculations and by experiment can be attributed to the fact that the rumpling structure is not taken into account in the calculation. This relaxation of the surface moves the Cl^- ions up by at least 0.15 Å [8]. In this case, it is more difficult to image the Na^+ lattice because of the increase in the interactions between the second layer of the apex (see figure 1) and the surface. Notice that at zero F_z force , the distance between a Cl^- ion of the surface and the last carbon atom of the apex is 3.325 Å (see table I).

TABLE I. Corrugation (Δz) and absolute altitude (Abs. z) between the tip apex and the Cl^- ion surface layer for the different scanning forces used.

Force/nN	-0.1	0.0	0.01	0.1	1.0
$\Delta z/Å$	0.390	0.353	0.353	0.346	0.380
Abs. $z/Å$	3.544	3.325	3.311	3.226	2.933

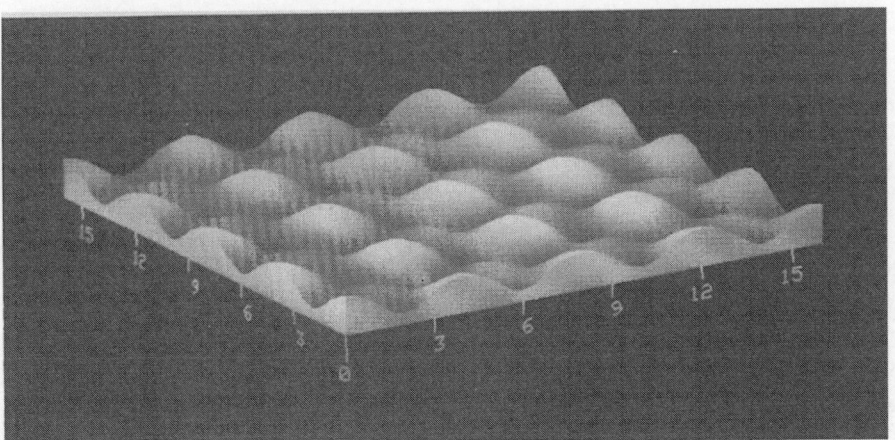

Figure 2: Constant force AFM image of a NaCl(100) surface for a 0.01 nN F_z scanning force. Scale: $\Delta z = 0.353$ Å from black to white, the scanned area is 15.96 x 15.95 $Å^2$.

2.2.   THE STABILITY OF A GOLD ATOM ON THE NaCl(100) SURFACE.

Since the seminal work of Barett [15], it is known that Au atoms preferentially adsorb onto a NaCl(100) surface at the edges of monoatomic steps when the surface is prepared by cleaving a NaCl substrate in UHV [16]. This property is used to decorate dislocation-induced steps on the NaCl(100) surface [11] as well as on surfaces of others materials such as KCl and MgO [17]. It is also known that gold particules can grow on atomically flat terraces of surfaces prepared in this way [18]. Models of the nucleation and the growth kinetics of these particules were used by Robinson and Robins to evaluate, using experimental data, the binding energy E_b and the surface diffusion barrier height E_d for an Au adatom [19]. These Experimental data were taken.in the flat regions of the NaCl(100) surface between the decorated steps. The authors determined $E_b = 0.68$ eV and $E_d = 0.27$ eV.

Such a binding energy clearly indicates that an Au atom is chemically bonded to the NaCl(100) surface as was studied theoretically by Fuwa and co-workers [9]. These authors found that Au mainly bonds, on top of a Na atom, but to the four nearest Cl neighbour via the $3p_z$ Cl atomic orbitals and to a less extend to the Na atom underneath. For our purpose, it is more convenient to use here a semi-empirical description of this binding with adjusted pairwise potentials between the Au atom and both the Na^+ and the Cl^- ions of the substrate [21]. It was demonstrated by J. Yanagihara that this approach leads to an E_b in good agreement with experiment [21]. Such an approach for NaCl(100) was first tackled by von Harrach [22] and by Chan and co-workers [20]. In this model, one must take into account three types of interactions: the dispersive van der Waals interactions, the core-core repulsive interactions and the electrostatic interactions [23]:

$$V(R) = \left(\sum_i C_{12} /||R\text{-}r_i||^{12} - C_6 /||R\text{-}r_i||^6 - C_8 /||R\text{-}r_i||^8 \right) - 1/2 \; \alpha_{Au} \; E_{loc}(R)^2 \quad (1)$$

where the repulsion constant C_{12}, the dipole-dipole C_6 and dipole-quadrupole C_8 constant are taken from [12,14]. R is the position of the Au atom on the NaCl(100) surface with respect to an origin located on the Na atom under the Au atom. α_{Au} is the dipolar polarizability of the Au atom and E_{loc} the local electric field near the NaCl(100) surface. This field is responsible for the relaxation of the NaCl surface because the gain in electrostatic energy is different for Na^+ and for Cl^-[7]. In (1), Eloc is only acting on Au because only the Au atom position will be optimized. In A full self consistent approach which would include a calculation of the rumpling structure of the NaCl(100) surface, the relaxation of the surface would change Eloc and therefore the R equilibrium distance. In this case, the summation in (1) must be extended to the polarization of both Na^+ and Cl^- ions.

Instead of calculating V(R) in the direct space as indicated in (1), the reciprocal lattice method was used in order to improve the computation time [12]. The V(R) curves are presented in figure 3 for different possible adsorption sites onto the NaCl(100) surface. The equilibrium position is found to be on top of the Na atoms, where $E_b = 0.61$ eV. This value is in good agreement with the experimental value [19]. Lateral diffusion of the Au atom from such a site to a neibouring one preferentially follows a path passing between two Cl^- ions, in a Na^+ - Na^+ direction. The diffusion barrier height is evaluated by the difference between E_b and the energy minimum at the Cl^- - Cl^- saddle point. Its estimated value is $E_d = 0.12$ eV, which is quite small compared with its experimental value [19]. This can again be attributed to the fact that the relaxation of the NaCl(100) surface is not considered here. A tenth of an angström difference in the height of the Cl⁻ atoms raises this barrier height to reach the experimental value [19]. The Au atom equilibrium height above the NaCl(100) surface is found to be 2.58 Å. This altitude is slightly too high compared with other theoretical results [9], which explains our small value for E_d.

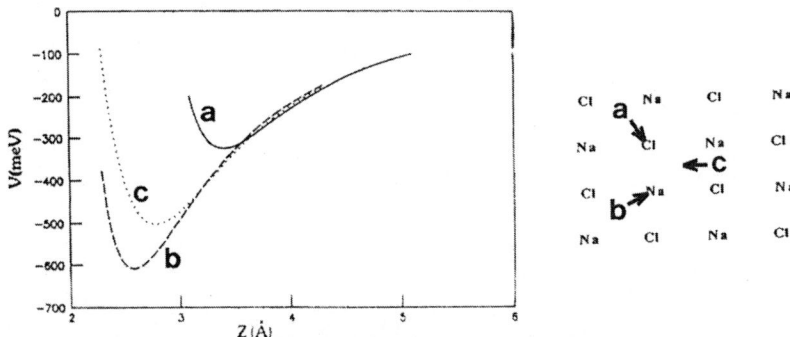

Figure 3: Potential energy curve of an Au atom adsorbed on an NaCl(100) surface for the three possible equilibrium position shown on this surface.

2.3.   THE CALCULATION OF AFM IMAGES OF A GOLD ATOM ON NaCl(100)

When the diamond tip (see figure 1) approaches the Au atom from on top at a distance $R_0 = (x_0, y_0, z_0)$, the total energy of this "tip apex - Au atom - NaCl surface" junction can be written:

$$U(R,R_0) = V(R) + V_{ts}(R_0) + V_{ta}(R,R_0) \qquad (2)$$

$V(R)$ is the semi-empirical potential (1). $V_{ts}(R_0)$ is the potential energy of the diamond tip apex interacting with the NaCl(100) surface, as approximated in section 2.1 by a Lennard-Jones potential. $V_{ta}(R,R_0)$ describes the interaction between the diamond apex and the adsorbed Au atom. The Parameters for $V(R)$ and $V_{ts}(R_0)$ have already been given. For $V_{ta}(R,R_0)$ we have preferred a Buckingham potential rather than a Lennard-Jones one:

$$V_{ta}(R,R_0) = \sum_i A_0 \exp\left(-a \, \|R-(R_0+R_i)\|\right) - C_6 / \|R-(R_0+R_i)\|^6 \qquad (3)$$

where R_i is the position of an apex carbon atom relative to the end atom of the tip. The first reason for this choice is that there are two parameters A_0 and a in (3) which can be used to tune the repulsive part of this potential. The second reason is that in molecular mechanic routines such as MM2 [24], A_0, a and C_6 have been optimized for many non-bonded couples of atoms using Buckingham potential. Generalized parameters for metals are also available. In (3), for each Au - C interaction , we have taken from the MM2 routine: $A_0 = 1.63 \, 10^6$ meV, $a = 2.796$ Å⁻¹ and $C_6 = 100950$ meV. Å⁶.

The force - distance $F_z(z_0)$ characteristic curve of the junction is presented in figure 4, when the end atom of the tip is centered on top of the Au atom and when this atom is frozen in its equilibrium position. Compared with the $F_z(z_0)$ characteristic curve calculated above the same Na^+ site with the Au atom removed, the attractive region is shifted away from 3 to 5 Å z_0. This is due to the 2.58 Å height of the Au atom in its equilibrium position. This attractive z_0 region is located at a lower z_0 on a Cl^- site than between two Cl^- sites because the facets of the apex are closer to the Au atom when the tip is between rather than on a Cl^- site.

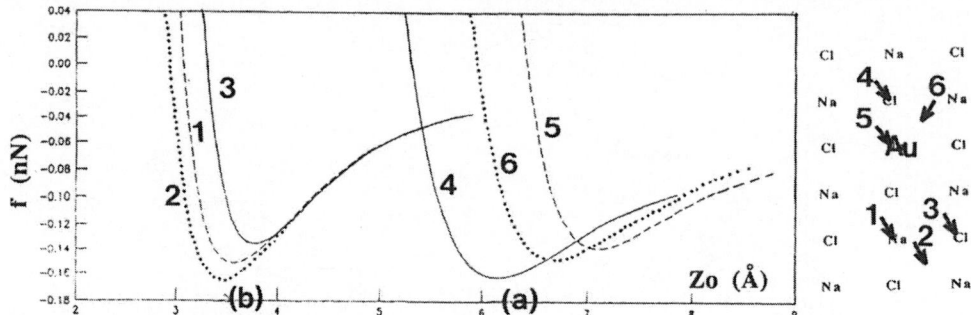

Figure 4: Force distance characteristic $F_z(z_0)$ on the Na^+ Au equilibrium site with (a) and without (b) this Au atom.

These lateral interactions and the shift towards large z_0 values of the $F_z(z_0)$ curves determine the apparent bump height produced by a single Au atom in an AFM scan obtained at constant force. Such scans are presented in figure 5. To calculate each scan, the Au atom is frozen in its 2.58 Å equilibrium position, even at very large repulsive scanning forces. In this case, the apparent bump height decreases when the scanning force increases.

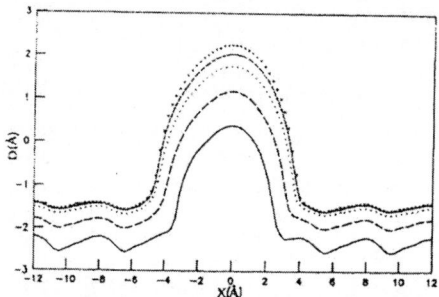

Figure 5: Constant force AFM scans along the Na^+ - Na^+ direction for different forces: (___) $F_z = 10$ nN, (-.-.-) $F_z = 1$ nN, (....) $F_z = 0.1$ nN, (-..-..) $F_z = 0.0$ nN and (....) $F_z = -0.03$ nN. Scale: $\Delta z = D + 4.5$ Å.

The background corrugation of the NaCl(100) surface increases until saturation is reached for forces larger than 50 nN. This behaviour is due to the fact that at small scanning forces (typically lower than 1 nN), the tip is far from the Au atom (z_0 is typically larger than 6 Å). At such heights, the van der Waals interactions (2) fall off very slowly when z_0 increases. Therefore, both the end carbon atom of the tip apex and the three carbons of its second layer interact with the Au atom. This lateral effect produces a very high apparent bump for the Au atom. On the contrary, for large scanning forces (typically larger than 10 nN), the tip height is small (z_0 is typically smaller than 5 Å). In this case, the van der Waals interactions (2) fall off very rapidly when z_0 increase. Therefore, the interactions occur principally via the last carbon atom of the tip. Consequently, the bump height is smaller in the repulsive than in the attractive mode. The variation of the NaCl(100) backgroung corrugation with z_0 can be explained in the same way. At large z_0, there is no difference in the interaction between the apex carbon atom and the Na^+ or Cl^- ions. Therefore, the background corrugation decreases when the scanning force deacreases.

A complete constant force AFM image of the Au atom adsorbed on a Na^+ site of the NaCl(100) surface is presented in figure 6. A zero force was chosen to minimize the Au - tip apex interaction, so that the Au atom is left in its equilibrium position, as discussed in section 3. The bump's height is 3.45 Å and its lateral width is 9 Å. These values are comparable with those obtained for an atomic adsorbate imaged by STM if it is physisorbed at 2.5 Å above the surface. Notice that the bump is not circular but displays a C3v symmetry. This is due to the interaction of the second layer of the apex with the Au atom.

Figure 6: Constant force AFM image of Au adsorbed on its equilibrium site for a zero F_z scanning force. Scale: $\Delta z = 3.5$ Å from black to white, the scanned area is 19.95 x 19.95 Å².

3. Moving a gold atom on a NaCl(100) surface with a diamond tip.

3.1. MOVING THE ATOM ?

To assemble an Au atomic wire on a NaCl(100) surface, gold atoms must be deposited on a freshly leaved NaCl crystal. Since Au atoms spontaneously accumulate at step edges to form clusters, isolated atoms, at low coverage, must be found on flat terraces between cleavage steps. An AFM image must be taken to locate these Au atoms using a very small scanning force so as to leave the atoms in their natural equilibrium position. After a specific Au atom to be moved has been chosen, the tip apex must be positionned above this atom. In order to have control over the displacement, there must be some attraction between the tip apex and the Au atom. This attraction can be exerted on top or laterally. If it is an on - top interaction, the displacement of the Au atom will occur via a sliding process. If it is a lateral attraction, the displacement will occur via a pushing process.

The sliding process works well for a Xe atom adsorbed onto a Ni(110) surface [25,26]. In this case, the Xe atom is trapped under the STM tip apex by a weak van der Waals attractive interaction when the apex approaches the Xe atom. Therefore, the Xe, held in this trap, follows the displacement of the tip [26]. The main difference between the Xe - Ni(110) system and the Au - NaCl(100) system is the amplitude of the diffusion barrier at the surface. This barrier is less than 20 meV for Xe on Ni(110) and it is larger than 0.12 eV for Au on NaCl(100). This means that a soft push on the top of the Au atom with a diamond tip will not be enough to create an attractive van der Waals well with a depth larger than the diffusion barrier height. For the Xe on Cu(110) system, this well can reach 35 meV in depth for a 15 meV barrier height if the STM tip apex is approched on top of the Xe atom at 6.4 Å above the surface [26].

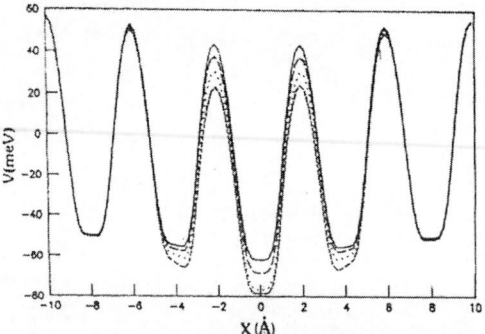

Figure 7: Plot of the potential energy minimum of the Au atom when this atom is displaced along a Na^+ - Na^+ direction with the tip apex fixed on a central Na^+ site for (___) $z_0 = 8$ Å, (-.-.-.) $z_0 = 7.5$ Å, (......), $z_0 = 7$ Å and (-..-..-) $z_0 = 6.5$ Å.

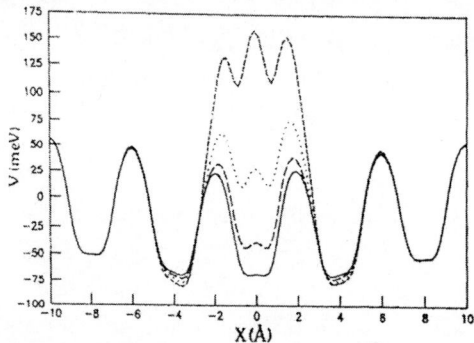

Figure 8: The same as figure 7 but for (___) $z_0 = 6.25$ Å, (-.-.-.) $z_0 = 6$ Å, (......), $z_0 = 5.75$ Å and (-..-..-) $z_0 = 5.5$ Å.,

Let us study this soft push on top of the Au atom. For the diamond apex positionned on a Na^+ site, we have calculated the variation of the potential energy minimum of an Au atom when it is moved along the Na^+ - Na^+ direction passing through the Na^+ site above which the tip is positionned. These curves are presented in figure 7 for different values of z_0. For large z_0 (typically between 6.5 and 8 Å), the apex creates an attractive depression on the chosen Na^+ site superimposed to the 0.12 eV corrugation of the Au minimum potential energy on NaCl(100). However, the stabilization energy is in all cases much lower than the 0.12 eV barrier height. If the z_0 height is further reduced, which is equivalent to the tip pushing more on the Au atom, the interaction becomes repulsive as presented in figure 8. In this case, even for z_0 around 6.25 Å, the Au atom is destabilized from its original equilibrium position. If one continues to press on the Au atom, the diffusion barrier decreases and for $z_0 = 5.5$ Å, the Au atom will escape from its Na^+ site to reach another one (figure 8).

Therefore because of the very large diffusion barrier height, a sliding process do not seem to allow a controlled displacement of a single Au atom on the NaCl(100) surface The Au atom can be forced to escape from its original equilibrium site when the tip presses on it. But the directionality of such an induced displacement is not controlable since the Au atom has an equal probability of jumping to any of the four nearest Na^+ sites surrounding the original one. Consequently, taking an AFM image at a zero scanning force will not change the equilibrium position of the Au atom because, at zero force, $z_0 = 6.53$ Å. This can be seen by plotting the variation of the F_z component of the force between the apex and the Au atom as the tip scans at a constant altitude (figure 9). If only this component of the force is taken into account, the energy gain for the Au atom due to its interaction with the apex is found to be only a few meV for a 6.53 Å constant height scan. This gain is smaller than the 0.12 eV barrier height.

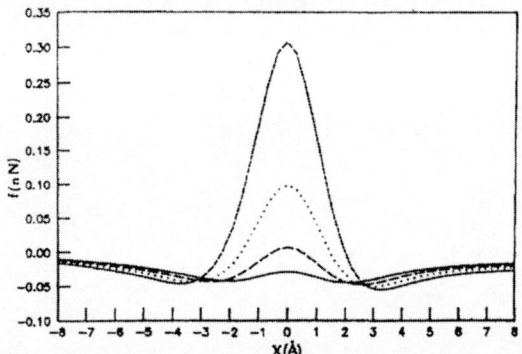

Figure 9: Variation of the F_z component of the tip apex - surface force when the tip is scanned at (___) $z_0 = 6.75$ Å, (----) $z_0 = 6.5$ Å, (.....) $z_0 = 6.25$ Å and (-..-..-) $z_0 = 6.5$ Å.

Let us now discuss the lateral process. In this case, instead of pressing on top of the Au atom, the tip is positionned near the Au atom but sideway. As presented in figure 10, the best choice is to put the apex in the middle between two Na^+ on the "c" site as defined in figure 3. In this case, the energy barrier created can be very large (see figure 10 a) but the two Na^+ sites around the "c" site are still possible equilibrium positions for the Au atom when a small 25 meV attractive lateral depression is added by the push exerted by the tip apex. If the Au atom is located on the Na^+ site to the right of this large barrier, a displacement of the tip toward this site destabilizes the Au atom. For example, in figure 10c and for $z_0 = 5.5$ Å, the barrier blocking the diffusion of the Au atom is now less than 30 meV. In this case, the next

Na$^+$ site can be occupied with a stabilization energy of more than 0.1 eV. By ending the tip displacement at the Na$^+$ site occupied by the Au atom before the push, this atom is now on the next Na$^+$ site as presented in figure 10d. Notice that this displacement can occur only in the direction of the tip mouvement since, as presented in figure 10b, there is still a large barrier preventing the Au atom from jumping in the direction opposite to the push. Moreover, the Au atom cannot escape laterally towards the two others Na$^+$ sites available. This is due to the fact that, by such a move, the tip apex becomes close to the two lateral "c" sites and that the barrier height on the two site is high enough at low temperature to prevent any lateral escape of the Au atom.

3.2. CALCULATION OF THE AFM IMAGE OF A LINE OF GOLD ATOM ON NaCl(100)

If we assume that the displacement process described in the prevoius section is reproducible, a chain of Au atoms may be assembled by pushing selected Au atoms one by one and bringing them onto Na$^+$ sites belonging to a Na direction on the surface. But the minimum possible distance between two Au atoms on the NaCl(100) surface is not known. What can be calculated from (1) is the potential energy of an Au atom on the surface which approaches another Au atom immobilized on a Na$^+$ site. As presented in figure 11, it is possible to bring two Au atoms onto two Na$^+$ sites separated by one or more Na$^+$ sites. But according to the van der Waals potentials used in (1), it is not possible to stabilize two Au atoms on two consecutive Na$^+$ sites. This is due to the fact that such potentials are not able to describe chemical bonding between two Au atoms.

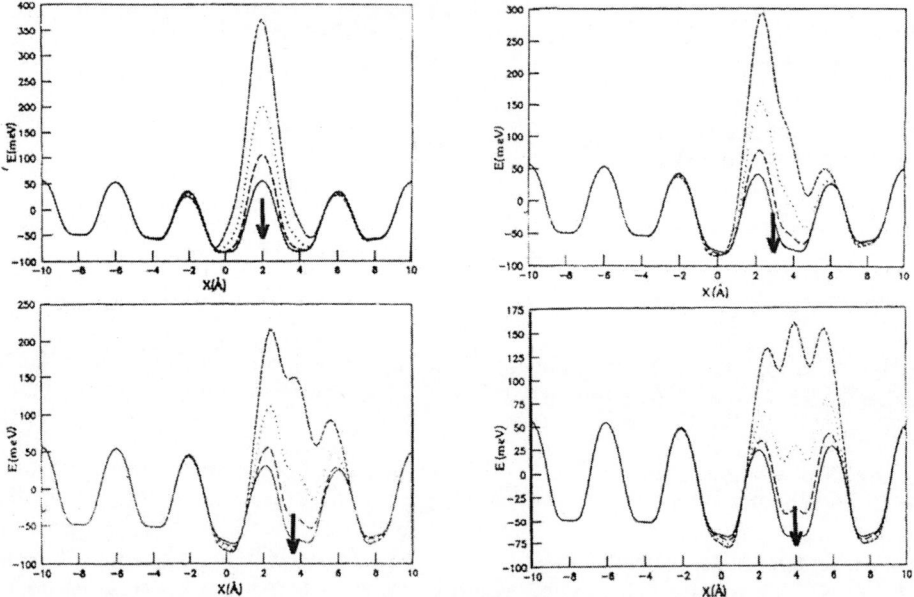

Figure 10: A possible sequence of tip apex position and displacement direction to push an Au atom from one Na$^+$ site the the next one in the direction of the tip displacement. The tip apex position is indicated on each figure with an arrow. The minimum energy of the Au atom is presented as a function of its position X along a Na - Na direction.

It is known that this bonding occurs on NaCl(100) surface since Au clusters are spontaneously formed on step edges. This bonding introduces an activation energy barrier as shown schematically in figure 11. This barrier must be overcome during the pushing of an Au atom in order to bring the Au atoms as close as possible in the chain. But we do not yet know the height of this activation barrier or the binding energy between two Au atoms on two consecutive Na$^+$ sites.

A force constant image of a chain of Au atoms assembled on consecutive Na$^+$ sites is presented in figure 12. A zero scanning force was used in order to leave each Au atom in its equilibrium position during the scans. The apparent bump height of each Au atom in the line is 3.35 Å. A small height variation is observed for the two terminal Au atoms of the chain. This is due to the facets of the tip apex which interact laterally with the first and the last Au atoms. Furthermore, each Au atom of the chain can be resolved when the structure is viewed from above. It was equally possible to resolve in the STM image each Xe atom of the Xe chain fabricated by sliding Xe atoms with an STM tip [25].

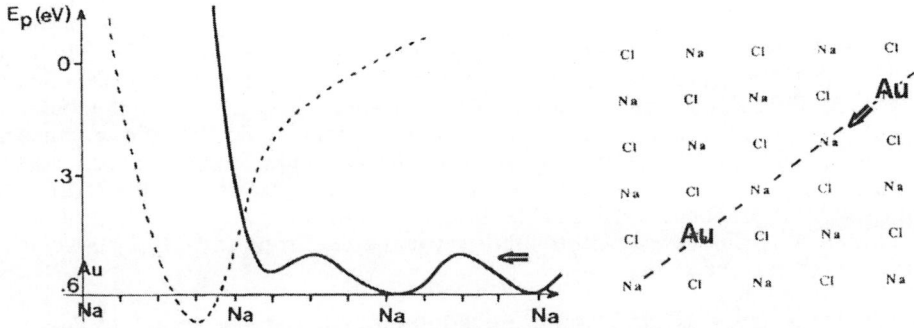

Figure 11: Potential energy of 2 Au atoms adsorbed on the NaCl(100) surface as a function of the distance between them when one of the two atoms is fixed on a given Na$^+$ site. (___) van der Waals contribution, (----) possible binding contribution.

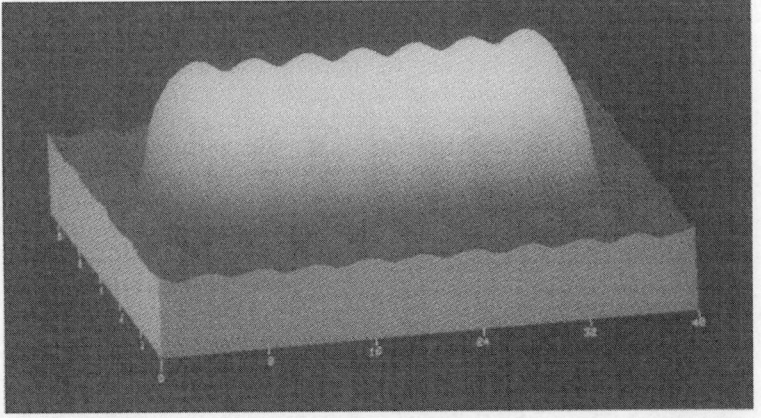

Figure 12: A constant force AFM image of a 7 in length Au atom chain adsorbed on NaCl(100) for a zero force. Scale $\Delta z = 3.5$ Å, The scanned area is 39.9 x 15.96 Å^2

4.   The elastic conductance of a gold atomic wire.

If the atomic wire described in the last section is connected to a permanent current source, electrons may flow through this wire. The intensity of this flow will be limited by many phenomena like (a) the diffraction of the Fermi electrons on the constriction formed by the connection between the atomic wire and the contact electrodes [27], (b) the energy difference between the Fermi level of the contact electrodes and the electronic levels of the atomic wire, (c) the displacement of the atomic wire atoms induced by the vibronic coupling between the flowing electrons and the vibrational modes of the wire or by Pieirls distorsion and (d) the electron - electron interactions between the flowing electrons and the electrons belonging to the wire certainly enhanced in the atomic wire compared to a bulk wire. This means that this wire will have a finite electric conductance G. Furthermore, if G << $2e^2/h$ and if the capacitance C of the "contact electrode - atomic wire - contact electrode" junction is such that $e^2/2C$ >> k_BT (T is the working temperature) the Coulomb blockade (e) can also limit the current flow through this atomic wire [28].

The (a) and (e) limitations to the current intensity are junction effects. This means that it is not possible to attribute an intrinsic resistance to an atomic wire independantly of its contacts [29]. Nevertheless, electronic structure of the atomic wire (b), vibronic coupling (c), Pieirls distortion (c) and electron - electron interactions (d) are characteristic of the material constituting the wire. Therefore, the conductance of an atomic wire has truly an intrinsic and an extrinsic origin. Both are necessarily related since the elastic contribution (b) to the conductance cannot be separated from the junction constriction effect as discussed by R. Landauer [29]. This is for example the case when a structural disorder in the atomic wire decreases the probability for the electrons to tunnel between the two contacts via the atomic wire.

The inelastic processes limiting the conductance can be evaluated using some generalization of the Joules law by calculating the kinetic energy of the atomic wire atoms as a function of the steady state current intensity in the wire. The elastic contribution can be calculated using the generalized Landauer formula with the scattering matrix calculated from the molecular orbitals of the atomic wire [30].

In a first approach, only the elastic scattering processes will be considered. This can be justified by the fact that the electron elastic mean free path in a metal like gold is 100 Å at room temperature to reach the micron range at low temperature [31]. Therefore, the inelastic processes will be of second order compared to the elastic processes in a metallic atomic wire of finite length like the one we can make with an AFM in a resonable amount of time. One elastic effect is the diffraction of the Fermi electrons on the constriction formed at the interface between the contact electrode and the atomic wire. For example, in a gold electrode, the Fermi wave length λ_F is 5.3 Å. Therefore, the Heisenberg relation $\Delta p.\Delta x$ > $h/4\pi$ leads to a minimum possible diameter for an atomic wire of $\lambda_F/4\pi$ = 0.4 Å. The diameter of an Au atomic wire is very close to this limit since the 6s Au atomic orbital radius is only 1.5 Å. Notice that in standard mesoscopic devices made of GaAs materials where resistance quantification have been first measured, the Fermi wavelength is around 400 Å [32].

Let us consider an atomic wire made of 7 gold atoms contacted to two gold electrodes each of 49 atoms in section .From the ESQC technique [33], the multi-channel scattering matrix of the Fermi electrons on the defect constituted by the gold electrode - Au wire - gold electrode junction can be calculated by describing each Au atom of the structure by all its valence atomic orbitals. Restricting each Au atom of the structure to its 6s atomic orbital to limit the computation time, the conductance G of this junction can be calculated using the generalized Landauer-Büttiker formula included in ESQC [33].

When there is no atomic wire between the two gold electrodes, the calculated conductance is only the conductance of a tunnel junction made of two small in section metallic wires. The tunneling current intensity through this junction is controlled by the through space electronic coupling between each atom of the two electrodes. Its resistance is very high, typically 100 GΩ for a 10 Å distance between the surface of the two electrodes, 49 atoms in section.

If now the atomic wire is built atom by atom, the electronic coupling between the two gold electrodes will change of character from a through space to a through bond coupling via the chain of Au atoms [34]. This is presented in figure 13a. When the atomic wire is longer than 3 atoms, the through space coupling becomes negligible for the section considered for the electrodes. The conductance is stabilized around the resistance quantum unit $2e^2/h$ (i.e. 12.9 KΩ). This is a constriction effect since there is no elastic or inelastic processes in the atomic wire limiting its intrinsic conductance. Furthermore, if the orbital description of the Au atomic wire is completed by its other atomic orbitals like the 5p and 5d ones, the conductance will increase because new electronic channels will be available for the Fermi electrons.

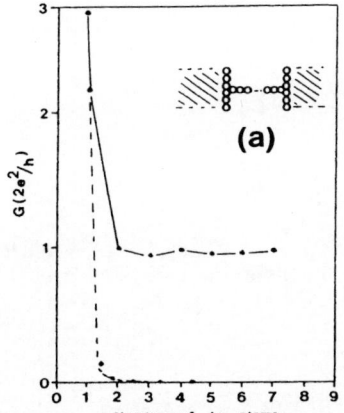 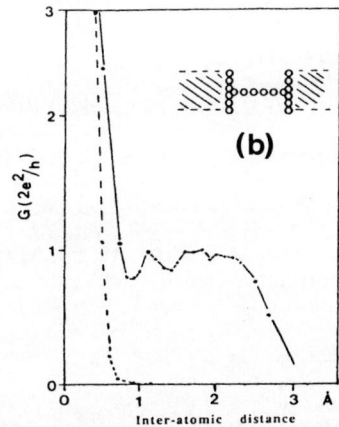

Figure 13: Variation of the conductance of a chain of Au atoms (a) as a function of the number of Au atoms in the chain for a 2.039 Å inter - atomic distance and (b) as a function of the inter-atomic distance for a given 7 atoms in length chain. For (a) and (b), (----) is a reference curve showing the conductance variation of a tunnel junction made only of the two contact electrodes without the atomic chain but with an equivalent distance separation.

Any departure from the atomic structure and geometry chosen to calculated G presented in figure 13a induces a change in G. This can be shown by varing the inter-atomic distance d in the atomic wire. The d = 2.039 Å distance was chosen in the previous calculation of G to respect the inter-atomic distance of the Au atoms in the contact electrodes. The variations of G as a function of d are presented in figure 13b. For very short distances (typically d < 1 Å), the distance between the two electrodes is very small and the through space coupling between these electrodes becomes larger than the through bond coupling via the atomic wire. Therefore, $G > 2e^2/h$ and follows the variation of the through space conductance. For an intermediate range of distances (typically $1 \, \text{Å} < d < 2.5 \, \text{Å}$), the through bond coupling is via the 6s electronic band of the Au atomic wire. This band is progressivelly formed by adding more atoms to the atomic wire and these atoms are enough coupled for the inter-atomic distance considered. For large distances (typically $d > 2.5 \, \text{Å}$), the interatomic electronic coupling in the atomic wire is not enough to ensure the formation of the 6s band. In this case, G is controlled by through bond coupling via discret Au levels. Then, G decayes very rapidly when d increases. G will only reached $2e^2/h$ if the Fermi level is in resonance with one of these atomic levels.

For the inter-atomic distance imposed by the NaCl(100) surface and if only the 6s orbitals are considered, the conductance of a the "contact electrode - Au atomic wire- contact electrode" will be well below $0.1 \, 2e^2/h$ (i.e. a resistance certainly largeur than one MΩ). A substrate with a smaller lattice constant like MgO or LiF will lead to a shorter Au inter-atomic distance in the wire than NaCl. Therefore, the resitance will be close to the maximum available $2e^2/h$ quantum resistance unit (rapported to the number of atomic orbitals used to described the electronic properties of an atom of the atomic wire).

5.   Aknowledgement

We would like to thank Fujitsu France for full computing support during this work

6.   References.

[1]: Strocio, J.A. and Eigler D.M. (1991) Science 254, 1319.

[2]: Garcia, R.G. (1992) Appl. Phys. Lett. 60, 1960.

[3]: Kobel, W. and Hanack, M. (1986) Inorg. Chem. 25, 103.

[4]: Baumgarten, M., Müller, U. and Müllen, K. (1992) in A. Aviram (ed.), Molecular Electronics, Sciences and Technology, American Institute of Physics Publishers, AIP262, 68.

[5]: Joachim, C. (1991) New Journ. Chem. 15, 223.

[6]: Benson, G.C. and Claxton, T.A. (1968) J. Chem. Phys. 48, 1356. Wolf, D. (1992), Phys. Rev. Lett. 68, 3315.

[7]: Welton-Cook, M.R. and Prutton, M. (1978) Surf. Sci. 74, 276.

[8]: Eichenauer, D. and Joennies, J.P. (1988) Surf. Sci. 197, 267.

[9]: Fuwa, K., Fujima, K., Adachi, H. and Osaka, J. (1984) Surf. Sci. 148, L659.

[10]: Rieder, R.H. (1982) Surf. Sci. 118, 57.

[11]: Meyer, G. and Amer, N.M. (1990) Appl. Phys. Lett. 56, 2100.

[12]: Girard, C. (1991), Phys. Rev. B 43, 8822.

[13]: Tang, H., Joachim, C. and Devillers, J. (1992), Surf. Sci., to be submitted.

[14]: Lakhlfi, A. and Girardet, C. (1991) J. Chem. Phys. 94, 688.

[15]: Bassett, G.A. (1958) Philosophical Mag. 3, 1042.

[16]: Bethge, H. (1982) Interfacial Aspect of Phase Transformation, Reidel, Dordrecht, p. 669.

[17]: Yanagihara, T. (1979) Surf. Sci. 86, 62.

[18]: Sato, H. and Shinozuki, S. J. (1970) Appl. Phys. 41, 3165.

[19]: Robinson, V.N.E. and Robins, J.L. (1974) Thin Solid Film 20, 155.

[20]: Chan, E.M., Buckingham, M.J. and Robins, J.L. (1977) Surf. Sci. 67, 285.

[21]: Yanogihara, T. and Yamaguchi, H. (1984) Jap. Journ. Appl. Phys. 23, 529.

[22]: von Harrach, H. (1974) Thin Solid Film 22, 305.

[23]: Steele, W.A. (1974) Interaction of Gases with Solid Surfaces, Pergamon Press, Oxford. Girard, C. and Girardet, C. (1987) J. Chem. Phys. 86, 6531.

[24]: Allinger, N.L., Kok, R.A. and Iman, M.R. (1988) J. Comp. Chem. 9, 591.

[25]: Eigler, D. and Schweizer, E. (1990) Nature 344, 524.

[26]: Bouju, X, Joachim. C., Girard, C. and Sautet, P. (1992) Phys. Rev. B submitted.

[27]: Tekman, E. and Ciraci, S. (1989) Phys. Rev. B 40, 8559.

[28]: Devoret, M.H., Esteve, D., Grabert, H., Ingold, G.C., Pothier, H. and Urbina, C. (1992) Ultramicroscopy 42, 22.

[29]: Landauer, R. (1989) J. Phys. C 1, 8099.

[30]: Sautet, P. and Joachim, C. (1988) Phys. Rev. B 38, 12238.

[31]: Schmid H., Rishton, S.A., Kern, D.P., Washburn, S., Webb, R.A., Kleinsasser, A., Chang, TH.P. and Fowler, A. (1988) J. Vac. Sci. Techno. B 6, 122.

[32]: van Wees, B.J., Kouwenhoven, E.M., Willems, E.M.M., Harmans, C.J.P.M., Mooij, J.E., van Houten, H., Beenakker, C.W.J., Williamson, J.G. and Foxon C.T. (1991) Phys. Rev. B 43, 12431.

[33]: Sautet, P; and Joachim, C. (1991) Chem. Phys. Lett. 185, 23.

[34]: Joachim, C. and Sautet, P. (1990) Scanning Tunneling Microscopy and Related Methods R. J. Behm, N. Garcia and H. Rohrer (eds.), Kluwer Academic Publishers, Dordrecht, Appl. Sci. Series Vol. 184, p. 377.

STRUCTURE, DYNAMICS AND ELECTRONIC PROPERTIES OF

MOLECULAR NANOSTRUCTURES OBSERVED BY STM

J.P. RABE
Max-Planck-Institut für Polymerforschung
Postfach 3148
DW-6500 Mainz
Germany

ABSTRACT. A scanning tunneling microscope (STM) has been used to characterize molecular nanostructures at surfaces and solid-fluid interfaces on length and time scales of 10 pm and 10 μs, respectively. We present the experimental set-up and describe imaging of solid surfaces under ambient conditions, including an organic conductor (difluoranthenyl-hexafluorophosphate), a layered semimetal (graphite) and a layered semiconductor ($MoSe_2$). We then discuss the determination of molecular structure and dynamics as well as electronic properties of molecular monolayers at solid-fluid interfaces. Finally, we give some perspectives for the modification of molecular nanostructures using an STM.

1.   Introduction

Nanostructures based on molecular materials have attracted considerable interest [1]. Molecules can be viewed as extremely well defined quantum structures. They also may have some internal degrees of freedom, which could give rise to switching properties. Due to the richness of organic chemistry these building blocks can be largely varied, in order to exhibit a particular property profile. Provided they are not too complex, chemists can make them rather pure and in large quantities, and living organisms are able to make them even more complex. In order to obtain still more complex structures one may assemble the individual molecules into supermolecular structures, where, in general, the inter-molecular interactions are smaller than the intra-molecular ones. If designed properly such "self-assembled" molecular structures can be rather perfect. We report in the following on a number of relatively simple molecular nanostructures, and what one may learn about them using a STM.

P. Avouris (ed.), Atomic and Nanometer-Scale Modification of Materials: Fundamentals and Applications, 263–274.
© 1993 *Kluwer Academic Publishers.*

2.   Experimental

An analysis of the capabilities and limitations of STM for a quantitative characterization of molecular materials [2], as well as a review of the application of STM to molecular materials have been given recently [3]. The results reported below have been obtained with a home-built instrument, which was designed with an emphasis on *low currents* and *high scanning speeds,* since particular difficulties encountered with molecular materials include low electronic conductivities and high molecular mobility. Of interest is also the *temperature controlled* sample stage [4]. Between room temperature and up to 120°C the instrument allows atomic scale resolution with currents as low as 200 fA and scan speeds up to 1 ms per line. The images are recorded on video tape in real time and digitized later for digital image processing on a graphic workstation or for static image display.

All images have been recorded under ambient conditions using Pt/Ir or W tips, which were either electrochemically etched or mechanically formed. A schematic of the experimental set-up is displayed in Fig. 1. The images of the holes in graphite were obtained in the "constant current mode" (displayed is a height profile), whereas all other images have been recorded in the "constant height mode" (displayed is a current profile).

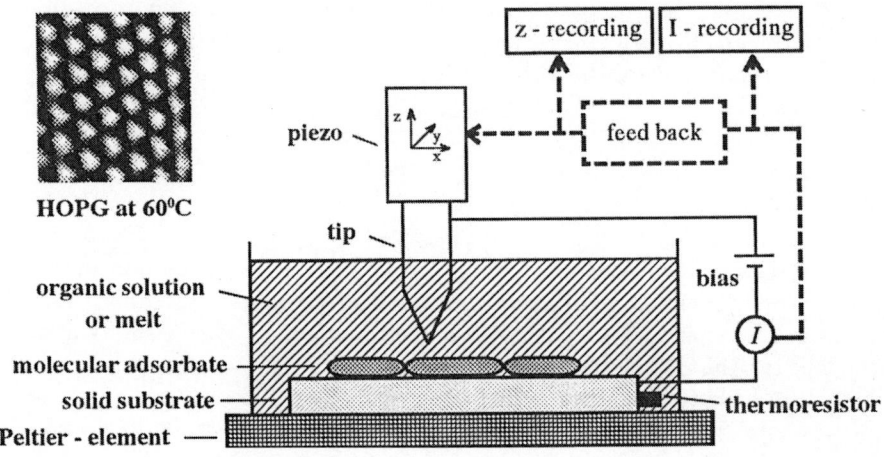

Fig. 1: Schematic of the experimental set-up [4].

3.   Solid surfaces under ambient conditions

Surfaces of conducting solids may be investigated under ambient conditions, provided they are sufficiently inert. Examples include thin films of gold or silver, which, if evaporated onto mica at some elevated temperatures, exhibit (111) surfaces [5]. Since STM, per se, does not require any order it allows also to study non-crystalline materials. Interesting examples include conductive polymers, like, e.g., iodine doped polydiacetylene, whose fibrillar structure has been visualized by STM [6]. However, for use as substrates for ultrathin organic layers single crystalline materials have the advantage that their local atomic structure is precisely defined.

3.1.   Organic Conductors

A class of molecular materials, whose surfaces can be directly investigated by STM are organic conductors. An example is given in Fig. 2, displaying an STM image and the corresponding model of the bc-surface of a single crystal of difluoranthenyl-hexafluorophosphate [7]. Even though no atomic resolution is achieved, the molecular columns are well resolved. Similarly, other single crystals of organic materials have been imaged before, including, e.g., tetrathiafulvalene-tetracyanoquinodimethane (TTF-TCNQ) [8] or β-(BEDT-TTF)$_2$I$_3$ [9]. A common problem is the fact that the limited chemical and/or mechanical stability of these materials, since it is observed that the outer layers may be altered during prolonged scanning.

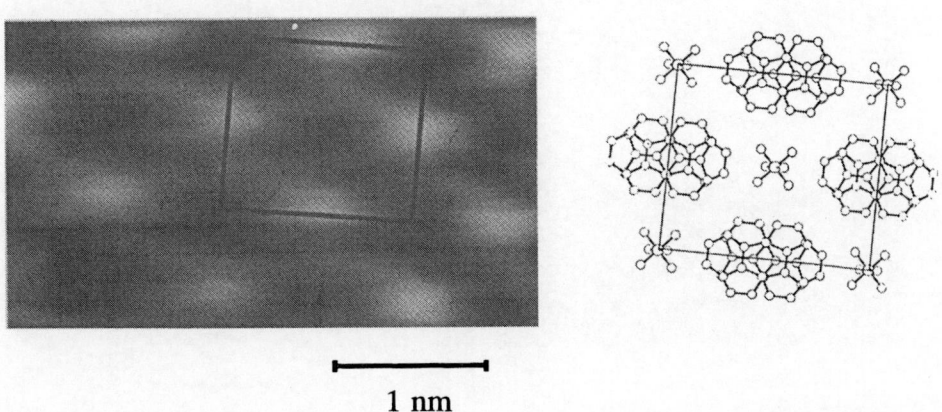

$\longmapsto\longmapsto$
1 nm

Fig. 2: STM image and model of difluoranthenyl-hexafluorophosphate [7].

3.2. Cleavage Planes of Layered Materials

A simple way to prepare atomically flat surfaces is to cleave a layered material. If the cleavage plane is chemically inert it can remain atomically well defined even under ambient conditions. A particularly interesting case is highly oriented pyrolytic graphite (HOPG), because it is both chemically as well as mechanically very inert. Moreover, highly perfect material is commercially available.

A number of rather stable layered semiconductors belong to the class of transition metal dichalcogenides. Materials, which have been investigated by STM include MoS_2 [10], $MoSe_2$ [11], and WSe_2 [12]. In Fig. 3 a typical image of a $MoSe_2$ basal plane is displayed. It indicates that the surface exhibits quite perfect terraces as far as atomical flatness is concerned. However, there are characteristic single atom defects, which are possibly due to other transition metal atoms substituting a molybdenum atom [11].

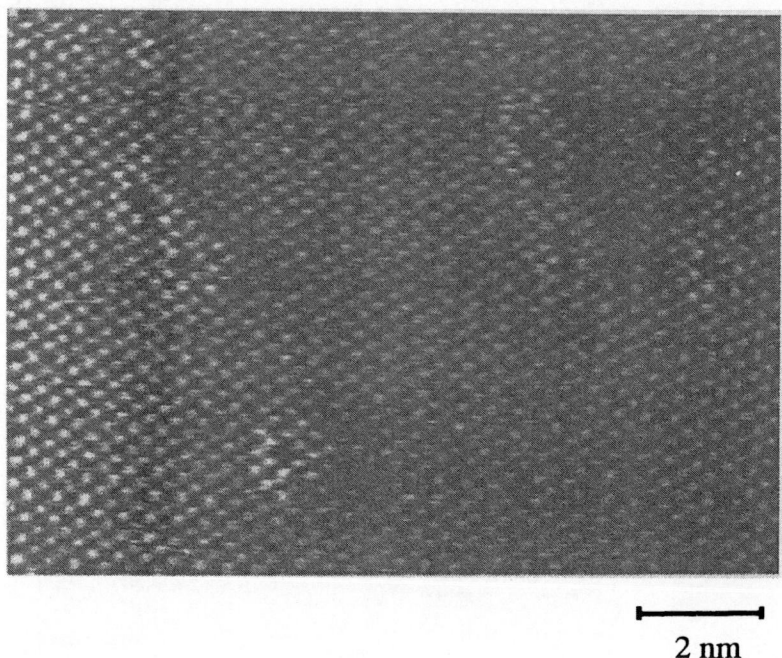

2 nm

Fig. 3: STM image of the basal plane of $MoSe_2$ [11].

4.   Molecular Monolayers at Solid-Fluid Interfaces

A class of molecular nanostructures, which have been investigated quite systematically by STM, are monolayers of alkanes and alkyl-derivatives, self-assembled at the interface between the basal plane of HOPG and a melt or an organic solution [4,7,13-18]. Fig. 4 displays the structure of a monolayer of a simple alkane, adsorbed from phenyloctane solution and imaged in-situ by STM. It reveals a highly ordered monolayer with lamellae, in which the main axes of the flat lying molecules are oriented perpendicular to the lamella boundaries [14]. Upon closer inspection one notices a change in contrast along an individual molecule, while across the image each molecule appears identical. The contrast change along a molecule has been attributed to a moiré pattern caused by the linear carbon chains in the graphite and the alkane, respectively, reflecting the incommensurability of the two linear lattices [14]. On the other hand, the identical appearance of all molecules within the two-dimensional molecular pattern reflects the commensurability of the molecular adsorbate lattice with the substrate [7,13,14].

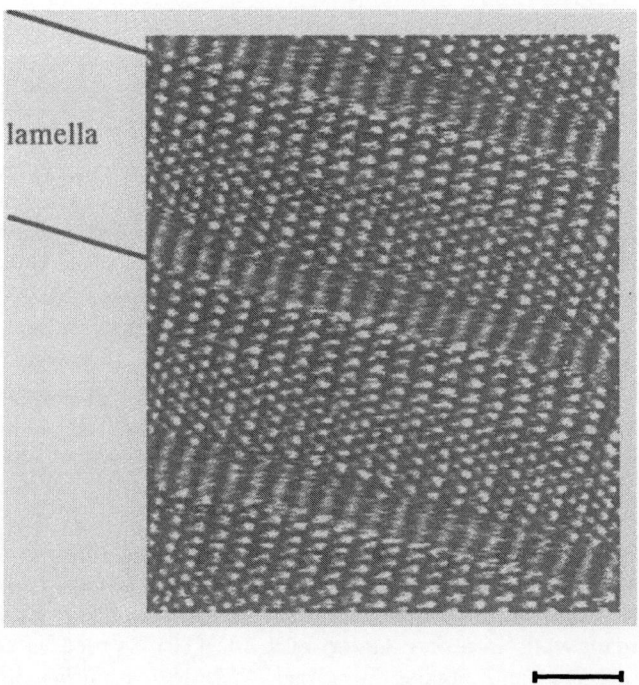

lamella

1 nm

Fig. 4: STM image of a $C_{27}H_{56}$ monolayer on HOPG [7].

Similarly, highly ordered monolayers form at the interface between alkane solutions and MoSe$_2$ surfaces (Fig. 5). Interestingly, however, the molecular structure is different: While on HOPG the alkane chains pack with their long axis perpendicular to the lamella boundaries, on MoSe$_2$ this angle is 60° [11].

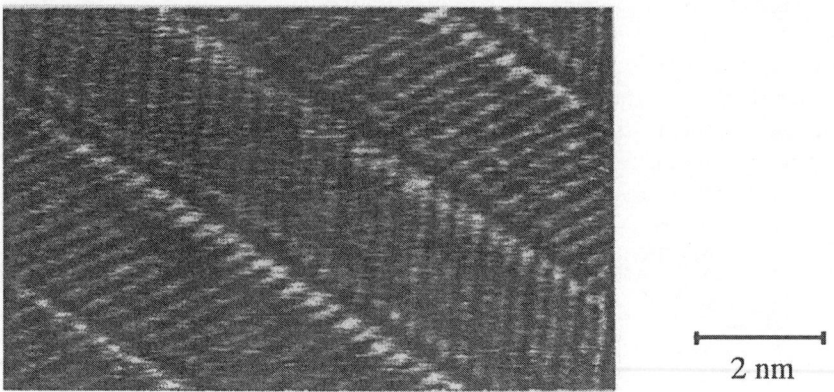

2 nm

Fig. 5: STM image of a C$_{32}$H$_{66}$ monolayer on MoSe$_2$ [11].

Many other long chain alkyl-derivatives have also been shown to form two-dimensional molecular patterns at the interface between HOPG and organic solutions. Particularly interesting are the images of the fatty acids [14], because they exhibit a superstructure along the lamellae. Since the lattice constants of the substrate are very well known, the superstructure allows a very accurate determination of the molecular distances within the adsorbate layer. In the case of the fatty acids the interchain distance is about 50 pm larger than in the case of the alkanes and alkanols. The reason is that the alkyl carbon skeleton planes are oriented parallel or perpendicular to the graphite plane, respectively [14,15]. Moreover, domain boundaries could be imaged at sub-molecular resolution [14,16,17], and different mixed monolayer phases have been observed [17,18].

So far the images have been interpreted as static, assuming that the time scales of molecular dynamics and image recording are sufficiently different. However, it is also possible to observe molecular reorientations directly with the STM. Examples include the motion of domain walls in a dialkylated benzene [16], as well as cooperative tilt flips within alkanol [14] and alkane lamellae [15]. Fig. 6 shows the progressive roughening of alkane lamellae at the interface between HOPG and an organic solution upon raising the temperature. The data indicate a continuous increase in the amplitude of a motion along the long molecular axis until the lamellar order disappears [4].

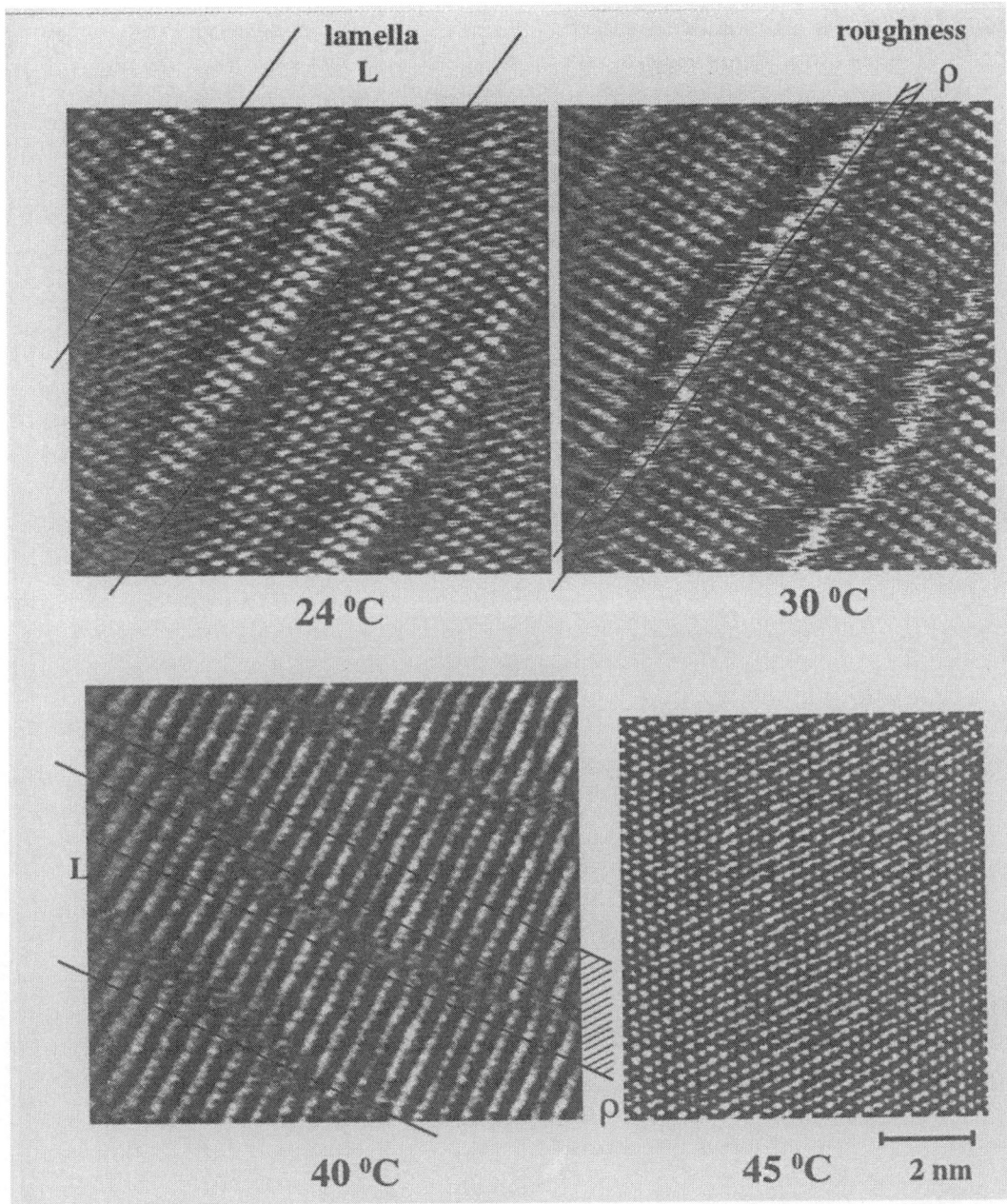

Fig. 6: Temperature induced roughening of adsorbed alkane lamellae [4].

STM imaging is also possible at the interface between a rigid conducting substrate and a "soft" and insulating organic solid on top [19,20]. In this case the STM tip may penetrate through the soft solid, and image just the first monolayer in contact with the rigid substrate. Provided the organic film is not too thick the forces exerted by the organic layer can be small enough to allow reproducible scanning.

The interpretation of the images in terms of structure and dynamics has been based to a large extent on symmetry and dimensions. Given the structure one may then use the contrast to learn something about the electronic properties of the molecules in their particular environments. Since the tunneling current will depend on the initial and final electronic states as well as on the barrier between them, it will contain information on tip, substrate and molecular adsorbate. For instance, in the case of xenon on nickel under ultrahigh vacuum conditions it has been argued that adsorbate states, which are electronically far away from the Fermi level, may contribute sufficiently to the electronic density of states at the Fermi level, in order to explain the contrast [21]. An interesting molecular example is given in Fig. 7. It displays a small

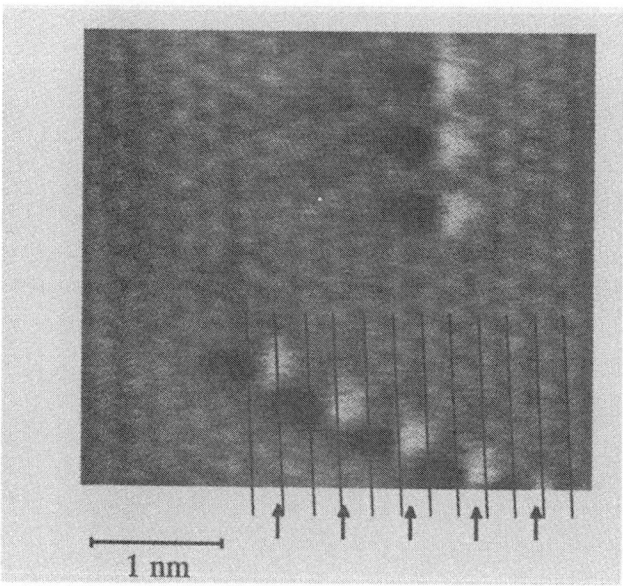

1 nm

Fig. 7: STM image at a domain boundary in a didodecylbenzene monolayer on HOPG. Under the given tunneling conditions the image is dominated by the underlying graphite lattice (vertical lines) and the bright benzenes on top (arrows) [3,22].

area within a didodecylbenzene monolayer on HOPG, recorded under imaging conditions, where the alkyl chains hardly contribute anything to the contrast, while the current through the benzene rings is particularly large [3,16]. Similarly, a considerable number of other alkylated molecular segments have been investigated [22]. It seems that, to a first approximation, the tunneling current increases as the energy difference between the molecular states and the Fermi level decreases. Quantum chemical ab-initio calculations of the electronic states of van der Waals complexes may be useful to quantify this [23].

5.   Perspectives for Modification of Molecular Materials

The STM allows not only the characterization but also the modification at interfaces. A comprehensive review has been given recently [24]. Several mechanisms may be invoked. So far, systematic work has been carried out primarily on surfaces of inorganic materials, but the underlying physics does not prevent the application to organic materials in the future. In order to use such processes for high density information storage purposes they need to fulfil some requirements: For instance, both writing and reading should be *fast*, the initial as well as the final structures should be sufficiently *stable*, the local change should *not perturb the molecular environment* much, and the change should be *clearly detectable* by STM. In the following we discuss some cases, which demonstrate that these requirements may be met in principle, even though the different requirements are not met equally well in all cases.

A fundamentally important process is field evaporation, which may be used to create holes and deposits on gold and silver [25,26]. Holes on Ag(111) have been fabricated, for instance, using bias pulses of about 5 V and as short as 10 nanoseconds at a base line current of 2 pA [26]. This may be attractive for high density data storage: On the hand, the primary writing process occurs at least on the 10 nanosecond time scale, and, on the other hand, depressions may be read out fast, since the height of the flying tip needs not to be adjusted.

Another interesting example is the local chemical modification of graphite under ambient conditions or at the interface with undried solvents. Metastable surface modifications of less than 1 nm diameter can be produced at a moderate bias of about 2 V [6,27], while long-term stable holes of a few nanometers in diameter could be etched with bias pulses of the order of 1 ms and 5 V [27-29]. The processes have been explained in terms of an anodic oxidation of graphite [29]. Moreover, holes typically one monolayer deep and between 5 nm and a few 100 nm in diameter have been made by gasification in air at temperatures around 650°C [30,31] (see also Fig. 8).

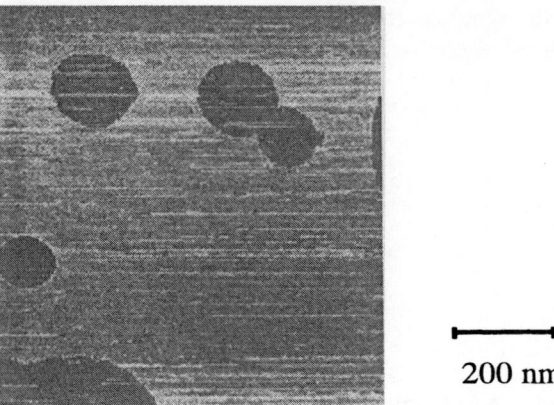

200 nm

Fig. 8: STM image of thermally etched holes in HOPG, which are typically circular
and one monolayer deep [31].

Interestingly, there are a number of modification mechanisms, which are particular to
molecular materials. An example may be the reorientation of molecules within a
molecular pattern. In two-dimensional patterns of long chain molecules several of the
prerequisites for information storage schemes described above can be fulfilled. Fig. 9
shows the image of a monolayer of long chain alkanes with a single square of
molecules, which are rotated by 90° with respect to the other molecules within this
domain. Obviously, the situation is metastable enough to be imaged, the rest of the
molecular pattern is not perturbed, and the square is easily detectable. While it is not
obvious, how to rotate the molecules in this square on purpose, one may imagine
related molecules, which should allow this process.

2 nm

Fig. 9: STM image of a metastable square of alkane ($C_{44}H_{90}$) molecules rotated by
90° with respect to the other molecules within this domain.

Another molecular property, which may be modified, is the conformation of a molecule. Azobenzenes, for instance, exhibit a stable trans- and a metastable cis-conformation. In bulk materials their conformation can be switched back and forth via the appropriate visible light. Fig. 10 shows that monolayers may be formed from an azobenzene derivative in its trans-conformation. Upon illuminating these patterns with light, which switches the molecules into their cis-conformation, the pattern can be destroyed [32]. A proper molecular design may allow that the cis-forms can also order well.

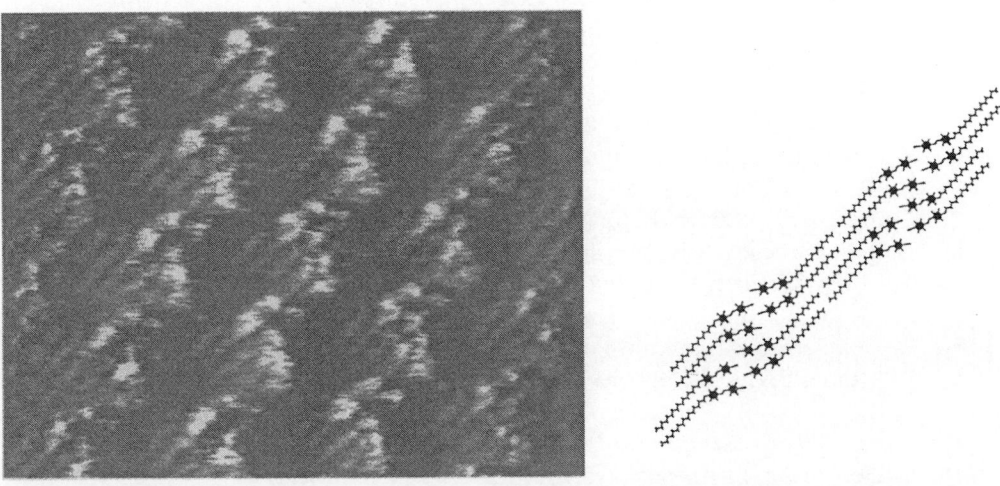

Fig. 10: STM image (12.7 nm x 10.2 nm) and model of a hexadecyl-cyanoazobenzene monolayer on HOPG [32].

Finally, chemical reactions, including complexations, dimerizations or polymerizations may be employed. While potentially reactive molecules (e.g., diacetylenes [22]) have been organized within molecular patterns, in which they may react, the demonstration of a controlled intermolecular reaction remains a challenge for the future.

6.   Acknowledgement

It is a pleasure to acknowledge my past and present coworkers S. Buchholz, L. Askadskaya, S. Cincotti, and A. Tracz, for their important contributions to the work reviewed here.

274

7. References

[1] *Nanostructures Based on Molecular Materials,*
W. Göpel, C. Ziegler (eds.), VCH Weinheim (1992).

[2] H.-J. Butt, R. Guckenberger, J.P. Rabe, *Ultramicroscopy* 46 (1992) 375.

[3] J.P. Rabe, *Ultramicroscopy* 42-44 (1992) 41.

[4] L. Askadskaya, J.P. Rabe, *Phys. Rev. Lett.* 69 (1992) 1395.

[5] S. Buchholz, H. Fuchs, J.P. Rabe, *J. Vac. Sci. & Techn. B* 9 (1991) 857.

[6] J.P. Rabe in [1].

[7] S. Buchholz, Dissertation, Mainz, 1991.

[8] T. Sleator, R. Tycko, *Phys. Rev. Lett.* 60 (1988) 1418.

[9] J.P. Rabe, S. Buchholz in: *Conjugated Polymeric Materials: Opportunities in Electronics, Optoelectronics, and Molecular Electronics*, p. 483, J.L. Brédas and R.R. Chance (eds.), NATO-ARW Series E, Kluwer, Dordrecht, 1990.

[10] M.G. Youngquist, J.D. Baldeschwieler, *J. Vac. Sci. Technol. B* 9 (1991) 1083.

[11] S. Cincotti, J.P. Rabe, *Appl. Phys. Lett.,* submitted.

[12] S. Akari, M. Stachel, H. Birk, E. Schreck, M. Lux, K. Dransfeld, *J. Microscopy (Oxford)* 152 (1988) 521.

[13] G.C. McGonigal, R.H. Bernhardt, D.J. Thomson, *Appl. Phys. Lett.* 57 (1990) 28.

[14] J.P. Rabe, S. Buchholz, *Science* 253 (1991) 424.

[15] R. Hentschke, B.L. Schürmann, J.P. Rabe, *J. Chem. Phys.* 96 (1992) 6213.

[16] J.P. Rabe, S. Buchholz, *Phys. Rev. Lett.* 66 (1991) 2096.

[17] S. Buchholz, J.P. Rabe, *Angew. Chem.* 104 (1992) 188; *Angew. Chem. Int. Ed. Engl.* 31 (1992) 189.

[18] R. Hentschke, L. Askadskaya, J.P. Rabe, *J. Chem. Phys.* 97 (1992) 6901.

[19] J.S. Foster, J.E. Frommer, *Nature* 333 (1988) 542.

[20] D.P.E. Smith, J.K.H. Hörber, G. Binnig, H. Nejoh, *Nature* 344 (1990) 641.

[21] D.M. Eigler, P.S. Weiss, E.K. Schweizer, N.D. Lang, *Phys. Rev. Lett.* 66 (1991) 1189.

[22] J.P. Rabe, S. Buchholz, L. Askadskaya, *Synth. Metals* 54 (1993) 339.

[23] G. Lambin, M.H. Delvaux, A. Calderone, R. Lazzaroni, J.L. Brédas, T.C. Clarke, J.P. Rabe, *Molecular Crystals Liquid Crystals*, in press.

[24] C.F. Quate, in: NATO Science Forum '90 *"Highlights of the Eighties and Future Prospects in Condensed Matter Physics"*, Plenum Press, 1991.

[25] H.J. Mamin, P.H. Guethner, D. Rugar, *Phys. Rev. Lett.* 65 (1990) 2418.

[26] J.P. Rabe and S. Buchholz, *Appl. Phys. Lett.* 58 (1991) 702.

[27] J.P. Rabe, S. Buchholz, A.M. Ritcey, *J. Vac. Sci. & Techn. A* 8 (1990) 679.

[28] T.R. Albrecht, M.M. Dovek, M.D. Kirk, C.A. Lang, C.F. Quate, D.P.E. Smith, *Appl. Phys. Lett.* 55 (1989) 23.

[29] S. Buchholz, H. Fuchs, J.P. Rabe, *Adv. Mater.* 3 (1991) 51.

[30] H. Chang, A.J. Bard, *J. Am. Chem. Soc.* 112 (1990) 4598.

[31] A. Tracz, G. Wegner, J.P. Rabe, *J. Chem. Phys.,* submitted.

[32] L. Askadskaya, C. Böffel, J.P. Rabe, *Ber. Bunsenges. Physikal. Chem.* (1993).

THE MODIFICATION OF SEMICONDUCTOR SURFACES BY MOLECULAR SELF-ASSEMBLY

D.L. Allara
Departments of Chemistry and Materials Science and Engineering
Pennsylvania State University
University Park, PA 16802
USA

ABSTRACT. The spontaneous adsorption of organic molecules at surfaces to form monolayer films has been a well known phemonenon for years. However, it is now widely recognized that specific molecule/substrate combinations under appropriate conditions can lead to stable, highly organized films, termed self-assembled monolayers (SAMs). These monolayers can be prepared as very uniform, robust structures of nanometer-scale thicknesses, with surfaces exhibiting a wide range of physical and chemical responses, distinctly different from the substrate upon which the film is attached. For these reasons a number of technological and scientific applications have been envisioned and recent efforts, predominantly in the latter area, have demonstrated the usefulness of these films, particularly for the system of alkanethiols on planar gold surfaces. In reality, very few examples of high quality SAMs exist. From a technological point of view it is important to explore applications to semiconductors. Two definitive examples of semiconductor-based SAMs are known, alkylsiloxanes on SiO_2/Si and alkanethiols on GaAs(100). General applications of these types of films are discussed, particularly for control of surface states and for use as ultrahigh resolution lithographic resist films. Finally, a brief discussion is presented of the need for development of new types of semiconductor/molecule chemistry to allow the development of new types of SAMs and the issue is raised of the requirement of interdisciplinary efforts in this type of research.

1. Review of General Concepts and Principles of Self-Assembled Monolayers

1.1 Definition and Description of Self-Assembled Monolayer Films

Self-assembled monolayers (SAMs) of organic molecules recently have evolved as a new class of thin film materials with exciting promises of new and novel applications, both scientific and technological[1]. While there are a number of ways to make extremely thin coatings on material objects, the term self-assembly generally is reserved for a process in which molecules, usually in the solution phase, spontaneously adsorb onto a surface to form a stable monolayer coating, often highly organized for specific types of

P. Avouris (ed.), Atomic and Nanometer-Scale Modification of Materials: Fundamentals and Applications, 275–292.
© 1993 *Kluwer Academic Publishers.*

molecules. Naturally, it has been known for many decades that molecules can adsorb to surfaces to make some sort of deposited surface films and there have been many applications involving this phenomenon, for example, the well-known inhibition of copper corrosion using benzotriazole. However, the term self-assembly has come to be associated with discrete monolayer films possessing a high degree of molecular organization, as opposed to the usual case of amorphous deposits of uncontrolled thickness. Because of intensive work carried out during the past decade, the general principles and the techniques (discussed in section 1.3) of the self-assembly process have become well known[2], although it should be pointed out that only a very limited number of film types have been discovered. The fact that the process is spontaneous and selective for only a monolayer implies that the free energy of adsorption is significantly negative for the first layer (high stability) but considerably less negative for further layers (low stability). In most cases, each molecule forms a specific chemical bond to the surface (chemisorption) and thus formation of a second layer is easily precluded by selecting an adsorbing molecule which provides an unreactive outer surface on the first layer. The premier example of this class of films is the system of organo-thiolates on gold surfaces[3,4] (discussed below). In other cases, the first layer is stabilized primarily through molecule-molecule interactions, invariably involving covalent crosslinking, which produces an insoluble, involatile macromolecular assembly (a gel membrane) incapable of removal by solvent or heat. In these cases, prevention of a second adsorbed layer must occur by using molecules which provide the outer surface of the first layer with a very low surface energy to discourage further adsorption relative to the initial substrate. The only important example of this system is that of alkylsiloxane assemblies[2,5], which because of the lack of substrate chemical specificity, can be formed on a variety of surfaces including SiO_2, Au, mica and Al_2O_3. However, on hydroxylic-group bearing surfaces such as hydrated SiO_2, an occasional attachment of the film to the substrate can occur via reaction of the film with a surface hydroxyl, a factor which can impart enhanced film adhesion. Because all of these films, regardless of the specific preparation method, can be controlled precisely to highly uniform monolayer thicknesses, a scale of 1-3nm usually, they represent a class of "nano" materials. Further, since the molecular assembly is confined to a surface plane, the film can be consisdered as 2-dimensional. In detail, of course, the structural complexities of the constituent molecules assembled at some given substrate surface will give rise to 3-dimensional properties (out of plane configurational states) but, in general, the major architectural features of the films are set by the constraints of the 2-dimensional pattern of assembly in the surface plane.

1.2 Potential Applications

In simplest terms, the films can be considered as a synthetic nanometer-scale thickness "skin" for a material object. Such objects can range from the macroscopic scale, with negligible surface/volume ratios, to nanometer-sized objects such as clusters, with significant surface/volume ratios. Whereas the presence of a SAM film can dramatically alter surface properties, bulk properties, such as mechanical strength or conductivity, will not change and so SAMs can provide, in principle, "designer" surfaces. SAMs are now being envisioned as a means to modify and control properties such as wetting, chemical reactivity, biological activity (e.g., compatibility with biological media), surface electronic states and mass transport, including electron and molecular transfer in and out of the material structure. Control of such properties can lead to a number of important applications[1,2] ranging from corrosion inhibition to biological sensors. The technologies expected to be impacted include energy conversion[1,6], transportation[6], microelectronics[1] and the health sciences[1]. As mentioned above, there are presently only two well-studied SAM systems, organothiolates/Au and alkylsiloxanes/SiO$_2$. Accordingly, most ideas for applications have stemmed from these two systems. The former is under current study primarily for applications to functionalized electrodes for electrochemistry[7] whereas the latter is commonly used as a means to make surfaces hydrophobic, particularly silicate glass surfaces. Gold has important applications as a corrosion resistant conductor, although cost limits the actual use, while glass (silicate) is an appropriate material for a wide variety of applications from optical devices to biological substrates. Since the field of surface self-assembly is just in its infancy, breakthroughs in major technological and/or manufacturing applications still await discovery.

Of particular interest in this paper are applications to semiconductor materials and processing. Applications that are envisioned here include chemically and biologically active transduction monolayers for molecule-specific sensors on semiconductor devices, etch resistant monolayers with molecular scale resolution for lithographic patterning of semiconductor devices and chemically bound monolayers which control device surface states via chemical interaction to the surface atoms of the semiconductor. As will be discussed below, a critical bottleneck in the discovery and development of applications is the discovery and development of appropriate SAMs for the semiconductors of interest. The technologically important materials are primarily the group IV elements, silicon and germanium; the III-V compounds, including gallium arsenide and indium phosphide and related alloys and oxides.

1.3 Preparation of SAMs

In one sense the preparation of self-assembled films is trivial. Once a viable molecule/substrate pair has been decided, the clean substrate is dipped directly into a solution, or alternatively exposed to a vapor, of the adsorbate compound and the surface adsorption continues spontaneously until all available surface sites are reacted (or covered) to form a complete monolayer. At this point, barring environmentally induced degradation, the sample is stable and can be removed for characterization or further processing for applications such as device fabrication. However, as will be seen with

GaAs(100), the preparation and stability must be considered as complex issues since elevated temperatures may be needed to drive the surface reactions and the films may be unstable over time to environmental degradation such as oxidation or hydrolysis. A schematic of processing methods is shown in Figure 1.

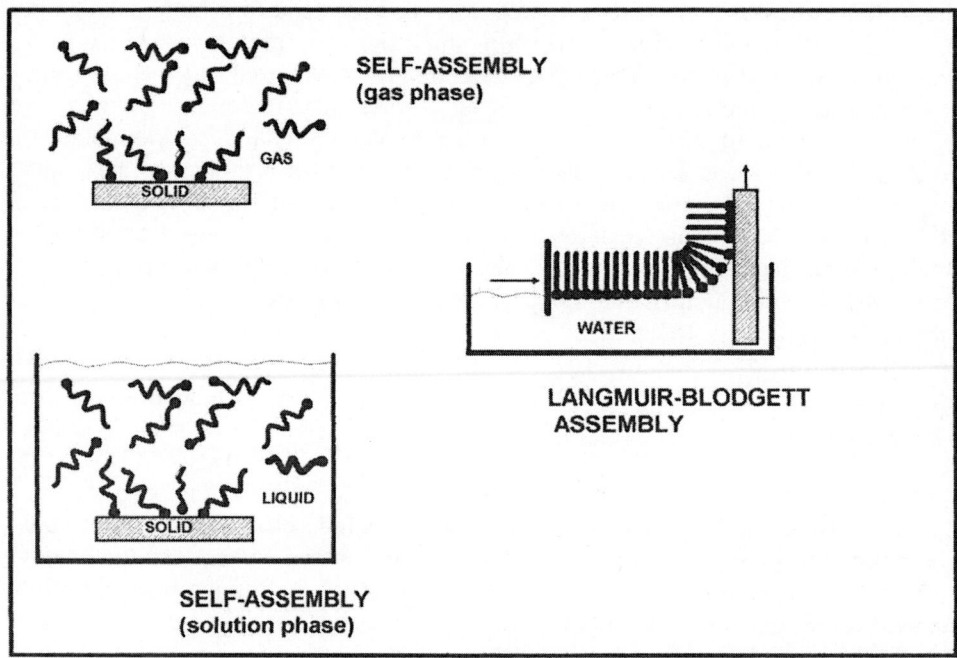

Figure 1. Methods for preparation of organized monlayer assemblies at surfaces.

The purpose of these pictorial schemes is to show the clear distinction between SAM and the popular Langmuir-Blodgett (LB) processing methods. The latter, as can be inferred from the figure, involves the imposition of a surface pressure on a water-surface bound monolayer of surfactant, followed by mechanical withdrawal of an immersed substrate which pulls a layer of film with it as it leaves the water phase. This process does not happen spontaneously but requires continuous external intervention. As such, the surface films which form exhibit structures which are direct functions of the processing variables. On the other hand, the SAM films are, in principle, driven to completion by free energy minimization and, as such, should always approach unique, equilibrium structures, provided, of course, that the kinetics will allow this. An excellent review of the above film assembly methods has been published[2]. The major class of molecules, by far, which has been used for film preparations consists of alkyl derivatives, which for the cases of polar, water-soluble derivative groups such as CO_2H, are also surfactants and thus useful for LB films. Surfactants also form a rich variety of related 3-dimensional "nano-structures" in the form of aqueous solution microphases and Figure 2 shows a variety of these which include bilayers. A large class of biologically significant self-assembled structures such as vesicles are formed from bilayers.

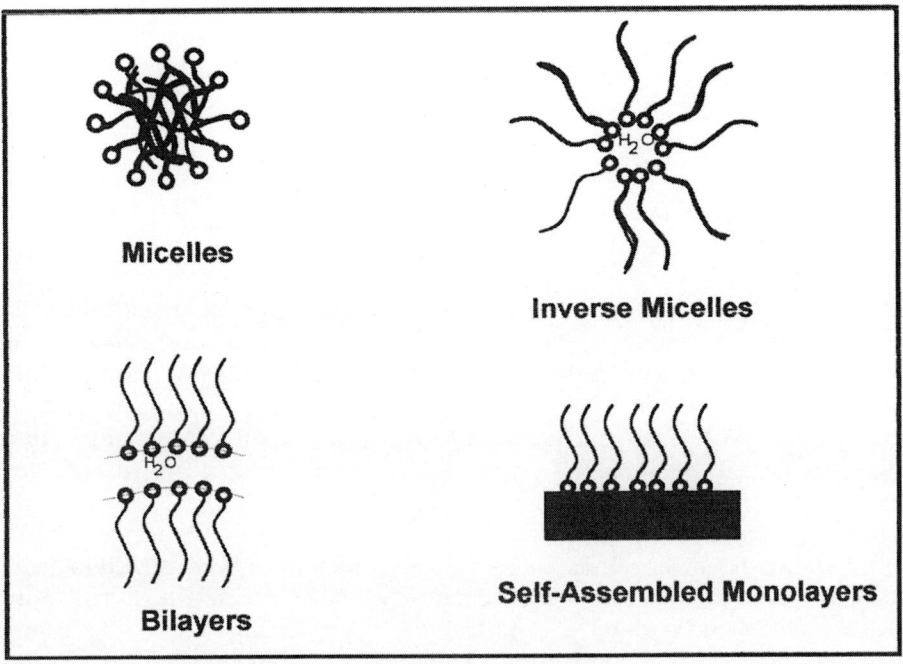

Figure 2. Related complex structures of amphiphilic molecules.

In the case of SAMs, the importance of the terminal group is in establishing substrate bonding or film crosslinking in order to provide a surface constrained phase. In contrast, for LB films and surfactant microphases, the terminal group provides water compatability (hydrophilicity). An example of this difference is given in the cases of SAMs which bond directly to substrate sites on zero-valent transition metals such as Au or Ag (coinage metals). In these cases polarizable (soft), easily reduced (acceptor) groups such as SH can provide strong bonding to the substrate but their weakly polar character makes them poor surfactant groups. On the other hand, the siloxy (trihydroxysilyl) group of alkylsixoanes should be a good surfactant polar group but it polymerizes quickly to form a network and thus cannot function as a typical monomeric surfactant. The important aspects of SAMs with respect to structure and substrate chemistry can be viewed in terms of conceptually selected regions of the films as shown in Figure 3.

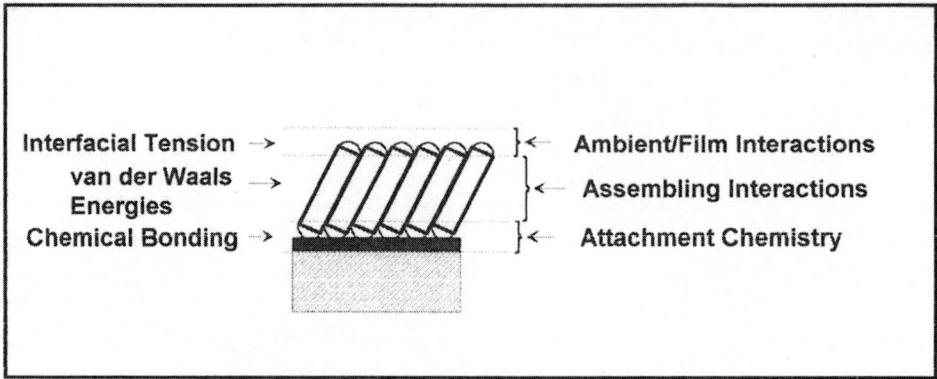

Figure 3. Important structural components of a self-assembled monolayer film.

For SAMs that react integrally with the substrate, the chemical interactions at the surface must be chosen to provide strong anchoring for thermal and solvent stability, as well as desired symmetry patterns in the case of "epitaxial" attachment. The selection of alkyl chains of length beyond approximately 12 carbons usually assures good molecular packing via van der Waals interactions and the presence of polar terminal groups on the non-anchoring end of the alkyl chains can be "tuned" to acheive specific wetting and chemical properties of the outer (ambient) film surface.

Table 1 shows the major self-assembly systems which have been studied to date, classified according to the attachment chemistry. One extremely important point for SAMs which depend upon direct chemical reaction with substrate sites for film integrity, is that in order to acheive a variety of terminal functionality at the ambient surface one must prevent these groups from competing with and/or interfering with the surface bonding. The strategy which has evolved for this purpose involves the selection of different polarizabilities and donor-acceptor properties, the so called hard-soft acid base character. This is pictorially shown in Figure 4 for a simple linear, terminally difunctional (telechelic) molecule.

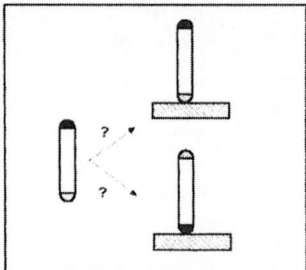

Figure 4. A representation of the selectivity of functional groups for anchoring difunctional molecules at surfaces in the self-assembly process.

For the case of the alkylsiloxane based SAMs which depend upon headgroup (SiO_x) crosslinking for stability, any terminal substituent groups must not interfere with the crosslinking or chain assembly processes. For this reason the production of high quality films seems only possible with very nonpolar terminal groups such as CH_3 and $CH=CH_2$. In this sense, the alkylsiloxane SAMs are extremely limited in chemical variability, although chemical reaction of the surface groups has been used to provide new functionality in several cases (for example, see reference 2). However, it must be pointed out that the formation of alkylsiloxane films is not very substrate dependent and as such they are much more universally applicable, at least in principle, to a wide variety of substrates (e.g., SiO_2, mica, Au, Al_2O_3, etc). This characteristic is in complete contrast to substrate-reacted SAMs which depend upon a specific adsorbate/substrate pair. For example, alkane thiols form strongly bound SAMs on the coinage metals or GaAs but are capable only of forming weakly bound, physisorbed layers on substrate materials such as SiO_2 or Al_2O_3 as no suitable adsorption chemistry is available. In order to access the envisioned technological applications of SAMs in semiconductor applications, it will be necessary both to develop new SAM/substrate chemistries and to develop innovative extensions of established ones. With regard to SAM/substrate interface chemistries, most of the knowledge in the field has evolved from studies of terminally substituted alkane thiol compounds on Au(111), a system which has proved invaluable for the development of the fundamental principles of self-assembly. This work will need to be extended to new types of surfaces.

Later in this paper, two SAM systems for semiconductor materials will be discussed, a newly developed system of alkane thiols on GaAs(100) and the well-known system of alkylsiloxanes which form high quality SAMs on the oxided silicon surface. To the author's knowledge, these are the only two definitive examples of SAMs on semiconductors. However, first the very intensely studied case of alkanethiolates/Au(111) will be discussed in order to demonstrate important points about structure and chemistry in SAMs.

2. Alkane Thiols/Au(111) - An "Epitaxial" Layer Example

The specific chemistry between the divalent S groups, SH and SS, and gold surfaces have been discussed above already. In this section several points will be introduced concerning the molecular structures of the alkane chains of alkanethiols on Au(111) and their arrangement on the surface plane. This system has been the most intensely studied SAM to date and it is instructive to examine the structural details which have been derived from experiment and to see their correlation with the nature of the substrate and the film constituent molecules. Some of the aspects discussed below can be found in detail in a recent review[8]. One of the important aspects of self-assembly at surfaces is the in-plane arrangement of the molecules on the substrate. Figure 5 depicts the effects of the lattice positions of rigid rod-shaped molecules on their geometry.

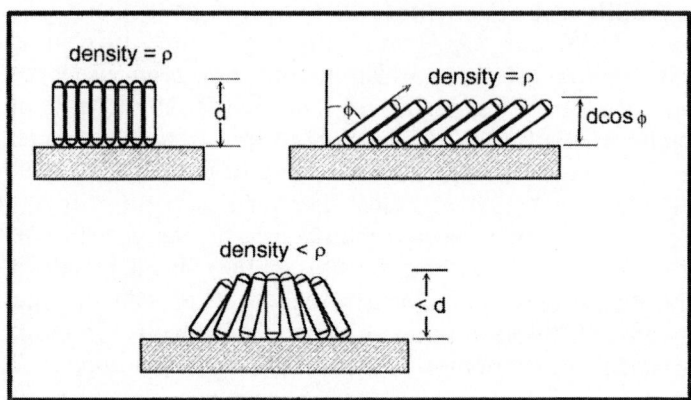

Figure 5. Representations of the effects of the surface lattice positions on the orientational arrangements of rigid rod molecules at surfaces.

For closest packed molecules only one structure is possible, an ensemble of identically arranged chains, each with vertical orientation to the surface, and a uniform film thickness (d). However, if the lattice spacings are opened up the chains can tilt in two limiting ways, uniformly with one angle (ϕ) or disorganized with a distribution of angles. Further, the film thickness will decrease below d (the closest-packed value) and the density will decrease.

Figure 6 shows a representation of a Au(111) surface with the hexagonally aranged Au atoms and the superimposed projections of several CH_2 chains positioned on the surface with their chain axes oriented perpendicular to the surface. It can be seen that the chain diameter of closest approach (~0.45nm) is larger than the lattice spacing of the Au atoms (0.29nm). Because of the steric constraints, each chain cannot place its S group at adjacent Au atoms but the next nearest neighbor atom, a ($\sqrt{3}$ x $\sqrt{3}$)R30° spacing, can accomodate the chains. The fact that this pattern has been been determined experimentally from a variety of data[8], including electron[9] and x-ray[10] diffraction, suggests that simple steric arguments suffice to make reasonable predictions of structure. However, it has been found for the case of alkane thiols on Ag(111) that significant reconstruction of the Ag surface takes place upon adsorption[11,12] and thus predictions based on the (111) lattice are incorrect. If the chains attached at the ($\sqrt{3}$ x $\sqrt{3}$)R30° surface sites on the Au(111) surface were in an all trans conformational sequence, they could be approximated as rods (of ~0.45nm diameter) and in order to pack tightly to minimize their interaction potentials (van der Waals dispersion interactions) the simple

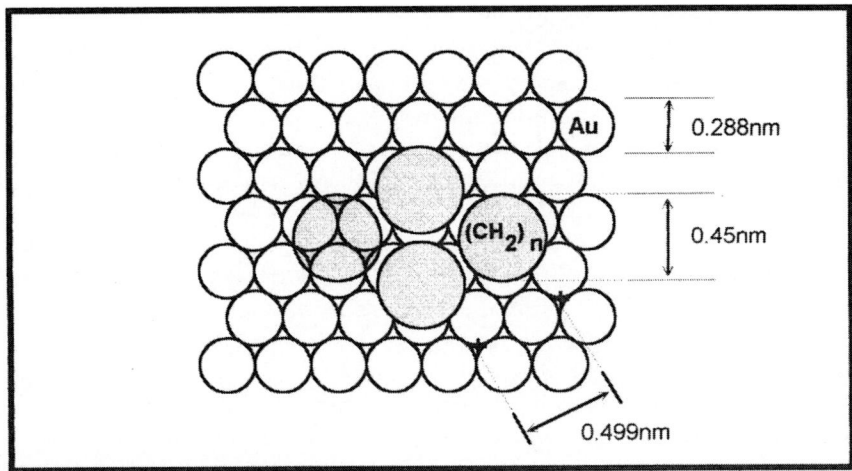

Figure 6. A schematic drawing of a Au(111) lattice with vertically oriented $(CH_2)_n$ chains located on three-fold Au lattice hollow sites.

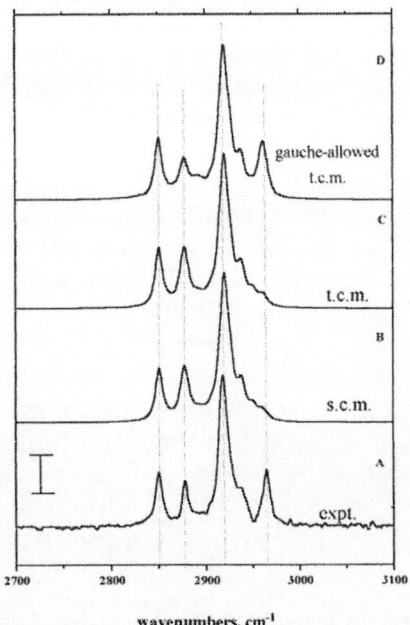

wavenumbers, cm⁻¹

Figure 7. Experimental and simulated infrared spectra for a monolayer of octadecane thiolate on a Au(111) surface. The simulations are for models of one chain (OCM) and two chains (TCM) per unit subcell and for fully extended, all-trans chains as well as a terminal gauche allowed chain model. The best match of simulation to experiment is for the two-chain model with terminal gauche conformations allowed. For details see text.

geometrical considerations implied in Figure 5 would make ϕ ~30°. Infrared (IR) spectroscopic evidence[11] has determined, in fact, that the chains exhibit nearly all trans conformations with average $\phi = 26\text{-}28°$ and thus the tilt-disordered structure in Figure 5 does not apply. With careful analysis[11] the IR spectra reveal that: 1.) the chains are not ideal rigid rods but rather the outer terminal CH_2 segments show some gauche disorder, a conclusion in excellent agreement with elegant temperature-dependent He scattering results[13] and 2.) the outer CH_3 group has an average orientation relative to the surface which depends upon whether the CH_2 chain is odd or even, a fact consistent only with a fixed Au-S-C bond angle with a typical value of ~110° for divalent sulfur. With respect to the first point, the "soft" film surface would seem to be in accord with the picture that thermal energy can provide the driving force to loosen the chains to a small extent at the outer surface where the intermolecular packing energies would be least disrupted. Other aspects of the structures of simple alkane thiols on gold surfaces have also been determined but will not be discussed further as details can be found elsewhere[8]. More important is the fact that a large variety of terminal functional groups have been placed at the surface of the films allowing wide variation of surface properties, particularly wetting. For example, films containing terminal OH, CO_2H, CO_2CH_3, $CONH_2$ and OR (where R=alkyl) have all been prepared[14]. Further variations can be made by changing the length of the chains, usually in the range of 10 - 24 CH_2 units and by mixing different molecules in the assembly process to provide a striking number of possible mixed composition films of variable chain lengths and terminal functional groups[8].

At this point, to the author's knowledge, almost all the applications of the above organothiolate/Au SAMs have been as scientific tools rather than technological or commercial. The major restriction of the thiol/Au system is that gold has very limited use as an engineering or technological material. Scientific applications, on the other hand are continually turning up and have included functionalized electrodes for selective electrochemistry[7], wetting/structure correlations[8,15], bioactive substrates for cell growth studies[16], chemically enhanced metal atom nucleation studies[17-20] and calibration samples for spectroscopy. Potential technological applications are indicated by studies of selective wetting pattern formation using "micromachining" techniques to inscribe the pattern and self-assembly to impart the wetting characteristics[21]. An actual technological application involves the use of thio-alcohol SAMs to promote adhesion of a polymer film used on a gold surface as a part of a commercial biosensor instrument[22].

3. Alkane Thiols/GaAs(100) - A Modified Semiconductor Surface

In contrast to the functionalization of gold by SAMs, which would appear to have limited technological application, there is a critical need to develop surface films and surface modifications for GaAs because of the importance of this material in device technology and, specifically, the importance of the surface both in processing steps and in device characteristics. According to the preceding discussion there are two strategies for developing a SAM for GaAs: develop a specific molecule/GaAs chemistry which will chemically attach an overlayer of molecules or develop conditions to prepare a high quality alkylsiloxane type of cross-linked surface membrane. The former has the advantage that chemical reaction with the semiconductor should lead to a well-defined termination of the surface and thus provide an opportunity to stabilize the surface properties in a favorable way. With this in mind, recently the former strategy has been explored and it has been discovered that alkane thiols will form a SAM on the GaAs(100) surface under conditions of immersing an HCl stripped GaAs surface in molten alkane thiol at ~100°C under an inert atmosphere[23]. Using a simple model based on steric interactions between chains, no surface reconstruction and a GaAs surface terminated by As atoms, one predicts that the CH_2 chains must bond at next nearest neighbor As atom sites, as for Au(111), as shown in Figure 8.

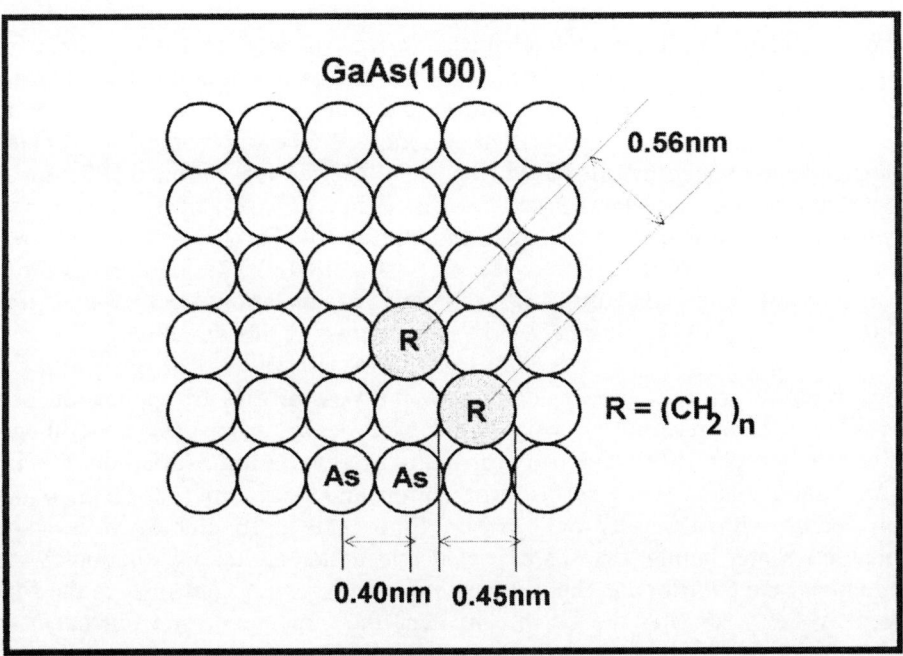

Figure 8. A schematic drawing of a GaAs(100) lattice with an As surface showing (CH_2) chains vertically oriented on top of next nearest As atom sites.

This model leads to a prediction of chains placed on a ($\sqrt{2}$ x $\sqrt{2}$),R45° (or c(2x2)) lattice with a spacing of 0.56nm, which results in specific film thicknesses for each given chain length and $\phi \sim 55\text{-}60°$ for maximum chain packing. A combination of ellipsometry and IR evidence confirms this prediction[23,24]. Further, x-ray photoelectron spectra[23-25] show that the molecule/semiconductor interface consists of S-As bonds with no apparent oxide (<several % coverage). The chain-chain spacings provide a significant fraction of potentially accessible free volume within the total film structure which in turn implies that thermal disordering should occur more easily than for the Au(111) SAMs which have smaller chain-chain spacings. IR evidence confirms this prediction[24].

In order to explore potential applications of these films, two separate sets of studies have been carried out recently in our laboratory. In the first, the SAMs have been incorporated in GaAs Schottky diode structures with a number of different metal electrodes and the I-V characteristics measured. The results show that the junctions exhibit good electrical characteristics with an ideality factor of 1.023 , showing near ideal thermioinic emission through the barrier, and with a modest increase of the barrier height of 40meV[26] over that obtained with just a stripped, uncoated GaAs surface. One can conclude that the SAMs lead to high quality device interfaces. An additional important issue associated with the nature of the film/semiconductor interface is whether or not the chemical bonding can affect the surface traps which in turn affect the pinning of the Fermi level and the electron-hole recombination lifetimes. Currently, there is no definitive evidence that the organized SAMs have a significant effect on surface trap population or character. However, it seems likely that improvements in the "perfection" of the SAM films would lead to reduction of surface traps based on two reasons. First, surface traps represent a small minority of surface sites (defects) and these may be the last to react in SAM formation. Second, adsorbed sulfide species ($S^=$ nominally)[27] and deposited films of small organothiols under conditions of unknown organization are known to improve recombination lifetimes[28]. Further work is in progress in this direction.

The second study involves the exploration of the possibilities of application of these SAMs as ultra-high resolution positive electron beam resists for nanolithography. Preliminary experiments[29] have been carried out by preparing SAMs from $C_{18}H_{37}SH$ on GaAs wafers with a (100) surface orientation and then exposing them to rastered electron beams with a variety of doses. Since this particular SAM is so highly hydrophobic ("water hating") it was expected that aqueous etching solutions would be unable to penetrate the film and thus only etch away GaAs in regions where the film was destroyed by the electron write beam. In fact, the expected result did occur and is somewhat striking in view of the fact that the film thicknesses were only ~1.5nm. In the initial experiments, it was demonstrated that trenches could be developed to a depth of a few tens of nanometers with a width of ~50nm, the limiting width of the electron beam. The lateral resolution exceeds that of the probing SEM resolution, ~50nm, and theoretically can be as high as the size of the SAM molecules, ~0.5nm, if the removal of the latter by the electron beam occurs on a molecule by molecule basis per initial collision, rather than removal of a region of molecules per collision, as occurs with typical polymeric resist films which are always much thicker. Measurement of resolution on the molecular scale requires tools such as AFM and these experiments are in progress. For practical applications to device fabrication there are a number of issues which must be addressed in detail. In addition to the resolution issue, the defect density is extremely important. The defect density should be related directly to missing chain (pinhole) defects in the SAM which would lead to development of a etched pit in the semiconductor surface. In order to minimize these types of defects it will be necessary to develop SAMs with extremely high and uniform coverage. One important question of current interest is what is the intrinsic lower limit of the defect population for SAMs in general. Since the chains are anchored on a surface plane, the system can be viewed as quasi-two dimensional and as such can be expected to exhibit higher intrinsic defect populations than 3-dimensional systems[30]. These defects could occur because of the pinning lattice imperfection and/or the details of the SAM formation process. In addition, there are questions related to grain boundary types of defects originating at points in the film where regions of differently aligned chains might meet. Significant scientific and technological questions remain to be answered in this area.

4. Alkylsiloxanes/SiO$_2$/Si - A 2-D Network

The class of SAMs made from alkylsiloxanes, $RSiO_x(OH)_y$, has been discussed in several sections above. In this section it is discussed in more detail from the point of view as a modification of a semiconductor surface. It is well known that solutions of octadecyltrichloro silane (OTS) in nonpolar solvents can lead to formation of monolayers of octadecylsiloxane films (termed OTS films) on the surface of oxided silicon[2,5]. This substrate usually leads to high quality films, in part because silicon can be obtained in the form of extremely smooth, planar wafers, an ideal type of substrate for preparing extremely uniform films. The quality of OTS films traditionally has been characterized by the wetting behavior with respect to liquids. Although the hydrophobic response is striking the most structurally sensitive behavior is observed for highly nonpolar liquids, particularly the oleophobic ("oil hating") behavior exhibited with n-alkanes. Recent work[31] has shown that the highest quality (highest oleophbicity) films can be produced by very careful temperature and solvent control during the preparation. Further, ellipsometry, forward recoil scattering and infrared spectroscopy[32] show that high coverages (close to theoretical) are obtained with chain structures approaching fully extended chains with very small, but yet finite, gauche defect content and average chain tilt angles of 6-8°. While the structural quality of these films is high, the substrate/film interface does not directly involve the bare semiconductor itself, but rather the oxide overlayer and as such the number of applications of the SAM is necessarily limited. For example, direct control of the Si surface states or carrier transport by the SAM is not possible without direct SAM/Si bonding. Of course, one may argue that the SiO$_2$/Si interface is of such high quality and the engineering so highly developed that a distinct change in device interface structure currently is unnecessary for commercial purposes. While this may be true, it is clear that the long term benefits of a wider materials science base for silicon surface modification is important because of the projected long term dominance of this material in microelectronics technology. Another limitation of the alkylsiloxane type of SAMs is the small variability possible for the terminal functionality of the adsorbing molecules (see section 1.3). In particular, hydrogen donor groups such as CO_2H completely interfere with the alkylsiloxane crosslinking process and prevent SAM formation. Other weaker donor groups will allow SAM formation but lead to poorer quality films. However, within these limitations there are potentially important applications to semiconductor technology in two directions. First, it is well known that one simple application is that of providing control of the adhesion of polymer resist films to silicon oxide surfaces. For this application, short chain (e.g., propyl) terminally functionalized alkyl siloxanes (e.g., NH_2) suffice with no need for organized films. More relevant to the present discussion of organized films is a possible application as nanolithographic resist films, similar to the application discussed in section 3 for GaAs. Attempts to use aromatic substituent-based siloxanes as deep-uv resists has been reported[33] and patterns have been generated on SiO_2/Si surfaces which show some selectivity for bio-adsorption. Other preliminary studies [34] have shown that patterns on the 50nm scale can be etched into SiO_2/Si using OTS films as electron beam resists combined with HF etching of the oxide. Further studies in these directions hopefully will lead to the development of new ultrahigh resolution (molecular scale) lithographic processing for nano-feature silicon devices.

5. Future Possibilities and Issues

Although the actual examples of SAM/semiconductor systems are extremely limited, it is clear that it is possible to develop such systems and that future applications seem sufficiently realistic to encourage further work. The most challenging opportunities would appear to be in developing SAMs which are directly bonded to sites at bare semiconductor surfaces as these structures inherently contain both the possibilities of indirect functions, such as protective films for resists, and of direct functions such as regulation of surface states and transduction of surface response for chemical sensors. There are two important types of issues imbedded in pursuing these opportunities. First there are the purely scientific and technological issues and second the issues of interdisciplinary efforts. As seen in the discussion above, development of the appropriate interface chemistry will require first an understanding of the chemical nature of the selected bare semiconductor surface. Much of this information comes from research in the area of surface physics. Once the bare substrate surface is understood and controlled, reaction chemistry can be tried until successful attachment mechanisms for obtaining semiconductor-molecule bonds are found. Much of the necessary chemistry is known in general, for example, reactions of silicon atoms with organic groups for developing Si/organic bonding, but a great deal of effort is needed to apply this general information to the specific task of surface reactions, and in particular of forming highly organized monolayer films. While the above discussion centers on new innovations in molecular-level chemistry of assembling surface films, there remain another class of innovations and opportunities to address in the "*supra*" molecular architecture of composite nanoscale material structures. These types of structures can include nanometer sized clusters linked together and/or to surfaces via functionalized molecular chains, such as depicted in Figure 9. These types of structures, and others not pictured as well, can include combinations of different types of materials, e.g., metals and semiconductors, incorporated in differently shaped objects and can lead to a variety of interesting electronic and optical properties associated with the unique structures, for example, sharp electromagnetic resonances caused by coupling between the indvidual resonances in arrays of clusters.

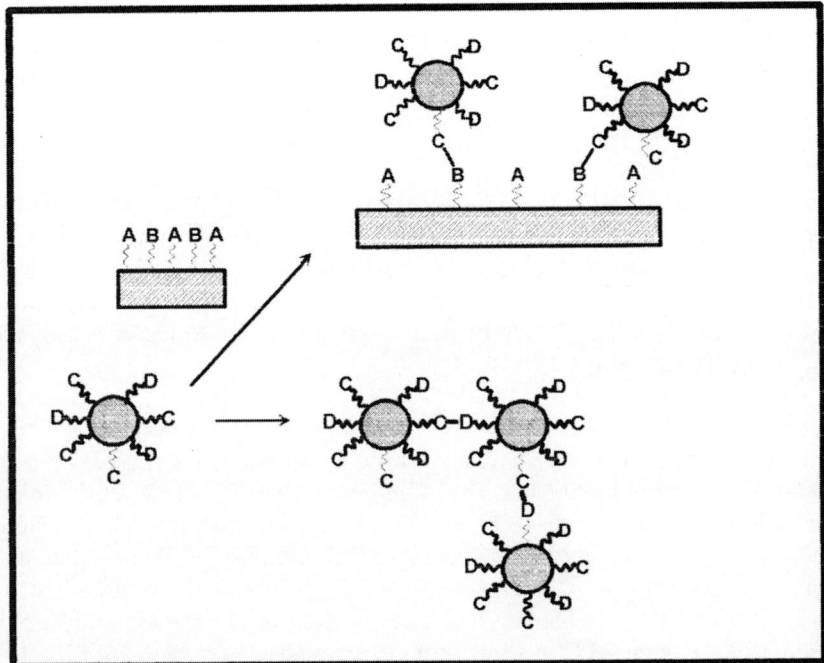

Figure 9. Nanometer-scale architectures using self-assembled monolayers.

An important point to be made with regard to the pursuit of the above opportunities is that these efforts will be highly multidisciplinary. From the chemistry point of view, each project easily can involve inorganic chemistry, organic chemistry, surface chemistry and surface analysis. The latter in particular is not trivial in that analysis of chemical features on the nanometer-scale is important and this will require development of advanced tools such as molecularly sensitive scanning probe microscopies. The subsequent characterization of electrical properties and development of processing applications involves electrical engineering and device (solid state) physics as well as the inclusion of other disciplines such as analytical chemistry and biological enginering for applications such as biosensors. In a broad sense, of course, the issue of interdisciplinary/multidisciplinary strategies is one of the most critical ones which lie ahead as the area of nano-technology evolves in the next decade.

References

1. J.D. Swalen, D.L. Allara, J.D. Andrade, E.A. Chandross, *et al*, *Langmuir*, 1987, *3*, 932
2. A. Ulman, *Ultrathin Organic Films*, John Wiley, NY, 1989.
3. R.G. Nuzzo and D.L. Allara, *J.Am.Chem.Soc.*, 1983, *105*, 4481; R.G. Nuzzo, F.A. Fusco and D.L. Allara, *J.Am.Chem.Soc.*, 1987, *109*, 2358.
4. C.D. Bain, E.B. Troughton, Y.T. Tao, J. Evall, G.M. Whitesides and R.G. Nuzzo, *J.Am.Chem.Soc.*, 1989, *111*, 321
5. J. Gun and J. Sagiv, *J.Coll.Interface Sci.*, 1986, *112*, 457; J. Gun, R. Isovici and J. Sagiv, *J.Coll.Interface Science,*1984, *101*, 201; R. Maoz and J. Sagiv, *J.Coll.Interface Sci.*, 1984, *100*, 465.
6. "Current Status, Research Needs and Opportunities in Applications of Surface Processing to Transportation and Utilities Technologies", A.W. Czanderna and A.R. Landgrebe, eds., Proceddings of a December, 1991 workshop, National Renewable Energy Laboratory, Golden, CO; to be published in *Critical Reviews in Surface Chemistry*, CRC Press, NY.
7. See for example: J.J. Hickman, D. Ofer, C. Zou, M.S. Wrighton P. Laibinis and G.M. Whitesides, *J.Am.Chem.Soc.*, 1991, *113*, 1128; C. Miller, P. Cuendet and M. Gratzel, *J. Phys.Chem.*, 1991, *95*, 877; M. Tarlov and E.F. Bowden, *J.Am.Chem.Soc.*, 1991, *113*, 1847; D.M. Collard and M.A. Fox, *Langmuir*, 1991, *7*, 1192; H.O Finklea, D.A. Snider and J. Fedyk, *Langmuir*, 1990, *6*, 371; J. Hickman, D. Ofer, P.E. Laibinis and G.M. Whitesides, *Science*, 1991, *252*, 688; C.E.D. Chidsey, C.R. Bertozzi, T.M. Putvinski and A.M. Mujsce, *J.Am.Chem.Soc.*, 1990, *112*, 4301; K.A.B. Lee, *Langmuir*, 1990, *6*, 709; T. Sagara, K. Niwa, A. Sone, C. Himmem and K. Niki, *Langmuir*, 1990, *6*, 254; H.C. Kelong and D.A. Buttry, *Langmuir*, 1990, *6*, 1319; C.E.D. Chidsey and D.N. Loiacono, *Langmuir*, 1990, *6*, 682.
8. L.H. Dubois and R.G. Nuzzo, *Annu.Rev.Phys.Chem.*, 1992, *43*, 437
9. L.S. Strong and G.M. Whitesides, *Langmuir*, 1988, *4*, 546
10. P. Fenter, P. Eisenberger, K.S. Liang , personal communication of a paper submitted for publication.
11. P.E. Laibinis, G.M. Whitesides, D.L. Allara,Y.T. Tao, A.N. Parikh and R.G. Nuzzo, *J.Am.Chem.Soc.*, 1991, *113*, 7152
12. P. Fenter, P. Eisenberger, J. Li, N. Camillone, S. Bernasek, *et al*, *Langmuir*, 1991, 7, 2013
13. C.E.D. Chidsey, G.Y. Liu, P. Rowntree and G. Scoles, *J. Chem. Phys.*, 1989, *91*, 4421; N. Camillone, C.E.D. Chidsey, G.Y. Liu, T.M. Putvinski and G. Scoles, *J.Chem.Phys.*, 1991, *94*, 8493.

14. R.G. Nuzzo, L.H. Dubois and D.L. Allara, *J.Am.Chem.Soc.*, 1990, *112*, 558
15. G.M. Whitesides and P.E. Laibinis, *Langmuir*, 1990, *6*, 87.
16. S. Ertel, B.D. Ratner, S.V. Atre and D.L. Allara, submitted for publication.
17. A.W. Czanderna, D.E. King and D. Spaulding, *J.Vac.Sci.Tech.*, 1991, *A9*, 2607
18. E.L. Smith, C.A. Alves, J.W. Anderegg, M.D. Porter and L.M. Siperko, *Langmuir*, 1992, *8*, 2707
19. D. Jung, P. Zhang, M. Chamberland and D.L. Allara, submitted for publication
20. M.J. Tarlov, *Langmuir*, 1992, *8*, 80
21. Abbott, J.P. Folkers and G.M. Whitesides, *Science*, 1992, *257*, 1380.
22. S. Lofas, M. Malmqvist, I. Ronnenberg, E. Sternberg, B. Liedberg and I. Lundstrom, *Sensors and Actuators*, 1991, *5*, 79.
23. C.W. Sheen, J. Shi, J. Martensson, A.N. Parikh and D.L. Allara, *J.Am.Chem.Soc.*, 1992, *113*, 1514.
24. C.W. Sheen, L. Chen, A.N. Parikh and D.L. Allara, to be submitted for publication.
25. B. Vannenberg, U. Gelius, C.W. Sheen and D.L. Allara, manuscript in preparation.
26. O.S. Nakagawa, S. Ashok, C.W. Sheen, J. Martensson and D.L. Allara , *Jap.J.Appl.Phys.*, 1991, *30*, 3759
27. C.J. Sandroff, R.N. Nottenber, J.C. Bischoff and R. Bhat, *Appl.Phys.Lett.*, 1987, *51*, 33; E. Yablanovitch, C.J. Sandroff, R. Bhat and T.G. Gmitter, *Appl.Phys.Lett.*, 1987, *51*, 439.
28. S.R. Lunt, G.N. Ryba, P.G. Santagelo and N.S. Lewis , *J.Appl.Phys.*, 1991, *70*, 7449.
29. R.G. Tiberio, H.G. Craighead, M. Lercel, T.Lau, C.W. Sheen and D.L. Allara, *Appl.Phys.Lett.*, in press.
30. D.R. Nelson and B.I. Halperin, *Phys. Rev. B.*, 1979, *19*, 2457.
31. J. Broszka and F. Rondelez , personal communication of paper submitted for publication.
32. A. N. Parikh and D.L. Allara, manuscript in preparation.
33. C.S. Dulcey, J.H. Georger, V. Krauthamer, D.A. Srenger, T.L. Fare and J.M. Calvert, *Science*, 1991, *252*, 551.
34. H.G. Craighead, R. Tiberio, C.W. Sheen and D.L. Allara, unpublished results.

Table 1
Major Self-Assembled Monolayer Systems

Type of Molecule	Substrate	Attachment Mechanism	Comments	Reference
$RSiO_x(H_y)$, alkysiloxanes	variable, usually SiO_2, also Al_2O_3, mica, etc.	2-D intermolecular crosslinking + limited surface crosslinks	surface chemical variability limited; very stable films	see text
RSH or RSSR	Au(111) primarily; also Cu, Ag	Metal-thiolate bonds	ordered films, films stable to $\sim <200°C$	see text
RCO_2H	Al_2O_3	RCO_2^- salt via surface induced deprotonation	surface chemical variability limited; similar to alkylsiloxanes	D.L. Allara and R.G. Nuzzo, *Langmuir*, 1985, *1*, 46 and 52
RCO_2H	Ag(oxided)	as above	as above; crystalline films can be formed; folded, doubly attached films known	N.E. Schlotter, M.D. Porter, T.B. Bright and D.L. Allara, *Chem.Phys.Lett*, 1986, *132*, 93; D.L. Allara, S.V. Atre, C.A. Elliger and R.G. Snyder, *J.Am.Chem.Soc.*, 1991, *113*, 1852
RNC	Pt	organo-metal bonding	low stability	J.J. Hickman, P.E. Laibinis, C. Zou, T.J. Gardner, *et al*, *Langmuir*, 1992, 8, 357
RSH	GaAs(100)	S-As bonding	thermally stable, some surface functionality variability	see text

MOLECULAR SELF-ASSEMBLY AND MICROMACHINING

N. L. ABBOTT, H. A. BIEBUYCK, S. BUCHHOLZ, J. P. FOLKERS, M. Y. HAN,
A. KUMAR, G. P. LOPEZ, C. S. WEISBECKER AND G. M. WHITESIDES
Department of Chemistry
Harvard University
12 Oxford Street
Cambridge MA 02138
USA

ABSTRACT. We are developing molecular self-assembly as a strategy for the fabrication of nanostructures. We have used self-assembled monolayer films formed by the chemisorption of organic molecules onto gold substrates, in combination with micromachining, (i) to pattern gold surfaces with well-defined regions of SAMs with contrasting properties, (ii) to transfer these patterns to the underlying gold film using SAMs as nanometer-thick chemical resists for wet etching, (iii) to construct microelectrodes by taking advantage of the dielectric barrier properties of SAMs, and (iv) to create contamination-resistant films on the surfaces of metals that are otherwise highly susceptible to contamination by adventitious molecular adsorbates and particles.

1. Introduction

Molecular self-assembly is the spontaneous self-organization of molecules into equilibrium supramolecular structures *(1-3)*. Although molecular self-assembly is ubiquitous in nature (for example, biological lipid membranes and crystallization), it is a new strategy in materials science for the fabrication of nanometer-scale structures. Using nature to provide the "existence theorem", we have designed and synthesized a range of organic molecules that spontaneously organize themselves into highly structured two-dimensional supramolecular arrays and we are currently developing applications of these new nanostructures.

Several characteristics of molecular self-assembly make it an attractive strategy for making nanostructures. First, molecular self-assembly is characterized by the spontaneous evolution to an equilibrium structure; intervention by the technologist, concurrent with the process of self-organization, is not necessary. Furthermore, because these structures exist at their thermodynamic minima, they are inherently self-repairing; a transient perturbation to the structure will be followed by a tendency to reorganize back to the equilibrium condition. Second, the process of molecular self-assembly can be remarkably free of errors; organization of as many as 10^{12} molecules can be routinely achieved with few defects in the resulting superstructures. Third, as demonstrated by the remarkably complex self-assembled structures found in nature, the use of clean room facilities is not a prerequisite for the self-assembly of molecules into structures with Angstrom-scale precision. Finally, when organic molecules are used, the properties of these molecules, and therefore, the functions of the

P. Avouris (ed.), Atomic and Nanometer-Scale Modification of Materials: Fundamentals and Applications, 293–301.
© 1993 *Kluwer Academic Publishers.*

resulting supramolecular assemblies, can be manipulated using well-known techniques of organic and biological synthesis. Recognition of these characteristics of self-assembly has prompted us to explore molecular self-assembly as a strategy for nanofabrication, using self-assembled monolayers of organic molecules as a model system *(4)*.

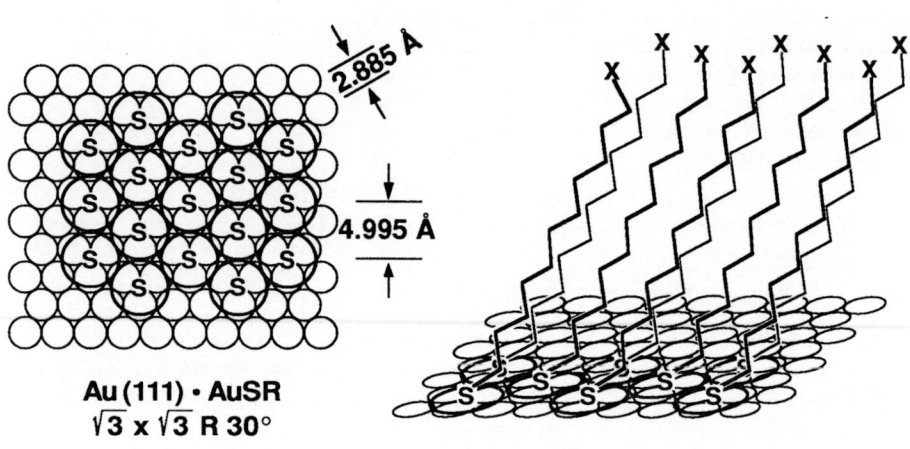

Figure 1. Schematic structure of a self-assembled monolayer formed by exposing a Au(111) surface to CH3(CH2)11SH.

Self-assembled monolayers (SAMs) are quasi-two-dimensional crystals formed by the chemisorption of organic molecules on the surfaces of metals (e.g. organic thiols on Au, Ag, Cu), metal oxides (e.g. alkane carboxylic acids onto Al_2O_3), and semiconductor oxides (e.g. organic trichlorosilanes on SiO_2) *(4)*. Exposure of the solid substrate to the organic molecules, either in solution or in the vapor phase, results in the spontaneous ordering of the organic molecules at the solid-solution or solid-vapor interface (Figure 1). Because these monolayer films are highly ordered and because the terminal moieties (X in Figure 1) of the organic molecules can be varied using techniques in synthetic organic chemistry, these systems offer a remarkable level of control over the properties of interfaces. As a result, SAMs have served extensively as model systems for the study of interfacial phenomena (wetting, adhesion, reactivity) in organic systems.

The best characterized SAMs are probably the organic thiolates on gold substrates *(4)*. Using this system we have explored four themes relevant to nanofabrication, including the development of self-assembly as a technique to complement existing methods of microfabrication, as well as a future technique for making nanostructures that could potentially serve as memory and logic devices.

2. Applications

2.1. CONTAMINATION-RESISTANT SURFACES

How molecular self-assembly can complement the already existing technologies used for making electronic devices can be illustrated with the idea of developing surfaces that are resistant to contamination. Current processes for microfabrication demand stringent clean room environments to reduce contamination (by particulates and vapors) of high free energy surfaces of electronic materials such as bare metals and silicon. Because the rate of contamination of a surface depends on the free energy of its surface (materials with high free energy surfaces adsorb and stick to contaminants more quickly and tenaciously than do low free energy surfaces), and because it is well established that nanometer-thick organic films formed from SAMs can be used to create low-energy interfaces on high energy substrates *(4)*, one promising application of molecular self-assembly is to passivate the exposed surfaces of materials during (and after) fabrication with appropriately chosen SAMs. For example, the surface free energy of bare gold, which is approximately 400 dynes/cm^2, can be reduced to 20 dynes/cm by forming a SAM that exposes methyl groups at the surface. Using scanning electron microscopy (SEM) to image films formed from molecular adsorbates, we have demonstrated that SAMs can be used to reduce the degree of contamination of a surface from airborne and solution-borne contaminants *(5)*. We have also observed that the electron beam-induced processes of contamination and desorption of carbonaceous films in the SEM can be influenced by the presence and type of SAM that is exposed to the electron beam *(5)*. Self-assembled monolayers formed from fluorocarbons (nanometer-thick "Teflon" films) appear to be the most effective in reducing the contamination of surfaces. We expect that issues such as the control of surface contamination and "stickiness" will become increasingly important with the development of contact lithography.

2.2. MICROMACHINING AND MOLECULAR SELF-ASSEMBLY TO PATTERN SURFACES

We have been developing a range of methods that combine molecular self-assembly with micromachining *(6)*, microwriting *(7)* and masking *(8)* to pattern surfaces with regions of SAMs with contrasting properties. The principle of region-specific deposition (or removal) of materials from a substrate broadly defines current microfabrication practices. Our approach, however, takes advantage of existing capabilities in organic chemistry to design and synthesize molecules with a variety of properties and to transfer these properties, using molecular self-assembly, to well-defined regions of a surface. We have combined molecular self-assembly with micromachining, microwriting and masking to pattern surfaces with features (a region of a surface with a property that is distinguishable from its surroundings) that range in size from centimeters to 100 nanometers and which, in principle, could extend down to the scale of molecular dimensions (nanometers) using AFM and STM. The potential ability to control structure with nanometer-scale precision contrasts with optical lithography which is limited using far field optics (by diffraction effects) to linewidths greater than 0.3 to 0.5 μm.

We have developed a process of patterning a surface by combining micromachining and molecular self-assembly in three steps (Figure 2): (i) formation of an initial SAM of an organic thiolate on gold; (ii) generation of regions of bare gold in the SAM by micromachining; and (iii) formation of a second SAM on these micromachined regions (6). With simple micromachining techniques, using either a surgical scalpel blade or the end of a carbon fiber, we have patterned a surface with micrometer and submicrometer (100 nm) resolution, respectively. We have demonstrated the control of surface properties by using this new type of microfabrication to generate well-defined regions of SAMs with contrasting wettabilities on gold and silver substrates. The way in which liquids wet these surfaces can be used as a convenient technique to image the patterns generated on such surfaces (these systems also act as important models for studying the fundamental properties of wetting and adhesion at organic interfaces: Figure 3). Although we have demonstrated the patterning of surfaces with regions of contrasting wettabilities, we believe that this principle can be generalized, for example, using organic molecules that have contrasting electroactive properties, rather than contrasting wetting properties. Indeed, SAMs formed from molecules containing electroactive moieties (for example, ferrocene) have been prepared (9).

In these examples of micromachining, we have used simple techniques in order to demonstrate the result of combining micromachining and molecular self-assembly. More recently, we have begun extending our use of micromachining techniques to include atomic force microscopy (AFM) and scanning tunneling microscopy (STM). Using STM with high bias voltages (2-10 V) we have removed well-defined regions of a SAM formed from hexadecanethiol on a gold substrate (10). We believe that the combination of AFM/STM and molecular self-assembly can be used to make patterns in SAMs that are (much) smaller than 100 nm. The ultimate limit to the resolution of this technique will probably be determined by the lateral mobility of the chemical species on the surfaces.

2.3. DIELECTRIC BARRIERS

Dielectric barriers are essential components of electronic devices. We, and others, have shown (with cyclic voltammetry) that a SAM formed from hexadecanethiol on a gold substrate can reduce the rate of transport of electrons between the gold surface and contacting electrolyte (0.1M H_2SO_4) by a factor about 10,000. Although there are many unanswered questions

Figure 2. Schematic illustration of the formation of 0.1 to 1 μm scale lines with property Y (for example, Y = COOH for a hydrophilic surface) in a surface of X (for example, X = CH3 for a hydrophobic surface) with micromachining and SAMs. We imply no asymmetry in the structure of the SAMs within the micromachined grooves. Au = evaporated film of gold; Ti = evaporated film of Ti used to promote adhesion of the Au to the silicon wafer (Si).

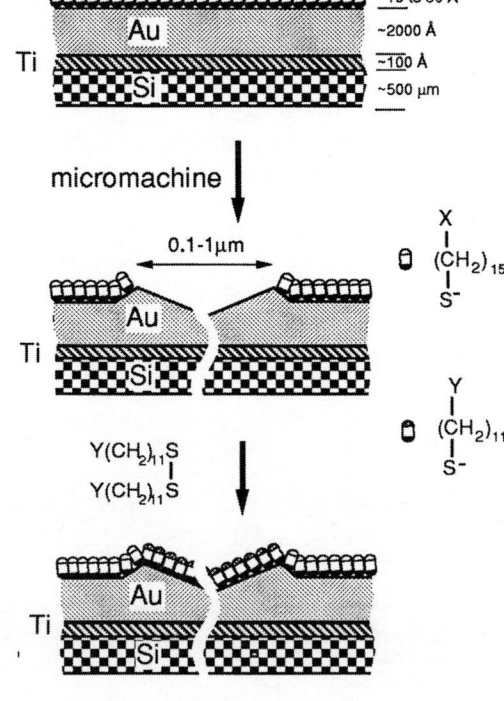

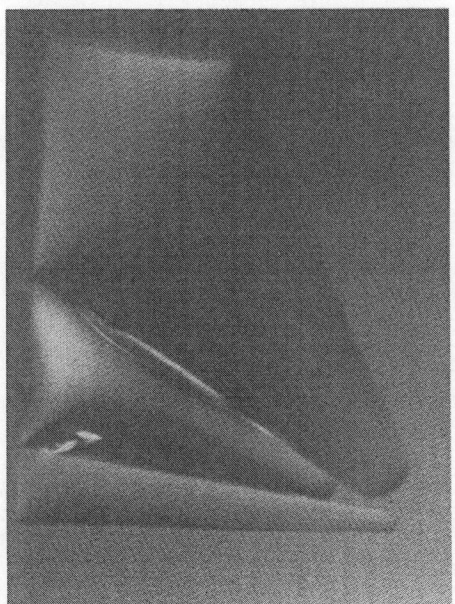

Figure 3. Drops of water on a hydrophilic SAM formed from HS(CH2)15COOH are pinned by micrometer-wide hydrophobic lines formed from [CH2(CH2)11S]2. The gold substrate was patterned with SAMs by combining micromachining and molecular self-assembly (see Figure 2).

surrounding the mechanism of electron transport across a surface covered with a SAM (e.g. pinhole defects, electron tunneling, thin spots), the ability of a SAM to serve as nanometer-thick dielectric barrier is a promising one. For example, the electrical function of metal oxides to act as barriers to electrons in tunneling junctions may also be achieved using SAMs. We have used the dielectric barrier properties of SAM to construct band microelectrodes, created by micromachining bare gold grooves in the surface of a gold substrate covered by a SAM. We have demonstrated microelectrode behavior (the absence of mass transport limitation during cyclic voltammetry) for an aqueous ferricyanide system (Figure 4). Because they have micrometer-widths, these microelectrodes are less prone to mass transport limitations than are macroscopic electrodes. The rapid response times of these electrodes and their ease of fabrication suggests that SAMs may serve as useful dielectric barriers for sensor systems.

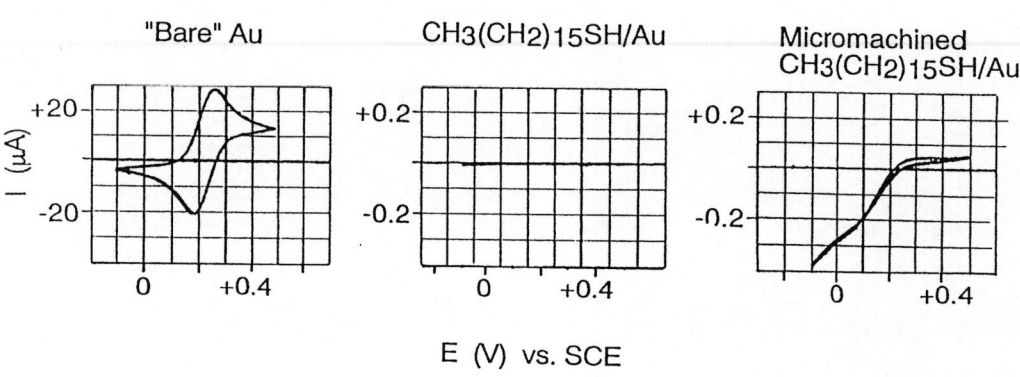

Figure 4. Rate of electron transfer to an aqueous solution of 1 mM $Fe(CN)_6^{3-}$ (0.1M $NaClO_4$, pH 7) using (a) a macroscopic bare gold surface, (b) a gold film covered by a SAM formed from $CH_3(CH_2)_{15}SH$, and (c) micrometer-wide band-shaped microelectrodes of bare gold formed by micromachining the surface from (b).

2.4. CHEMICAL RESISTS

SAMs can serve as nanometer-thick resists. In addition to being good dielectric barriers, SAMs can also be good barriers to certain ions in aqueous solution, including ionic species that act as wet chemical etchants *(11)*. We have demonstrated that SAMs formed from hexadecanethiol on a gold film (200 nm thick) can protect the gold film from an aqueous solution of alkaline potassium cyanide (a well known etchant of gold) for a sufficient time that unprotected gold areas can be etched to the silicon substrate *(11)*. The ability of a SAM to act as a barrier to etchants of metals has allowed us to develop simple techniques to transfer patterns created in SAMs on a gold substrate into the underlying gold film. For example, using micromachining to expose a bare gold line in a SAM formed from hexdecanethiol and then immersing the entire sample in an etching solution of 0.1M KCN (in 1M KOH), we have created micrometer wide trenches in a gold film (Figure 4). With an alternative procedure, we have used neat liquid hexadecanethiol as an "ink" to write (using micropens) micrometer-wide lines of hexadecanethiol onto a gold film. By etching the entire film we have formed conducting microwires of gold from the regions protected by the SAM (Figure 4).

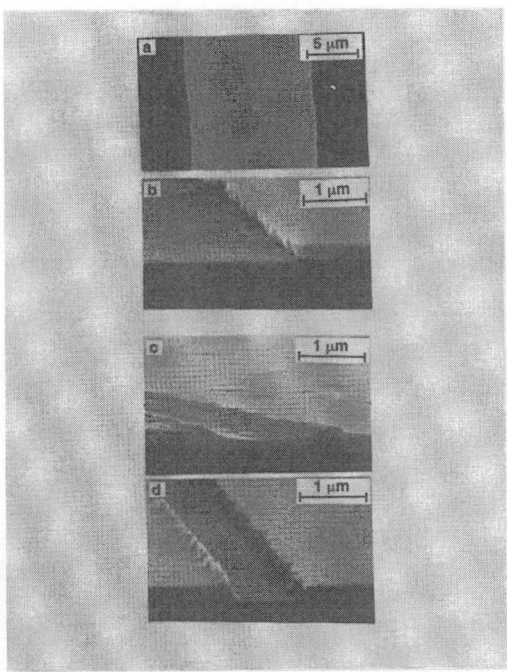

Figure 4.   Scanning electron micrographs of a 10 μm gold wire (a and b) and a 1 μm trench (c and d) in a gold film. Both microstructures were made using a SAM of $CH_2(CH_2)_{15}S^-$ as a resist to an etching solution of 0.1M KCN and 1M KOH (see text and *(11)* for details).

3. Future Prospects and Conclusions

If molecular self-assembly is to serve as a useful strategy for nanofabrication, we believe that some of the important, and yet unanswered questions, are (i) what degree of perfection is required of a self-assembling technique?; (ii) how do we measure the degree of imperfection in a SAM?; and (iii) how far are we from what is required? To address these types of questions, we are developing assays for the detection of defects in SAMs. In one approach we use wet etching to detect the defects because etch pits (which can be observed by SEM) form at defects sites. The results of this assay indicate that (in a wet chemistry laboratory) we are able to produce SAMs on gold with fewer than 1 defect per mm^2 (or 1 defect per 10^{12} molecules)(12).

Many other interesting challenges will be faced before molecular self-assembly becomes a routine tool for nanofabrication. For example, our current forms of micromachining and microwriting are serial processes, and therefore, these techniques may be too slow for industrial processes. It is possible, however, that these techniques can be made to be parallel, for example, by stamping-type processes. Finally, we note that our recent demonstrations of forming SAMs on colloidal gold suggest exciting possibilities for nanofabrication (13). For example, colloids covered with SAMs (dielectric barriers) and assembled onto a patterned surface could form the basis of single electron transistors.

Molecular self-assembly as a principle for the fabrication of small-scale structures is a new strategy that is at the stage of "concept" rather than "development". Using SAMs as model systems, we are exploring the potential of molecular self-assembly to serve both as a supplement to existing technologies and, perhaps, further in the future, as the foundation of new types of electronics devices.

ACKNOWLEDGMENTS

Supported by the Office of Naval Research and the Defense Advanced Research Projects Agency.

REFERENCES

(1) Swalen, J.D.; Allara, D.L.; Andrade, J.D.; Chandross, E.A.; Garoff, S.; Israelachvili, J.; McCarthy, T.J.; Murray, R.; Pease, R.F.; Rabolt, J.F.; Wynne, K.J.; Yu, H. (1987) "Molecular Monolayers and Films", Langmuir, 3, 932-950.
(2) Whitesides, G.M., Mathias, J.P. and Seto, C.T. (1991) "Molecular Self-Assembly and Nanochemistry: A Chemical Strategy for the Synthesis of Nanostructures", Science (Washington, D.C.), 254, 1312-1319.
(3) Ulman, A. (1991) An Introduction to Ultrathin Organic Films: From Langmuir-Blodgett to Self-Assembly, Academic Press, San Diego CA.
(4) Whitesides, G.M. and Laibinis, P.E. (1990) "Wet Chemical Approaches to the Characterization of Organic Surfaces: Self-Assembled Monolayers, Wetting and the

Physical-Organic Chemistry of the Solid-Liquid Interface", Langmuir, 6, 87-96; and references therein.

(5) Lopez, G.P., Biebuyck, H.A. and Whitesides, G.M., unpublished results.

(6) Abbott, N.L., Folkers, J.P. and Whitesides, G.M. (1992) "Manipulation of the Wettability of Surfaces on the 0.1 to 1 Micrometer Scale Through Micromachining and Molecular Self-Assembly", Science (Washington, D.C.), in press.

(7) Biebuyck, H.A., Kumar, A. and Whitesides, G.M., unpublished results.

(8) Lopez, G.P., Biebuyck, H.A. and Whitesides, G.M., unpublished results.

(9) Hickman, J.J., Ofer, D., Laibinis, P.E., Whitesides, G.M. and Wrighton, M.S. (1991) "Molecular Self-Assembly of Two-Terminal Voltammetric Microsensors with an Internal Reference", Science (Washington, D.C.), 252, 688-691.

(10) Abbott, N.L., Buchholz, S. and G.M. Whitesides, unpublished results.

(11) Kumar, A., Biebuyck, H.A., Abbott, N.L. and Whitesides, G.M. (1992) "The Use of Self-Assembled Monolayers and a Selective Etch to Generate Patterned Gold Features", J. Am. Chem. Soc., in press.

(12) Han, M.Y. and Whitesides, G.M., unpublished results.

(13) Weisbecker, C.S. and Whitesides, G.M., unpublished results.

CHARACTERIZATION OF THE INTERACTION OF C_{60} WITH AU(111)

ERIC I. ALTMAN and RICHARD J. COLTON
Chemistry Division, Code 6177
Naval Research Laboratory
Washington, DC 20375-5000, USA

ABSTRACT. The nucleation, growth and structure of C_{60} films on Au(111) was studied using UHV-STM/S, LEED, AES, and TOF-SIMS and the results were compared with previous results for a mixed fullerene film (E.I. Altman and R.J. Colton, Surface Science, in press). The presence of the larger fullerenes was found to have no significant effect on either the growth or the structure of the fullerene films. In both cases, close-packed layers with a lattice constant of 1.0 nm grow out from the steps on both the upper and lower terraces. No LEED pattern due to C_{60} was observed, however, STM indicated that two ordered structures predominate on the Au(111) surface: (1) a layer with a periodicity of approximately 38 Au spacings; and (2) a layer with a $2\sqrt{3} \times 2\sqrt{3}$ R30° unit cell. In both structures, STM images show an apparently "inhomogeneous" film with some molecules brighter than others. The presence of brighter molecules was not due to larger fullerenes as in the earlier study; TOF-SIMS indicated the film contained only C_{60}. Time lapsed STM images, intramolecular contrast, and STS results showed that the "inhomogeneity" of the film can be attributed to electronic differences between molecules bound in different rotational orientations on the surface. Tunneling spectra indicate that the variation in bonding is primarily due to the efficiency of coupling between the C_{60} LUMO and the Au surface. A number of internal structures were observed that appear to be due to a convolution of the structure of the C_{60} molecule and the underlying Au(111).

1. Introduction

Recently there have been numerous theoretical and experimental studies aimed at characterizing the physical and chemical properties of C_{60} and larger fullerenes [1-10]. These studies have shown that: 1) C_{60} forms face-centered cubic (fcc), simple cubic (at low T), and hexagonal close-packed (hcp) (with C_{70}) crystalline phases [1,3-5]; 2) at room temperature C_{60} molecules in the crystal spin freely [6,7]; 3) fcc C_{60} is a semiconductor with a band gap of approximately 1.6 eV [8]; and 4) alkali doped C_{60} films are superconducting [9]. Numerous reactions and derivatives of fullerenes have also been reported [10]. In addition, many applications for these materials are being investigated including coatings, lubricants, and non-linear optical devices [11].

Since many of the proposed applications of the fullerenes depend on surface interactions and the use of thin films, and because the catalytic chemistry and electrochemistry of fullerenes depends upon their interaction with metal surfaces, we have been studying the nucleation, growth, structure, and thermal stability of fullerene films on Au(111) using scanning tunneling

P. Avouris (ed.), Atomic and Nanometer-Scale Modification of Materials: Fundamentals and Applications, 303–314.
© 1993 *Kluwer Academic Publishers.*

microscopy (STM), Auger electron spectroscopy (AES), low energy electron diffraction (LEED), and time-of-flight secondary ion mass spectrometry (TOF-SIMS). Au(111) was selected for these studies because this surface undergoes an unusual $23x\sqrt{3}$ long-range reconstruction which has been shown to act as an ordered array of nucleation sites for the growth of Ni, Fe, and Co films [12-15]. The influence of the film on the Au surface reconstruction can also be used as a measure of the strength of the interaction between the film and the gold surface. Previous studies have shown that a strong interaction is required to lift the reconstruction [13-15] and that ordinary hydrocarbons do not lift the reconstruction [16].

Previously, we reported a systematic study of the structure of mixed fullerene films on Au(111) as a function of coverage (from 0.01 monolayers (ML) to in excess of a monolayer) and annealing temperature [17]. In this earlier paper it was shown that the fullerenes grow in a layer-by-layer manner on Au(111) with the nucleation of the first layer strongly influenced by the Au reconstruction. The fullerenes were found to nucleate preferentially at locations along step edges corresponding to regions of bulk fcc termination. At higher coverages, close-packed fullerene layers with a lattice constant of 1.0 nm were observed to grow out from the steps on both the upper and lower terraces. Adsorption on the terraces was found to lift the Au reconstruction indicating that the fullerenes bond strongly to the Au surface. No LEED pattern due to the fullerenes was observed; however, STM indicated that two ordered phases predominated: (1) an "in-phase" layer with a periodicity of approximately 38 Au lattice spacings in which the crystallographic directions of the overlayer match those of the substrate and the lattice mismatch is large; and (2) a $2\sqrt{3}x2\sqrt{3}$ R30° layer with a nearly perfect lattice match. It was also found that multilayers could be desorbed at 300°C but that the monolayer did not begin to desorb until heated to 500°C; another indication that fullerenes strongly adsorb on Au(111). The strength of the fullerene-Au interaction was very surprising since Au is generally considered catalytically inert. Therefore, other metals, such as Pt and Ni, would be expected to interact even more strongly with the fullerenes suggesting that the reactivity of the fullerenes can be controlled through catalytic chemistry. Several different intramolecular structures were observed including multiple lobes and striped structures with the striped structures more prevalent.

The study of mixed fullerene films on Au(111), however, left several questions unanswered and raised several new issues. These questions include: (1) What is the role of the larger fullerenes in the ordering of the films?; (2) Is the observed variation in internal structure of the mixed fullerene film due to the different molecules on the surface or is it due to different adsorption site geometries?; and (3) Why do the fullerenes adsorb strongly on Au(111) while ordinary hydrocarbons only physisorb on Au? In addition, the lattice mismatch of the 38x38 structure should result in a periodic arrangement of molecules bound in three-fold hollow sites, bridge sites, and on-top sites. The height differences between molecules bound in these different sites should be measurable by STM. However, this height variation could not be resolved in STM images of mixed fullerene films. To address these issues we have studied, and report here, the interaction of pure C_{60} with Au(111) using tunneling spectroscopy and barrier height imaging in addition to STM, LEED, AES, and TOF-SIMS.

2. Experimental

Experiments were performed in an UHV system equipped with a scanning tunneling microscope, LEED optics, an Auger spectrometer, a sputter-ion gun, a quadrupole mass spectrometer and a fullerene evaporation source. The base pressure of this system is $1x10^{-10}$ torr. TOF-SIMS was performed in a separate system.

The tunneling microscope was designed and built at the Naval Research Laboratory and has been described previously [17]. Images were obtained at positive and negative biases between 1 and 2.5 V and currents of 0.1-1.0 nA. In this range the bias had little effect on the images except where noted. Spatially resolved tunneling spectra were obtained by interrupting the feedback loop and measuring the I-V characteristics of the tunnel junction at certain points (every fourth pixel) while a normal constant-current image is obtained. Spectroscopic images are created from these data by mapping the current measured at each point for a particular voltage. Similarly, barrier height measurements were performed by interrupting the feedback loop and measuring the I-s characteristics of the tunnel junction. The barrier height at each point is then calculated from the slope of a ln(I) vs. s plot. More details on the tunneling spectroscopy and barrier height measurements will be provided in a forthcoming publication [18].

Epitaxial gold films deposited on mica were used as substrates in this study. Preparation of these gold films has been described previously [17]. The gold films were prepared in a separate vacuum chamber. After transferring the sample to the UHV chamber, the gold surface was cleaned by Ar ion sputtering at 500 eV and annealing to 600°C until AES revealed no traces of impurities and a sharp LEED pattern characteristic of Au(111) was obtained. STM images of these films showed terraces as wide as 100 nm. Images of the terraces revealed a herringbone pattern characteristic of the 23x√3 reconstruction [12,13]. The herringbone pattern was used as a convenient marker for the crystallographic orientation of the Au surface.

Pure C_{60} was obtained from SES Research, Houston, TX. C_{60} powder was loaded into a tantalum foil basket and sublimed onto the sample by heating to 400°C. During evaporation, the sample was held at room temperature.

3. Results and Discussion

The presence of the larger fullerenes had no noticeable effect on either the growth or structure of the fullerene films. For both a pure C_{60} film and a mixed fullerene film containing over 40% of C_{70} and larger fullerenes, at the lowest coverages (0.02 ML) the fullerenes adsorbed almost exclusively at step intersections and at steps separating narrow terraces (less than 2.0 nm wide, too narrow to reconstruct). As the coverage was increased, the fullerenes populated steps separating wider terraces at locations where the steps intersect fcc terminated regions of the reconstructed surface. At higher coverages, close-packed layers with a lattice constant of 1.0 nm grow out from the step edges as shown in the STM images of a mixed fullerene film in Figure 1.a and a pure C_{60} film in Figure 1.b. As shown in Figure 1, STM indicates no measurable difference between the mixed fullerene film and the pure C_{60} film in either the spacing or the packing geometry of the molecules even though the mixed fullerene film contains significant amounts of fullerenes over 15% larger in size than C_{60}. It has been suggested that the presence of the larger fullerenes may stabilize certain structures in thick fullerene films and bulk crystals [3,5,19]. However, these results indicate that the structure of the monolayer is not strongly affected by the presence of larger fullerenes.

STM also indicated that the orientation of the fullerene film with respect to the Au(111) substrate was not influenced by the presence of larger fullerenes. Again, no LEED patterns due to the pure C_{60} overlayer were observed, but STM indicated two preferred orientations of the overlayer: (1) an in-phase structure with a periodicity of approximately 38 Au lattice spacings; and (2) a layer with a 2√3x2√3 R30° unit cell. An STM image of the in-phase structure is given in Figure 2.a, and an image of the $2\sqrt{3}x2\sqrt{3}$ R30° structure in Figure 2.b. The most striking feature in both Figures 2.a and 2.b is the presence of brighter, or apparently higher (by 0.04 nm) molecules. Figure 2 also shows that there is a tendency to form linear chains of bright molecules along crystallographic directions. Again, these features are not due to the presence of larger fullerenes in the film. TOF-SIMS analysis of the film showed only trace amounts (<1%) of larger fullerenes. In mixed fullerene films, bright or high spots were observed that were attributed to larger fullerenes [17]. However, in the mixed fullerene films these features were considerably higher (0.1-0.2 nm) than the other molecules and appeared randomly distributed.

Height variations on the order of 0.04 nm were expected for the in-phase structure. A consequence of the lattice mismatch of the in-phase structure is a periodic variation in adsorption site geometry from three-fold hollow sites, to bridge sites, to on-top sites [17]. If no distortion from the bulk spacing found for fcc (111) C_{60} occurs, then the periodicity would be 38 Au lattice spacings. Assuming hard spheres, a C_{60} molecule in an on-top site would be 0.016 nm higher than a bridge bound C_{60} and 0.033 nm higher than a three-fold hollow bound C_{60} -- height variations measurable by STM. The pattern observed in Figure 2.a, however, cannot be reconciled with a pattern expected from the lattice mismatch. The lattice constants of the C_{60} layer (1.0 nm) and the Au surface (0.288 nm), makes it impossible for chains of adjacent molecules to be bound in equivalent sites. However, the images show chains containing as many as 8 bright molecules. Also, similar patterns of brighter and dimmer C_{60} molecules was also observed in the $2\sqrt{3}x2\sqrt{3}$ R30° domains. Since $2\sqrt{3}$ times the Au spacing (or 0.996 nm) is the same as the C_{60} spacing (1.0 nm) within experimental error, all the molecules in this structure can lie in equivalent adsorption sites. Therefore, the observed variations in the height of the C_{60} molecules cannot be unambiguously attributed to molecules sitting in on-top sites rather than three-fold hollow or bridge sites.

Insight into the cause of the apparent height variations was gained by studying STM images of the same sample area as a function of time. Figure 3 shows two STM images of the same area taken 3 minutes 45 seconds apart. As shown in Figure 3, the molecules marked with a " + " have changed from either bright to dim or vice-versa without any measurable shift in position indicating that the same C_{60} molecule in the same position on the surface can appear different in STM images. Because the molecule may be able to rotate with respect to the surface, the appearance of an adsorbed C_{60} molecule in STM may also depend on the orientation of the molecule. This suggests that electronic differences between molecules of different rotational orientations could be responsible for contrast between otherwise identical C_{60} molecules. (It should be noted that rotation of napthalene molecules on Pt(111) over a similar time scale has been observed with STM [20].)

Spectroscopic measurements indicated that differences in electronic state densities (and not height differences) do in fact cause certain molecules to appear either brighter or dimmer. Figure 4 shows an STM topographic image in 4.a and the corresponding spectroscopic image at 0.85 V (sample positive) in 4.b. Tunneling spectra obtained over different molecules are shown in Figure 5. The black areas in Figure 4.b clearly correspond to the location of the three dimmer molecules in the center of Figure 4.a indicating lower density of the unoccupied states at the positions of these molecules. The tunneling spectra shown in Figure 5, also indicates lower state density at the dimmer molecules at negative sample biases (occupied states), although this difference is much less pronounced than for positive sample biases (unoccupied states). This suggests that greater contrast between the molecules would be observed at positive rather than negative sample biases. Greater contrast was in fact observed at positive sample biases.

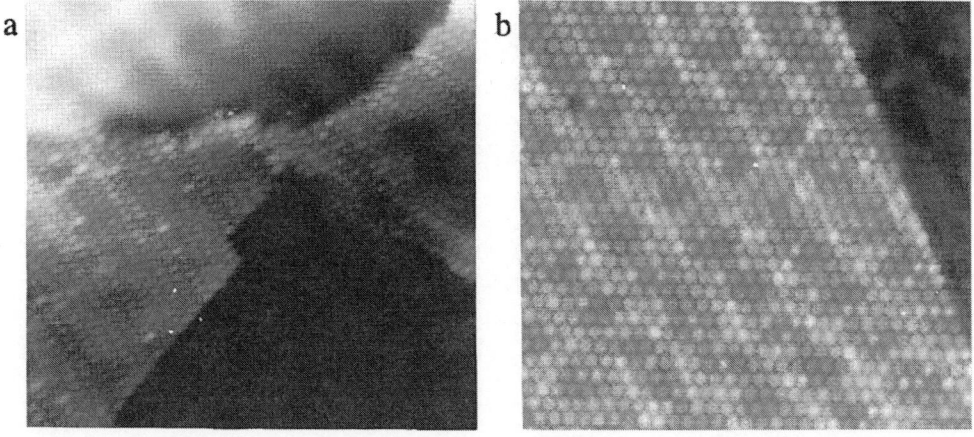

Figure 1)   STM images of (a) mixed fullerene film (30 x 30 nm) and (b) pure C_{60} film (29 x 29 nm).

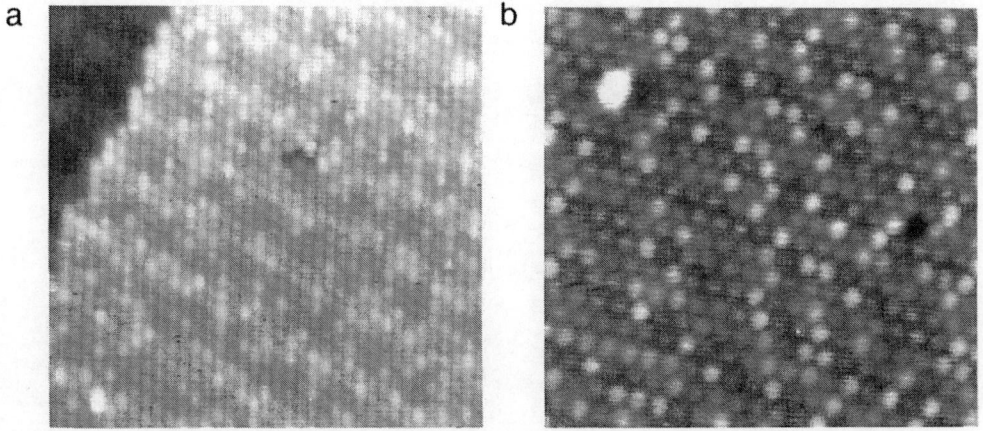

Figure 2)   STM images of (a) an in-phase 38x38 C_{60} domain (29 x 29 nm) and (b) a $2\sqrt{3}x2\sqrt{3}$ R30° C_{60} domain (19 x 19 nm).

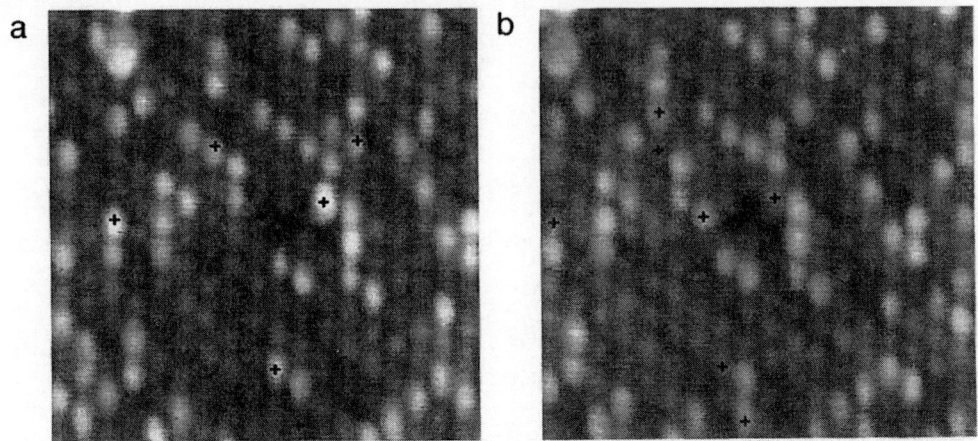

Figure 3)   STM images (14.5 x 14.5 nm) of an in-phase domain in a pure C_{60}
             film taken 3 minutes 45 seconds apart.   The "+" symbols mark
             molecules that have changed contrast.

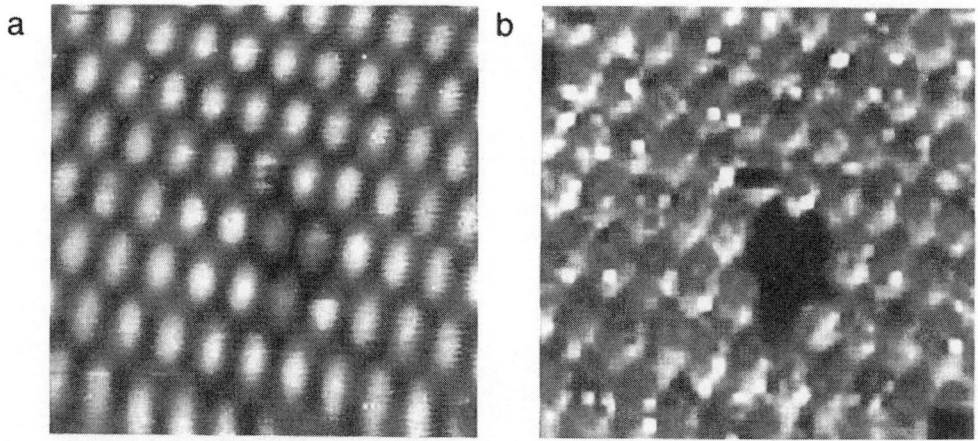

Figure 4)   (a) Constant-current image recorded at -1.25 V (sample negative)
                 and 1 nA (10 x 10 nm).
             (b) Spectroscopic image for 0.85 V (sample positive) obtained
                 simultaneously with the constant current image in (a).

Barrier height measurements also indicated differences between the molecules. These measurements indicated a barrier height 0.9 eV lower over the brighter molecules compared to the rest of the molecules. A reduced barrier height means a greater tip-sample separation is required to maintain the same current resulting in an apparent height increase. The barrier height and spectroscopic results clearly show that variations in surface-adsorbate bonding changes the contrast of C_{60} molecules adsorbed on the surface. Since the patterns shown in the STM images in Figures 2 and 3 cannot be solely attributed to molecules bound at different surface sites, the rotational orientation of the molecule must be responsible for the observed contrast.

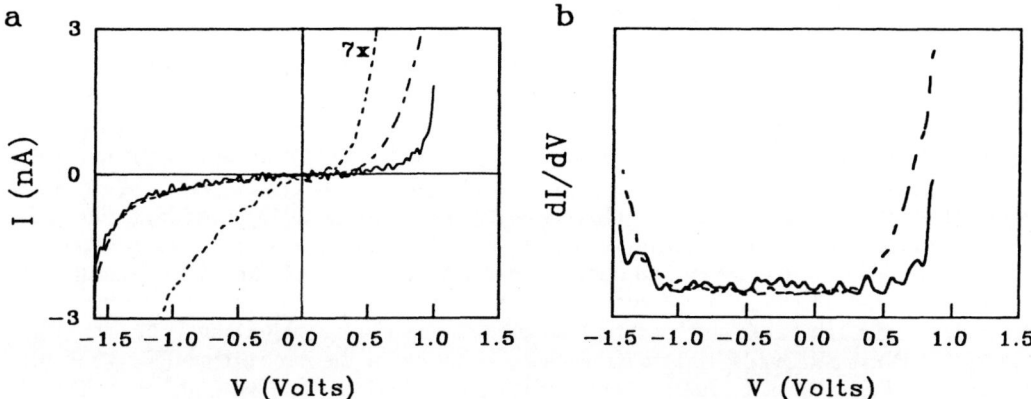

Figure 5)   Tunneling spectra recorded simultaneously with images in Figure 4 plotted as (a) I vs. V and (b) dI/dV vs. V. The solid line indicates spectra taken over dimmer molecules and the dashed line indicates brighter molecules.

The tunneling spectra shown in Figure 5 offer additional insight into the bonding of C_{60} to the Au surface. The I-V curves in Figure 5.a show a band gap of less than 0.2 eV, considerably smaller than the band gap expected for bulk C_{60}. The narrower band gap can be explained in terms of state broadening due to adsorption on the Au surface and to a contribution of the underlying Au to the tunnel current [21]. Below the Fermi level (E_F), the spectra taken over the brighter and dimmer molecules (Figure 5.b) are similar, both showing the onset of a peak at -1.3 eV. Conversely, above E_F the spectra are very different. The spectrum obtained over the brighter molecules displays an intense shoulder beginning at 0.4 eV, while this feature shifts to higher energy over the dimmer molecules. These results are similar to ultraviolet photoelectron spectra (UPS) and inverse photoelectron spectra (IPS) of C_{60} on polycrystalline Au reported by Ohno, et. al. [22]. IPS results for submonolayer coverages showed a peak at less than 1 eV above E_F, a shift of 0.5 eV towards lower energy compared to thick films. In contrast, UPS results showed little difference between monolayer and thicker films indicating that bonding occurs mostly through charge transfer to C_{60} and hybridization of the LUMO (lowest unoccupied molecular orbital). The tunneling spectra and STM images demonstrate that certain orientations of the molecule on the surface give rise to stronger bonding and that this stronger bonding occurs primarily through more efficient interaction of the unoccupied states of the molecule with the metal surface. The shifts in the LUMO derived levels may also explain why C_{60} bonds much more strongly to Au than do other hydrocarbons such as benzene. IPS results for benzene adsorbed on Au show only very small shifts in the LUMO upon adsorption on Au. However, the LUMO level of benzene is 4.8 eV above E_F [23] while the LUMO of C_{60} is less than 2 eV above E_F [22], suggesting that unoccupied states near E_F are required for chemisorption on Au.

High resolution images showing intramolecular contrast also indicate that different rotational orientations of the molecules on the surface cause some molecules to appear brighter or dimmer. A series of STM images showing intramolecular structure within an in-phase domain are given in Figure 6. Here most of the molecules appear to be the same height and display no internal structure; but the molecules that appear brighter and dimmer display internal structure. As many as ten different internal structures can be picked out of Figure 6 including "x" shaped molecules, molecules with three stripes with different intensities of the stripes, and molecules with a central bright spot surrounded by three, four, or five lobes. While these structures cannot be easily related to the truncated icosahedral structure of the C_{60} molecule, several important points can be drawn from these results. First, structures that give rise to bright spots or brighter molecules appear distinctly different from those that give rise to dimmer molecules. Also, in structures with stripes, the same structures always have the stripes oriented along the same crystallographic directions of the Au(111) surface. For example, the "x" shaped structures shown in Figure 6.a have one stripe oriented in the [110] direction and the other in the [1$\overline{1}$2] direction. This suggests that the observed structures are a convolution of the electronic structure of the C_{60} molecule and the underlying Au. The "x" shape can be rationalized in terms of a bridge bound C_{60}. However, Figure 6.d shows adjacent "x" shaped molecules with the "x" in the same orientation. The lattice matching of the overlayer and the substrate makes adjacent bridge bound molecules with the same crystallographic orientation impossible. Further, repeated imaging again showed that a molecule could change internal structure without any measurable lateral shift. The structures then cannot be rationalized solely in terms of the adsorption site geometry. Therefore, the observed appearance of an adsorbed C_{60} molecule depends on both the rotational orientation of the molecule on the surface and on the geometry of the surface site.

The images in Figure 6 also show that most of the molecules display no resolvable internal structure. Similar results were obtained in previous studies of mixed fullerenes on Au(111) [17] and pure C_{60} on polycrystalline Au. On GaAs(110), however, the C_{60} appeared uniform with no resolvable internal structure [25]. The C_{60} bonding to GaAs is weak, and it was suggested that internal structure was not observed because the molecules are spinning on the surface [22,25]. In crystalline C_{60}, NMR and neutron scattering data indicate that the molecules spin with rotational period on the order of 1 ns at 260 K [7], many orders of magnitude too fast to be imaged by STM. Therefore, intramolecular structure can only be observed if the interaction with the surface is strong enough to freeze the rotation of the molecule. This suggests that the molecules on Au(111) that do not display internal structure may be weakly bound to the surface and possibly spinning. However, this explanation is not consistent with tunneling spectra and STM images that show internal structure within the most weakly bound molecules. Also, AES, which averages over a large area, demonstrates that C_{60} desorbs from Au(111) at 500°C, indicating that most of the molecules on the surface are strongly, not weakly, bound. Finally, STM indicates changes in orientation of the molecules on the time scale of seconds to minutes. The intramolecular corrugation will also depend on the degree of electron localization, and therefore lack of internal structure can also be an indication of electron delocalization and symmetric charge distribution. If this is the case, intramolecular corrugation may be present,

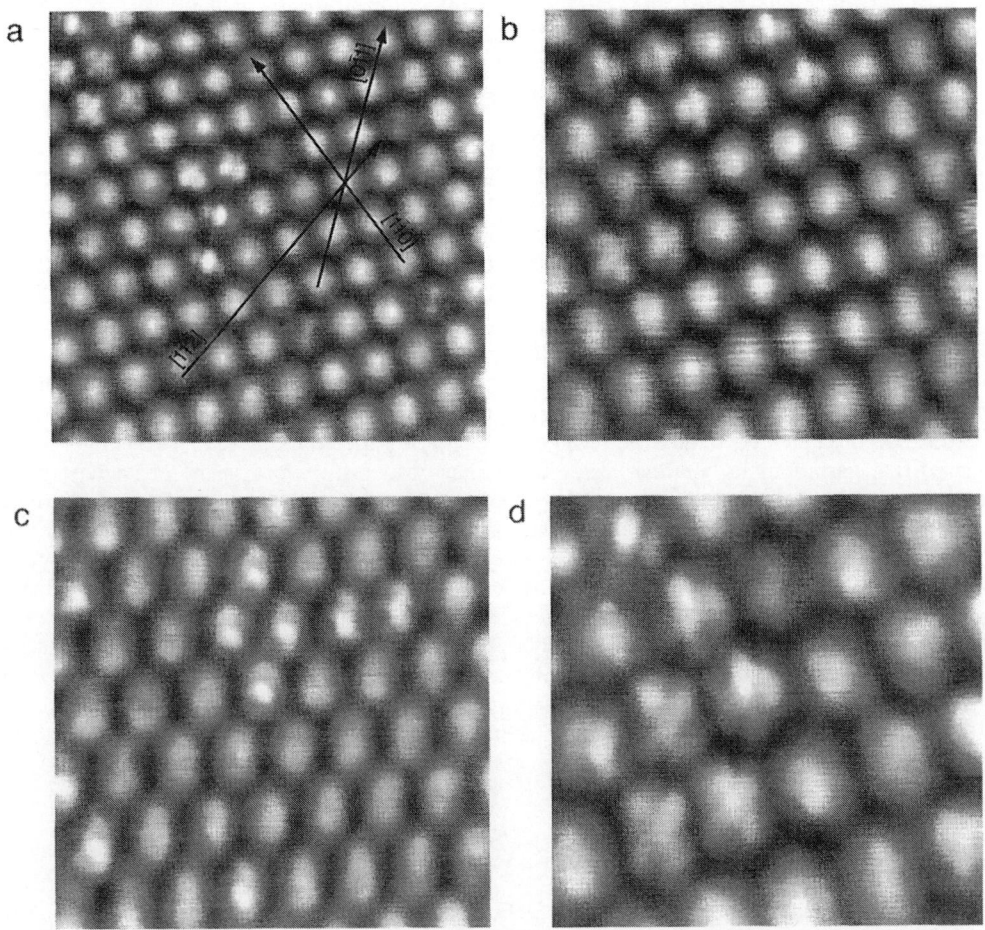

Figure 6)   STM images displaying intramolecular contrast. (a) 10 x 10 nm, (b)
7.3 x 7.3 nm, (c) 7.3 x 7.3 nm, and (d) 4.8 x 4.8 nm.

but the STM does not have sufficient resolution to resolve the structure. The lowest corrugations measured with the STM used in this study is approximately 0.005 nm.

4. Summary

STM results of mixed fullerenes and pure C_{60} indicate that both grow in a layer-by-layer manner on Au(111) with nucleation occurring at specific locations along the steps defined by the 23x$\sqrt{3}$ reconstruction. The presence of larger fullerenes does not affect the packing, structure, or orientation of the film with respect to the Au substrate. In both cases close-packed structures with a spacing of 1.0 nm were observed. Two predominant ordered phases were observed: (1) a layer with a periodicity of approximately 38 Au lattice spacings; and (2) a layer with a $2\sqrt{3}$x$2\sqrt{3}$ R30° unit cell. For pure C_{60} films, apparent height variations in the film were observed in both phases. STS and barrier height measurements indicated that the apparent height differences were due to electronic variations caused by differences in adsorbate bonding depending on the surface site and the rotational orientation of the molecule. Tunneling spectra indicated that variations in bonding were due primarily to differences in the efficiency of the interaction of the Au with the unoccupied states of the molecule. A number of different intramolecular structures were observed. The observed structures were aligned along specific crystallographic directions of the underlying Au. The intramolecular contrast therefore appears to be a convolution of the molecular structure and the structure of the underlying Au. Time-lapsed STM images suggest that C_{60} adsorbed on Au can rotate on the surface over the time scale of seconds to minutes compared to 1 ns for crystalline C_{60}. The observation of internal structure, the tunneling spectra, and the observed rotation times along with our previous report of a desorption temperature 200°C higher for monolayer fullerenes compared to thick films and that the fullerenes lift the Au(111) 23x$\sqrt{3}$ reconstruction all indicate an unusually strong interaction between fullerenes and Au(111).

5. Acknowledgements

E.I. Altman acknowledges the support of ASEE as an ONT Postdoctoral Fellow. The authors also thank the Office of Naval Research for financial support of this work.

6. References

1. Krätschmer, W. , Fostiropoulos, K. and Huffman, D.R. (1990) 'The Infrared and Ultraviolet Adsorption Spectra of Laboratory Produced Carbon Dust: Evidence for the Presence of the C_{60} Molecule', Chem. Phys. Lett. 170, 167-170.
2. Krätschmer, W., Lamb, L.D., Fostiropoulos, K. and Huffman, D.R. (1990) 'Solid C_{60}: A New Form of Solid Carbon', Nature 347, 354-358.
3. Van Tenedeloo, G., Op De Deeck, M., Amelinckx, S., Bohr, J. and Kräthschmer (1991) 'Phase Transformation in Solid C_{60}/C_{70}: an Electron Microscopy Study', Europhys. Lett. 15, 295-300.
4. Heiney, P.A., Fischer, J.E., McGhie, A.R., Romanow, W.J., Derenstein, A.M., McCauley Jr., J.P., and Smith III, A.B. (1991) 'Orientational Ordering Transition in

Solid C_{60}', Phys. Rev. Lett. 66, 2911-2914.

5. Tong, W.M., Ohlberg, D.A.A., You, H.K., Williams, R.S., Anz, S.J., Alvarez, M.M., Whetten, R.L., Rubin, Y and Diederich, F.N. (1991) 'X-Ray Diffraction and Electron Spectroscopy of Epitaxial Molecular C_{60} Films', J. Phys. Chem 95, 4709-4712.

6. Yannoni, C.S., Johnson, R.D., Meijer, G., Bethune, D.S., and Salem, J.R. (1991) '^{13}C NMR Study of the C_{60} Cluster in the Solid State: Molecular Motion and Carbon Chemical Shift Anisotropy', J. Phys. Chem. 95, 9-10.

7. Neumann, D.A., Copley, J.R.D., Cappelletti, R.L., Kamitakahara, W.A., Lindstrom, R.M., Creegan, K.M., Cox, D.M., Romanow, W.J., Coustel, N., McCauley Jr., J.P., Maliszewskyj, N.C., Fischer, J.E. and Smith III, A.B. (1991) 'Coherent Quasielastic Neutron Scattering Study of the Rotational Dynamics of C_{60} in the Orientationally Disordered Phase', Phys. Rev. Lett. 67, 3808-3811.

8. Saito, S. and Oshiyama, A. (1991) 'Cohesive Mechanism and Energy Bands of Solid C_{60}', Phys. Rev. Lett. 66, 2637.

9. Stephens, P.W., Mihaly, L., Lee, P.L., Whetten, R.L., Huang, S.M., Kaner, R., Deiderich, F. and Holczer, K. (1991) 'Structure of Single-Phase Superconducting K_3C_{60}', Nature 351, 632-634.

10. Hammond, G.S. and Kuck, V.J. (eds.) (1992) Fullerenes - Synthesis, Properties, and Chemistry of Large Carbon Clusters, ACS Symposium Series 481, ACS, Washington, DC.

11. Most of the fullerene applications reported to date have been speculative and appear mainly in various science news articles. A paper by Brenner, D.W., Harrison, J.A., White, C.T. and Colton, R.J. (1991) 'Molecular Dynamics Simulations of the Nanometer Scale Mechanical Properties of Compressed Buckminsterfullerene', Thin Solid Films 206, 220-224 discusses the mechanical properties of C_{60}; and a recent paper by Wang, Y. (1992) 'Photoconductivity of Fullerene Doped Polymers' Nature 356, 585-587 demonstrates the photoconducting properties of a fullerene-doped polymer.

12. Wöll, Ch., Chiang, S., Wilson, R.J. and Lippel, P.H. (1989) 'Determination of Atom Positions at Stacking Fault Dislocations on Au(111) by Scanning Tunneling Microscopy', Phys. Rev. B39, 7988-7991.

13. Chambliss, D.D. and Wilson, R.J. (1991) 'Relaxed Diffusion Limited Aggregation of Ag on Au(111) Observed by Scanning Tunneling Microscopy', J. Vac. Sci. Technol., B9, 928-932; Chambliss, D.D., Wilson, R.J. and Chiang, S. (1991) 'Ordered Nucleation of Ni and Au Islands on Au(111) Studied by Scanning Tunneling Microscopy', J. Vac. Sci. Technol. B9, 933-937; Chambliss, D.D., Wilson, R.J. and Chiang, S. (1991) 'Nucleation of Ordered Ni Island Arrays on Au(111) by Surface-Lattice Dislocations',Phys. Rev. Lett. 66, 1721-1724.

14. Stroscio, J.A., Pierce, D.T., Dragoset, R.A. and First, P.N. (1992) 'Microscopic Aspects of the Initial Growth of Metastable fcc Iron on Au(111)', J. Vac. Sci. Technol. A10, 1981-1985.

15. Voightländer, B., Meyer, G. and Amer, N.M. (1991) 'Epitaxial Growth of Thin Magnetic Cobalt Films on Au(111) Studied by Scanning Tunneling Microscopy', Phys. Rev. B 44, 10354-10357.

16. Haiss, W., Lackey, D., Sass, J.K. and K.H. Besocke (1991) 'Atomic Resolution Scanning Tunneling Microscopy Images of Au(111) Surfaces in Air and Polar Organic Solvents', J. Chem. Phys. 95, 2193-2196.

17. Altman, E.I. and Colton, R.J. (1992) 'Nucleation, Growth, and Structure of Fullerene

Films on Au(111)', Surf. Sci., in press.

18. Altman, E.I., DiLella, D.P., Lee, K.P., Ibe, J.P. and Colton, R.J. (1992) 'Data Acquisition and Control Systems for Atom Resolved Tunneling Spectroscopy', to be published.

19. Dietz, P., Fostiropoulos, K., Krätschmer, W. and Hansma, P.K. (1992) 'Size and packing of Fullerenes on C_{60}/C_{70} Crystal Surfaces Studied by Atomic Force Microscopy', Appl. Phys. Lett., 60, 62-64.

20. Hallmark, V.M., Chiang, S., Brown, J.K. and Wöll, Ch. (1991) 'Real Space Imaging of the Molecular Organization of Napthalene on Pt(111)', Phys. Rev. Lett. 66, 48-51.

21. Eigler, D.M., Weiss, P.S., Schweizer, E.K. and Lang, N.D. (1991) 'Imaging Xe With a Low-Temperature Scanning Tunneling Microscope', Phys. Rev. Lett. 66, 1189.

22. Ohno, T.R., Chen, Y., Harvey, S.E., Kroll, G.H., Weaver, J.H., Haufler, R.H. and Smalley, R.E. (1991) 'C_{60} Bonding and Energy-Level Alignment on Metal and Semiconductor Surfaces', Phys. Rev. B 44 13747-13755.

23. Frank, K.H., Dudde, R. and Koch, E.E. (1986) 'Electron Affinity Levels of Benzene and Azabenzenes on Cu(111) and Au(110) Revealed by Inverse Photoemission', Chem. Phys. Lett. 132, 83-87.

24. Chen, T., Howells, S., Gallagher, M., Yi, L., Sarid, D., Lichtenberger, D.L., Nebesney, K.W. and Ray, C.D. (1992) 'Internal Structure and Two-Dimensional Order of Monolayer C_{60} Molecules on Gold Substrates', J. Vac. Sci. Technol. B10, 170-174.

25. Li, Y.Z., Patrin, J.C., Chander, M., Weaver, J.H., Chibante, L.P.F. and Smalley, R.E. (1991) 'Ordered Overlayers of C_{60} on GaAs(110) Studied with Scanning Tunneling Microscopy', Science 252, 547-548; Li, Y.Z., M. Chander, Patrin, J.C., Weaver, J.H. Chibante, L.P.F. and R.E. Smalley (1991) 'Order and Disorder in C_{60} and K_xC_{60} Multilayers: Direct Imaging with Scanning Tunneling Microscopy', Science 253, 429-433.

MOLECULAR AND CELLULAR ORANIZATES ON THE ELECTRODE SURFACE FOR ELECTRONIC CONTROL OF THEIR FUNCTIONS

M.AIZAWA
Department of Bioengineering
Tokyo Institute of Technology
Nagatsuta, Midori-ku, Yokohama 227
Japan

ABSTRACT. Monolayers of redox enzymes such as glucose oxidase, fructose dehydrogenase and alcohol dehydrogenase are molecularly organized on the electrode surface and are facilitated with electronic communication with the electrode through molecular wire of conducting polymer. The enzyme activity of these molecularly interfaced enzymes are electronically modulated, which provides us with a new design principle of biomolecular electronic devices. Furthermore monolayer of cellular organizates of neurons has been formed on the solid material surface under potential-controlled direction of neurite outgrowth.

1. INTRODUCTION

Our current efforts have been concentrated on building up either biomolecular organizates or cellular organizates on the surfaces of electronic solid materials with retaining their biological functions.The purpose of this research is to implement biomolecules, specifically proteins, into biomolecular electronic devices. Although there have long been endeavors to realize biomolecular electronic devices composed of proteins, the protein molecules find difficulties in electronic communication with electronic conductive materials, which could be essential in input/output of information.

Biomolecular electronic devices are designed on the basis of the use of single molecule properties rather than bulk properties of molecules. One of key issues in fabricating biomolecular electronic devices is how to communicate between the molecular size features of a chip and the molecular device to be achieved. Another difficult problem is associated with interfacing a single molecule with larger features. It is our basic approach to implement molecular wire of conducting polymer as molecular interface between a protein molecule and electronic solid material.

This paper describes the electrochemical fabrication of monolayer of a redox enzyme on the electrode surface, the electrochemical fabrication of molecular interface for the enzyme, and the electron transfer (electronic communication) between the enzyme and the electrode through the molecular interface.

P. Avouris (ed.), Atomic and Nanometer-Scale Modification of Materials: Fundamentals and Applications, 315–325.
© 1993 *Kluwer Academic Publishers.*

2. MOLECULAR ORGANIZATES ON ELECTRODE SURFACE

One of the key technologies for fabricating biomolecular electronic devices is to make protein molecules organized in monolayer scale on the electrode surface. Since most proteins are soluble in water and fragile in conformation, a limited method is applicable to make protein organizates in monolayer scale. With due regard to these properties of proteins the methodology should be based on making use of self-assembly of proteins on the electrode surface. In this paper a potential-assisted self-assembly of proteins is proposed to make protein molecular organizates on the electrode.

The potential-assisted self-assemble is carried out in an electrolytic cell equipped with a platinum or gold electrode (working electrode) on which protein molecular organizates are formed, a platinum counter electrode, and a Ag/AgCl reference electrode. The potential of the working electrode is precisely controlled with a potentiostat with referring the Ag/AgCl electrode. A protein solution should be prepared with taking account of protein isoelectric point because protein charges negatively in the pH range above its isoelectric point.

Fructose dehydrogenase (FDH) is an oxidoreductase which has PQQ (pyrrolo-quinoline quinone) as prosthetic group. Upon enzymatic oxidation of D-fructose, the prosthetic group (PQQ) is reduced to $PQQH_2$ and then an electron acceptor reoxidizes $PQQH_2$ to PQQ, liberating two electrons. FDH has an isoelectric point of pH 5.0.

FDH molecular organizates was formed on the platinum or gold electrode surface by potential-assisted self-assembly. The potential-assisted self-assembly is schematically illustrated in Fig.1. FDH was dissolved in a pH 6.0 of phosphate buffer to make its charge negative. FDH molecules adsorb on the electrode surface primarily due to electrostatic interaction. Rate of adsorption depended on electrode potential. In the potential range from 0 to +0.5 V vs. Ag/AgCl, adsorption rate sharply increased with potential. Furthermore FDH molecules may be organized on the electrode surface in such a manner as the negatively charged site of FDH molecule faces to the positively charged surface of the electrode. As was expected, the biological function of FDH, enzyme activity, was fully retained even in an electrode-bound form. The enzyme activity of electrode-bound FDH was determined after the potential-assisted self-assembly. Distinctive potential dependency of the enzyme activity was obtained as shown in Fig.2. With a good correlation to the potential dependency of the enzyme adsorption, the enzyme activity increased with electrode potential at which the potential-assisted self-assemble was performed.
It is noted that the amount of protein adsorbed is precisely regulated by potential-controlled time. One can easily obtain a monolayer of electrode-bound protein in either a full surface coverage or less surface coverage with retaining its biological function.

In addition to FDH, the potential-assisted self-assembly was applied to several redox enzymes such as glucose oxidase and alcohol dehydrogenase.

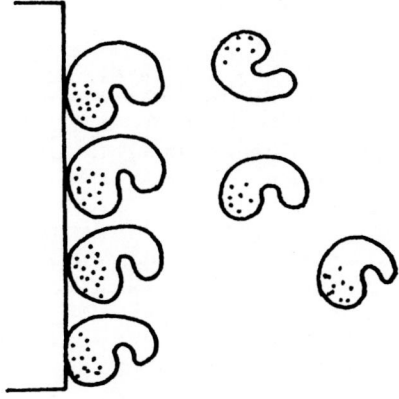

Fig.1   Potential-assisted self-assemble of proteins

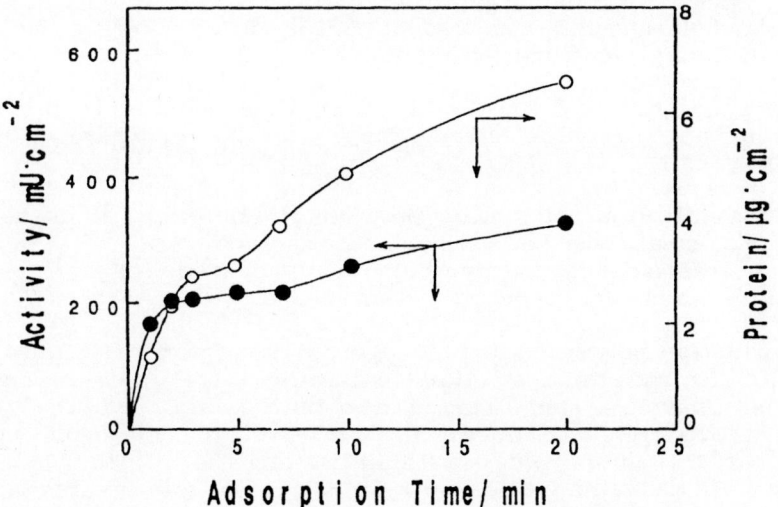

Fig.2   Time dependence on the adsorbed enzyme activity (●) and on the the amount of protein adsorbed (O). FDH was adsorbed on Pt electrode at 0.5 V (vs. Ag/AgCl) in a phosphate buffer solution of pH 6.0 containing 5 mg/ml of FDH. Acquisition of each datum was repeated at least two times, and the resulting values were averaged.

3. MOLECULAR WIRE OF MOLECULAR INTERFACE

Another critical issue in fabricating biomolecular electronic devices is how to electronically communicate between an electrode-bound protein and an electrode. There are two distinctive examples of ideal electron transport systems composed of electron transferring proteins. The one is an electron transport system in the inner mitochondrial membrane. The other falls in an electron transport system in the photosynthetic thylakoid membrane. In these systems intermolecular electron transport efficiently occurrs among different redox potentials of electron tranferring proteins in accordance with potential gradient. However, these electron transferring proteins find difficulty in electron transfer on the electrode surface due to steric hindrance of the electron transferring site of proteins from the electrode surface.

To solve the problem of insufficient electron transfer of electrode-bound proteins, the molecular interface has been incorporated between the electrode surface and electron transferring proteins[1,2].

There are several molecular interfaces for redox enzymes to promote electron transfer at the electrode surface [3-6].
    1) Electron mediator : Either electrode or enzyme is modified with an electron mediator in various manners.
    2) Molecular wire : The redox center of an enzyme molecule is connected to an electrode with such a molecular wire as conducting polymer chain (Fig.3).
    3) Organic salt electrode, and conducting polymer electrode : The surface of an organic electrode may provide enzymes with smooth electron transfer.

In this investigation conducting polymer of polypyrrole has been used as the molecular wire of molecular interface. As an example, the molecular interface of polypyrrole was prepared for the electrode-bound FDH which was made by potential-assisted self-assembly[7].

Electrooxidative polymerization of pyrrole was performed on the FDH-adsorbed electrode in a solution containing 0.1 M pyrrole and 0.1 M KCl under anaerobic conditions at a potential of 0.7 V. The thickness of the polypyrrole membrane was controlled by a polymerization charge. After washing with distilled water, the PP/FDH/Pt electrode was kept in Mallvaine buffer of pH 4.5 at 4C for several hours (8-20h) until further experiments commenced.

Almost all the FDH molecules on the FDH-adsorbed electrode surface seemed to demonstrate their activity because of the mild immobilization at less extreme potential and the easy diffusion of the substrate, although the employment of a polymer matrix to cover the enzyme layer seemed to affect the diffusion process of the substrate, causing the decrease in apparent enzyme activity. Therefore, it was very important to make the polymer membrane as thin as possible to minimize the effect of the membrane on substrate diffusion and to ensure the complete coverage of the enzyme layer. The effect of PP coating on the enzymatic activity of the PP/FDH/Pt electrode is shown in Fig. 4 by changing the polymerization charge from zero to 20mC. Four

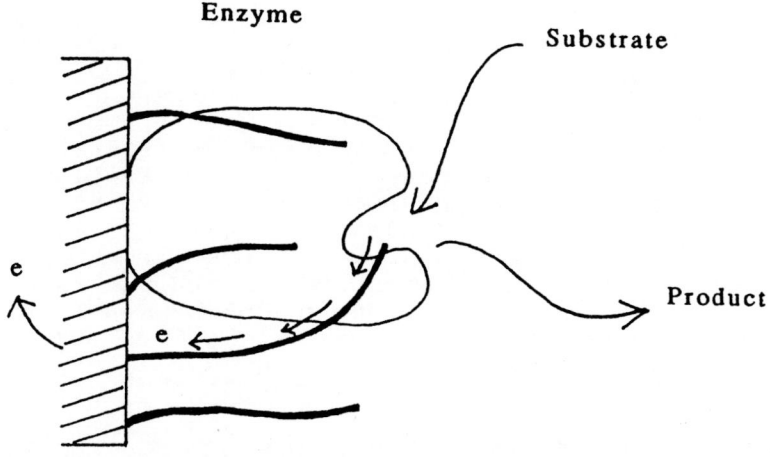

Fig.3 Molecular wire of conducting polymer for molecular interface

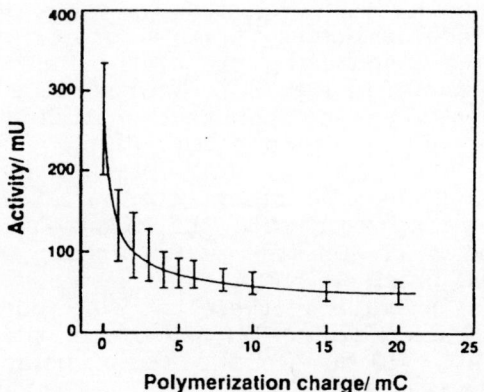

Fig.4 Effect of PP thickness on the enzymatic activity of the PP/FDH/Pt electrode. Activity was measured within 2-3 h after preparation.

polypyrrole-interfaced electrodes were prepared at each polymerization charge, and the mean value was compared. The activity was
determined by the optical method described previously. In general,
the thickness of the polypyrrole membrane is directly proportional to
the polymerization charge. Fig. 4 shows that the apparent activity of
FDH sharply dropped when the polymerization charge reached to 2-
3mC charge and then it fell gradually with the increase in polymerization charge. The PP membrane thickness prepared by passing a
polymerization charge of 2-3mC corresponds roughly to the monomo-
lecular thickness of FDH. The thickness is estimated to be 50-80 A,
provided that the enzyme is spherical. In this study, a polymerization
charge of 4 mC was passed to polymerize pyrrole, as it seems likely
that this electricity is enough to cover the enzyme completely and to
retain sufficient enzyme activity.

4. ELECTRONIC COMMUNICATION THROUGH MOLECULAR INTERFACE

Electronic communication between electrode-bound proteins and an
electrode is confirmed by such electrochemical characterizations as
differential pulse voltammetry.

Differential pulse voltammetry of the FDH/Pt electrode was performed
immediately after the enzyme adsorption because the enzyme on the Pt
surface was easily desorbed from the surface. On the other hand,
electrochemical measurement of the PP/FDH/Pt electrode was carried
out after 20h of storage in the buffer, since it took more than 6-8h
incubation to observe repeated electrochemical response. After 6-8h of
storage the electrode demonstrated a stable response for a few days.
The differential pulse voltammograms of the FDH/Pt and the PP/FDH/
Pt electrodes are shown in Fig. 5. In both cases a pair of anodic and
cathodic peaks were observed which were attributed to the electrochemical oxidation and reduction of the quinoprotein at redox potentials of 0.08 and 0.07 V(vs Ag/AgCl), respectively. In our previous
study[5,6], we found that the redox potential of PQQ in a solution of
pH 4.5 is 0.06 V(vs Ag/AgCl) with the polypyrrole electrode. Therefore, the redox peaks of FDH are related directly to the electrochemical process of the prosthetic PQQ. The important difference between
the electrochemical activities of these two electrodes is as follows : by
the employment of a conductive PP interface, the redox potential of
FDH shifted slightly (10 mV) to the redox potential of PQQ, which
indicates more smooth and easier electrical shuttling between the
prosthetic group of FDH and the electrode through the PP interface.
In addition, the anodic and cathodic peak shapes and peak currents
of the PP/FDH/Pt electrode were identical, which suggests reversibility of the electron-transfer process. One further point requires
emphasis : the peak current of the PP/FDH/Pt electrode is about 8
times greater than that of the FDH/Pt electrode. The significant
increase in redox peaks strongly supports our concept of the molecular interface.
Due to the incorporation of FDH molecules in the π-conjugated
conductive polymer on the surface of the electrode, the
prosthetic PQQ electrically communicates with the base
electrode through the molecular wiring, since the elec-

tron transfer between PQQ and the electrode was enhanced significantly in the presence of the PP film. These results clearly indicate that conductive PP works as an ideal molecular interface.

5. POTENTIAL-MODULATED ENZYME ACTIVITY

Redox enzymes catalyze either oxidation or reduction of corresponding substrate. However, these enzymes cannot be regenerated by themselves if neither electron acceptor nor donnor is associated. An oxidase, for instance, accepts an electron from corresponding substrate to be oxidized. The enzyme remains in a reduced form when the electron cannot be transfered to such an electron acceptor as oxygen. Redox enzymes are thus generally associated by either electron acceptor or donnor for their regeneration.

In absence of any electron acceptors the reduced form of an oxidase could transfer an electron to an electrode of which potential is appropriately controlled. The electrode may work as not only electron acceptor to regenerate the oxidase but signal transducing device to quantitate the enzyme reaction.

These indicate a possible potential dependency of enzyme reaction rate (enzyme activity of electrode-bound enzyme.

Potential dependency of the biocatalytical activity of the PP/FDH/Pt electrode is shown in Fig. 6.
The dependency was investigated by adding 5 mM fructose, and the resulting current response was compared. The applied potential-current response curve was divided into three parts. First, the potential range was less than the redox potential of FDH (0.07 V) where negligible response current was generated because in this potential range less oxidation of fructose occurred since very little FDH was in the oxidized form. Second, the potential range extended from the redox potential of FDH to the rest potential of the PP/FDH/Pt electrode (0.35 V), where a sharp increase of responese current at around 0.1 V and then a gradual increase were observed. Third, the potential range was at a potential higher than the rest potential. The sharp increase in response current was observed up to 0.6 V. At a potential higher than this potential the response current fell sharply, probably due to the irreversible deactivation of FDH. The possible reasons for the deactivation at higher potential may be (1) higher electrical field causes the conformational change of the FDH in such a way that the enzyme loses its prosthetic group, or it may change the structure of the enzyme into an inactive form, and (2) higher potential drastically chenges the pH inside the membrane interface in a manner such that the enzyme loses its activity. However, enzyme activity at the

322

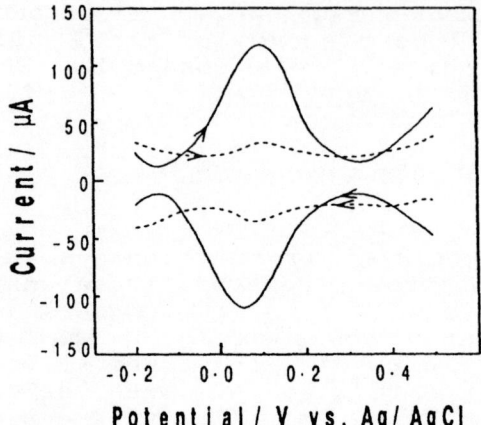

Fig.5   Differential pulse voltammograms of FDH/Pt (- - -) and PP/
FDH/Pt (—) electrodes in McIlvaine buffer solution of pH 4.5.

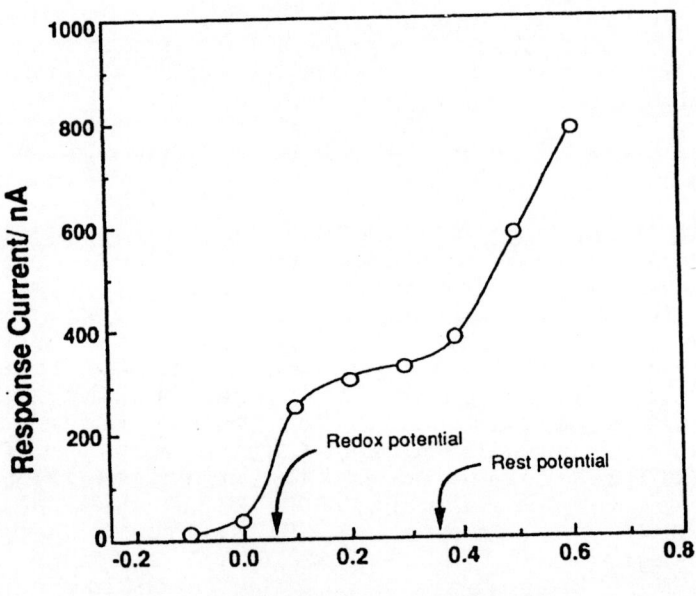

Fig.6   Dependency of the enzymatic activity of the PP/FDH/Pt
electrode on applied potential. Potentials were applied to the same
PP/FDH/Pt electrode. When the residual current became steady state,
fructose solution was injected to a final concentration of 5 mM, and
the resulting increase in anodic current was measured as the response
current.

PP/FDH/Pt electrode can be controlled reversibly in the potential range from 0.1 to 0.6 V. The increase in anodic current upon the addition of fructose can be explained by considering the following three processes ; (1) diffusion of substrate into the PP interface ; (2) electron transfer from substrate to the enzyme upon the enzymatic oxidation of fructose (i.e.,FDH-PQQ is reduced to FDH-PQQH$_2$) ; (3) reoxidation of reduced FDH to oxidized FDH-PQQ by releasing an electron to the electrode through the PP wiring. All of these three steps may be accelerated by the application of higher potential. However, we speculate that the last step is mainly responsible for the sharp increase in response current as described above. In such a case, the last step, i.e., the electron-transferring step from FDH to the electrode, is the rate-limiting step. As the enzyme activity was reversibly controlled, the PP interface electrode was controllable as far as the applied potential was in the range from 0.1 to 0.6 V without any deterioration of the enzyme. Since PP has a positive charge in this potential range, the doped π-conjugated polymer works as a molecular wire which carries electrons between a biomolecule and an electrode.

6. ARTIFICIALLY DESIGNED CELLULAR ORGANZATES

One of challenging endeavors is to implement nerve cells into biomolecular electronic devices. Multiple input signals are processed by a single nerve cell to transmit sole output signal to adjacent nerve cell through the axon. Furthermore the signal is processed at the synapse which is a junction of three dimennional neuronal network.

Our serial investigation have shown that the proliferation of mammalian cells are cultured on the electrode surface. In the line of the investigations we attempted to construct 2-dimentional in vitro neural network cultured on the potential-controlled electrode surface[8].

Rat pheochromocytoma (PC12) cells were plated on the surface of an optically transparent electrode coated with collagen. The electrode was placed at the bottom of a culture vessel with a counter electrode and a Ag/AgCl reference electrode. The electrode potential of the optically transparent electrode was controlled with a potentiostat and a function generator.

A patterned electrode of In_2O_3 was fabricated on a glass plate. The pectinated In_2O_3 electrodes were patterned in parallel at a distance of 20μm. PC12 cells were placed on the surface of the patterned electrode plate and cultured in the presence of NGF. The electrode potential was controlled during culture.

PC12 cells were cultured on the optically transparent electrode in the presence of NGF to initiale differentiation. In the potential range below 0.3 V vs. AG/AgCl, the cells differentiated to neurons with extending neurites. In sharp contrast differentiation was markedly inhibited above 0.35 V vs. Ag/AgCl.

The findings have encouraged us to construct an electrically controlled 2-dimensional _in_ _vitro_ neuronetwork model.

The PC12 cells were placed on the surface of the patterned electrode plate. Potential controlled culture was continued for 14 days in the presence of NGF. The neurites were grown in the direction of the glass stria. The electrode prevented the neurites from growing. This should be the first step to control the direction of neuronetwork formation.

7.CONCLUSION

Protein molecular organizates have successufuly been assembled by the potential-assisted self-assemble on the electrode surface, while the biological activity is fully retained. The potential-assisted self-assemble seems feasible to fabricate protein molecular orgaizates on the electrode surface. Molecular wire of conducting polymer finds effective on facilitating electronic communication of electrode-bound protein molecules with the electrode. The enzyme activity of electrode-bound proteins is modulated by electrode potential.

Cellular organizates of neurons are formed on the solid material surface by potential-controlled direction of neurite outgrowth.

These should provide us with new technology in fabricating biomolecular electronic devices.

REFERENCES

1. Aizawa, M. "Protein Molecular Assemblies and Molecular Interface for Bioelectronic Devices" in Molecular Electronics-Science and Technology (ed., A. Aviram), Engineering Found. (1989) pp. 301-308.
2. Aizawa, M., S.Yabuki, H.Shinohara, "Biomolecular Inter face" in Molecular Electronics (ed., F.T.Hong), Plenum Press (1989) pp. 269-276.
3. Yabuki, S.; Shinohara, H.;Ikariyama, Y.; Aizawa, M. J.Electroanal.Chem.Interfacial Electrochem. 1990, 277, 179.
4. Yabuki, S.; Shinohara, H.; Aizawa, M, J.Chem.Soc.,

Chem.Commun. 1989, 14, 945.
5. Shinohara, H.; Khan, G.F.; Ikariyama, Y.; Aizawa, M.,
   J.Electroanal.Chem.Interfacial Elactrochem. 1991, 304,
   75-84.
6. Khan, G.F.; Shinohara, H.; Ikariyama, Y.; Aizawa, M,
   J.Electroanal.Chem. Interfacial Electrochem. 1991, 315,
   263-274.
7. Khan, G.F., Kobatake, E., Shinohara, H., Ikariyama, Y.,
   Aizawa, M., " Molecular Interface for an Activity Controlled Enzyme Electrode and its Application for the
   Determination of Fructose", Anal. Chem., 1992, 64,
   1254-1258.
8. Aizawa, M., Motohashi, N., Shinohara, H., Ikariyama,
   Y., Furukawa, S., "Electrically Controlled Formation of
   Neuronetwork", Proc. 12th Ann.Int.Conf. IEEE Engineering in Medicine and Biol. Soc., 1990, 1740.

CHARACTERIZATION AND APPLICATION OF NANOSCALE ARTIFACTS IN SCANNING TUNNELING MICROSCOPY

R. BERNDT* and J.K. GIMZEWSKI
IBM Research Division
Zurich Research Laboratory
8803 Rüschlikon, Switzerland

ABSTRACT: We describe various experiments in which nanostructures fabricated with the tip of a scanning tunneling microscope are utilized as tools. These artifacts serve as geometric and electronic reference structures and as model systems for photon emission experiments to probe local electromagnetic fields and inelastic tunneling processes. We apply bias-dependent imaging and photon emission from the STM to deduce chemical information on nanostructures and the STM tip.

I. Introduction

The utility of scanning tunneling microscopy to locally study geometric arrangements of crystal surfaces and adsorbed materials and also aspects of their electronic structure is well documented [1]. In studies of this kind the interaction of tip and sample is often considered as a disagreeable complication. These interactions, however, are the basis of a related area of research: the STM may be used to modify surfaces and fabricate nanoscopic objects [2, 3]. The fascination with this ability has stimulated a large variety of experiments starting from the early days of scanning tunneling microscopy [4] and has recently lead to the manipulation of single atoms [5] and the construction of simple devices [6]. However, the development of methods to construct small objects has so far not been accompanied by a similar improvement in their characterization.

Recently, we have demonstrated that intense photon emission from the gap of an STM operated on metal surfaces is due to inelastic tunneling (IET) excitation of tip-induced plasmon (TIP) modes [7]. Therefore, by detecting photons emitted from an STM, investigation of the electromagnetic interaction of two nanometer-sized objects at nanometer distances is feasible, which is relevant for properties of nanostructured materials or clusters and appears to be involved in surface enhanced Raman scattering [8]. Here we discuss simultaneous measurements of constant-current STM topographs and of the integral photon intensity (photon maps) as well as fluorescence spectra of the photons emitted from the STM. We point out that a thorough understanding of such experiments requires consideration of tip and sample as an intimately coupled system.

In the context of this workshop we present a variety of experiments conducted in our laboratory in which structures deliberately induced by the STM tip have been used to

---

*Present address: Université de Lausanne, Institut de Physique Expérimentale, 1015 Lausanne, Switzerland

P. Avouris (ed.), Atomic and Nanometer-Scale Modification of Materials: Fundamentals and Applications, 327–335.

study local properties of the structures themselves, the surrounding surface or the tip. We demonstrate the use of nanoscale structures as markers and test objects in STM experiments. The topic of chemical characterization of small objects on a surface and of the probe tip, which can be of comparable importance, is addressed.

II. Reference Structures

A detailed analysis of a time lapse series of STM images is often hampered by considerable drift, which may occur, in particular, after large tip displacements. Moreover a clear assignment of changes in images to changes caused by deliberate variation of external parameters such as the tip voltage V_t is impeded by possible changes in tip state which are frequently observed. Using the tip of an STM, reference structures may be generated, which help to circumvent these difficulties.

An application of surface modification to this end is presented in Fig. 1. It shows an

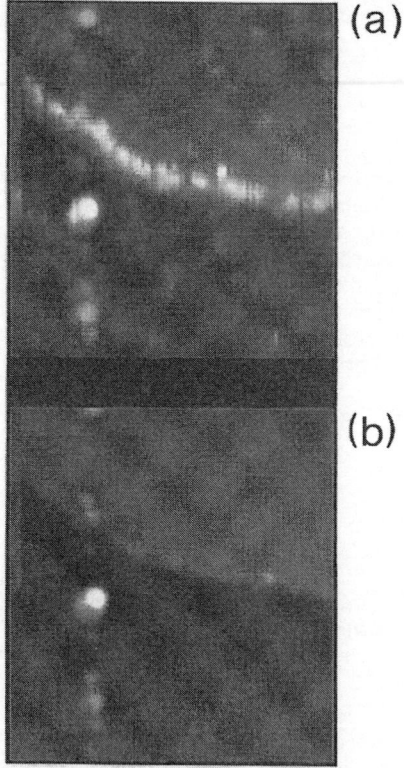

Figure 1. Bias dependence of STM topographs of a 250 Å × 250 Å area of a Ti film that has been exposed to 20 L of molecular oxygen. Topographs rendered as gray-scale images are shown for $i_t = 1$ nA and $V_t = -3.5$ V in (a) and $V_t = +3.5$ V in (b). Height scale = 7 Å. A row of protrusions running down the left-hand side of the image was written with the tip as described in the text.

STM topograph of a 250 Å × 250 Å area of a Ti film evaporated onto Si(111) in ultra-high vacuum. The film has been exposed to 20 L of molecular oxygen. Characterization of evaporated films by STM prior to oxygen exposure reveals large well-ordered areas separated by steps consistent with the growth of (0001) Ti layers. After exposure to oxygen a high density of oxygen-induced structures is observed, which exhibit an intriguing bias dependence. In order to produce a reference structure we have applied a series of short positive voltage pulses to the tip. This procedure was found to create a line of protrusions (see left-hand side of Figs. 1a and b). At negative tip voltage $V_t = -3.5$ V (Fig. 1a), the oxygen induced structures appear as protrusions of ~1.5 Å in height and ~20 Å in width. Steps (for an example see the step running approximately horizontally through the surface area shown in Fig. 1) appear to be covered with structures of similar dimensions. Reversal of the bias polarity results in the image shown in Fig. 1b. The oxygen induced structures appear as shallow depressions (depth ~0.2−0.5 Å). The images of steps change similarly. The voltage dependence of the artificial structures, which have dimensions on a similar scale as the oxygen-induced features (at $V_t = -3.5$ V), is much less pronounced. These features appear as protrusions over a variety of tunnel voltages at different polarities, which is indicative of a metallic character. The change in apparent size of the artificial features is dominantly related to the change in the tip-surface distance with the bias voltage at constant current. This permits us to exclude changes of tip state as a possible origin of the observed bias dependence of the oxygen-related structures. Instead, these observations are typical of electronic effects that play an important role in the imaging mechanism. In the framework of Lang's theoretical model for jellium surfaces with various adsorbed atoms [9], the larger features imaged at $V_t = -3.5$ V should reflect an increase in the unoccupied density of states (DOS) in the sample, whereas at low or positive bias a decrease in occupied DOS would be expected. A detailed analysis [10] shows that our STM observations are indeed consistent with electron spectroscopic data of the oxidation of Ti [11].

III. Model Systems for Photon Emission Experiments

In addition to the static electric field between tip and sample of an STM, there exist dynamic electromagnetic modes induced by the close proximity of tip and sample [12]. We have recently demonstrated that these localized, tip-induced plasmon modes couple strongly to tunneling electrons [7]. On metal surfaces the lowest-energy mode, which has an approximately dipolar symmetry (cf. Fig. 2), gives rise to emission of visible

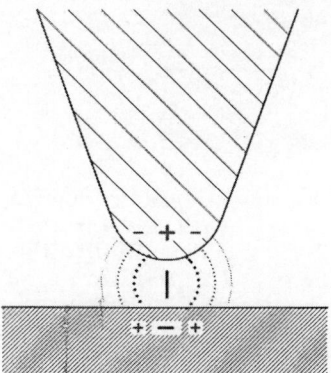

Figure 2. Schematic model of a tip induced plasmon mode in the STM configuration, which is localized in the tip-surface gap region. The + and − symbols indicate the charge density oscillations frozen at a point in time.

light from the tunneling gap. The resonance frequency of these modes and also the resulting photon intensity are affected by the dielectric properties of tip and sample. In order to visualize the localization of these modes it is desirable to prepare single metallic particles with characteristic radii in the 100 angstrom range. Based on theoretical considerations [13, 14], enhanced emission is expected from such particles.

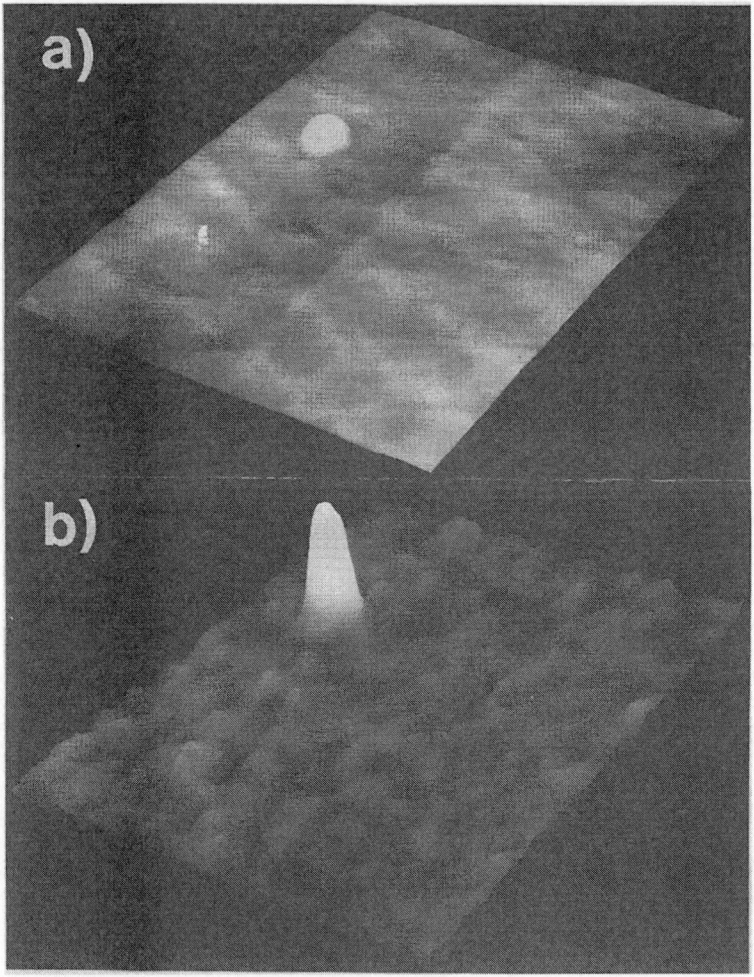

Figure 3. Topograph and photon map measured simultaneously in constant current mode on a 2500 Å × 2500 Å area of a Cu(111) surface with $V_t = 2.2$ V and $i_t = 2$ nA (height = 50 Å). (a) Topographic image. The surface had been modified prior to this measurement by applying short negative voltage pulses to the tip. This resulted in the formation of a protrusion and the creation of a dislocation network. (b) Photon map observed simultaneously. The light intensity data was used for shading with a grey-scale corresponding to 0 to 4000 counts per second. From Ref. [20].

Using metal evaporation and annealing we have prepared three-dimensional Ag particles on Si [13]. However, on these samples strong coupling occurs between Ag particles, which severely complicates image interpretation. With the STM it is feasible to create an individual particle on a flat surface area where no interaction with neighboring particles is present. Figure 3 is an STM topograph of a 2500 Å × 2500 Å area of a Cu(111) crystal surface which has been modified by applying a negative voltage pulse to the tip. By this procedure a protrusion with approximate lateral dimensions of ~250 Å and a height of ~50 Å was generated. For a structure of such dimensions the observed image may differ from the actual shape due to convolution with tip shape. Comparison of the topographic image (Fig. 3a) and the map of integral photon intensity measured simultaneously (Fig. 3b) reveals fairly constant photon emission for tip positions above flat portions of the surface. Above the protrusion light emission is increased by an order of magnitude as expected [13, 14]. Furthermore, the sharp contrast occurring in the photon map clearly indicates that the emission is due to a mode localized to the tip and the protrusion. It should be noted that the observed increase in photon intensity reflects a local enhancement of the electromagnetic field of TIP modes. Similar field enhancements are believed to be relevant in surface enhanced Raman scattering [8, 15]. In this context the STM experiment may be regarded as a model system for investigating the electromagnetic modes arising between nanometer-sized objects.

IV. Deduction of Chemical Information from Photon Intensity Maps

In the present section photon emission data is used to deduce chemical information on surface structures. The chemical identity of materials affects the observed photon intensity on two conceptually different levels. Firstly, the dielectric properties of a material determine strength and frequency of TIP modes [7]. Secondly, the local density of final states for inelastic tunneling processes, which give rise to photon emission, is sensitive to chemistry on an atomic scale [7, 16]. Sample and tip materials affect the photon emission characteristics in a similar manner. Subsequently it is demonstrated that the dielectric properties of the tip have a significant effect on photon spectra which may in turn be used to characterize the tip. Furthermore, photon maps are used to deduce the chemical identity of nanostructures on surfaces.

IV.A THE TIP

Figures 4(a) and (b) display fluorescence spectra obtained on a Au(110) surface with a W tip and a Au-covered tip, respectively. Figures 4(c) and (d) show analogous data obtained on Ag(111) using W and Ag-covered tips. Significant changes occur in spectral distribution and total intensity when W tips are covered with thick layers of noble metal. The overall intensity increases by an order of magnitude and the emission maximum is shifted to the red. Emission peaks become sharper and in the case of Ag a multiple peak structure develops. These observations are directly related to the different dielectric properties of W and noble-metal tips. A detailed discussion of this data and theoretical model calculations will be reported elsewhere [17]. The results shown in Fig. 4 indicate that photon emission spectra may be utilized to characterize the tip material on a scale that is relevant for TIP modes. This probably does not permit chemical identification on an atomic scale. However, it should be noted that we have repeatedly observed changes from emission spectra typical of a W tip to those associated with noble metal tips when the STM was operated at elevated tunneling currents ($i_t \gtrsim 100$ nA) on noble metal surfaces. This suggests that sufficient material transfer between tip and sample occurs under such conditions to significantly alter photon emission spectra.

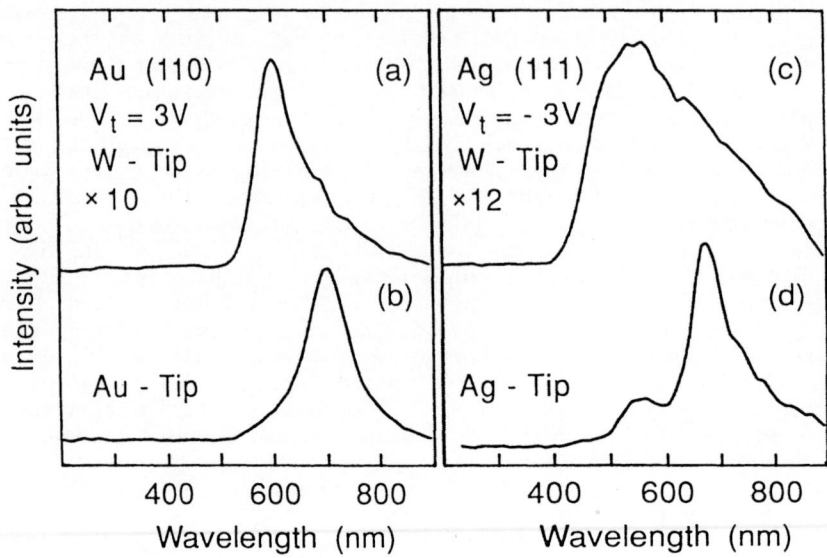

Figure 4. Effect of tip material on fluorescence spectra. (a) and (b) display experimental results measured on Au(110) surfaces for W and Au tips, respectively. (c) and (d) show spectra from Ag(111) surfaces accumulated using W and Ag tips, respectively. Spectra have been normalized using the factors given in the figure.

IV.B SURFACE STRUCTURES

Figure 5 shows (a) a topographic image and (b) a photon map of a Cu(111) surface observed simultaneously. From flat surface areas a fairly constant photon intensity is observed in agreement with model calculations for a W tip on Cu [7]. Intensity variations found at steps have been interpreted in terms of a geometry-dependent variation of the matrix element for inelastic tunneling processes [18]. Here we focus on the contrast observed in the photon map at a line of structures written using positive voltage pulses, which were applied to the tip (see top of Figs. 5a and b). These pulses have generated protrusions of similar dimensions as observed previously on Ti (cf. Fig. 1). Most of these structures reduce the photon intensity, although an enhancement is also observed for few of the protrusions. The data presented in Figs. 1, 3, and 5 is consistent with the following reasoning. When creating the artificial structures only few materials are present: Cu and Ti, respectively, from the sample, W from the tip, and possibly O or C from tip contamination [19]. From previous measurements on clean and oxygen-exposed Ti [16] we conclude that transition metals such as W and adsorbates such as O or C reduce the photon intensity with respect to that of a clean Cu surface. Therefore the observed increase in photon intensity for the structure in Fig. 2 suggests that application of negative pulses to the tip has created a structure consisting dominantly of Cu. This is indicative of a thermally assisted process. On the other hand, positive pulses often result in smaller structures (cf. Figs. 1 and 5). We have fabricated structures of similar dimensions on Au[110] surfaces [20]. The reduction of photon intensity found at these structures suggests that they mainly

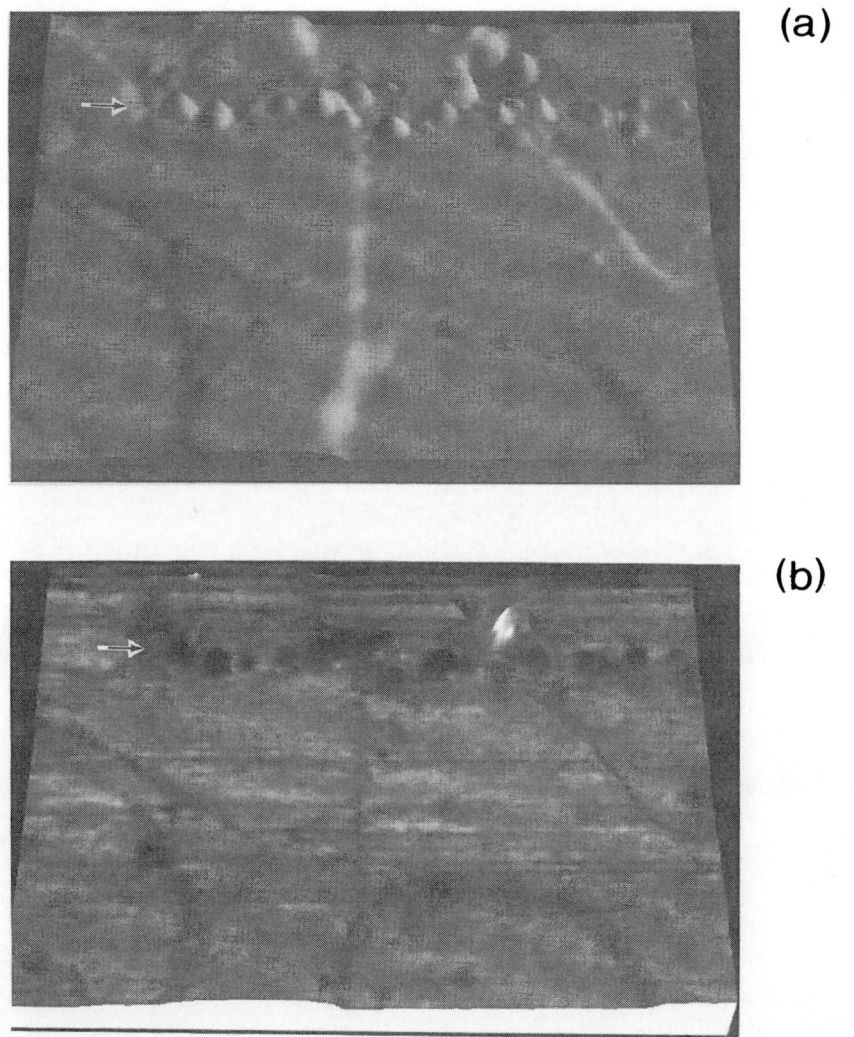

(a)

(b)

Figure 5. (a) Constant current topography and (b) photon STM images of a Cu(111) surface exhibiting terraces separated by steps. A horizontal line of small structures indicated by arrows was created by a series of electric field pulses at positive tip polarity. $V_t = 2.2$ V, $i_t = 5$ nA. Area $= 500$ Å $\times$ 300 Å, intensity scale $= 6000$ cps and height scale $= 15$ Å.

contain tip material transferred by field or current-assisted desorption. The dissimilarity of these structures with those induced by oxygen adsorption, which is apparent in bias-dependent images (cf. Fig. 1), indicates that they are not induced by O (or C). Consequently, our observations lead to the conclusion that the protrusion contains W, which is transferred to the sample from a tip apex consisting of W.

V. Concluding Remarks

I-V tunneling characteristics and bias-dependent imaging can provide some useful information on the metallicity or non-metallicity of surface structures. Yet a more detailed analysis of their chemical identity is desirable. In the above examples we have presented some applications of surface modification with the STM to perform local experimentation and we have attempted to make the first step of a chemical analysis of the structures obtained. We anticipate that the unique properties of small clusters of matter will permit a variety of interesting experiments involving local probe microscopies and photons.

Acknowledgments

We wish to thank E. Courtens, H. Rohrer and B. Reihl for their support. R. Schlittler provided invaluable experimental assistance.

References

[1]   Behm, R.J. (1990) "Scanning Tunneling Microscopy and Related Methods," in R.J. Behm, N. Garcia, and H. Rohrer (eds.), NATO Advanced Studies Institute Series E, Vol. 184, Kluwer, Dordrecht, p. 173 and Feenstra, R.M. (1990) *ibid.* p. 211.

[2]   Staufer, U. (1992) "Surface Modification with a Scanning Proximity Probe Microscope," in H.J. Güntherodt and R. Wiesendanger (eds.), Scanning Tunneling Microscopy Vol. II, Springer Verlag, Berlin, Heidelberg, in press.

[3]   Quate, C.F. (1990) "Manipulation and Modification of Nanometer Scale Objects with the STM," in Proc. NATO Science Forum '90 "Highlights of the Eighties and Future Prospects in Condensed Matter Physics," Biarritz, France, Sept. 16-21, 1990, Plenum Press, (1992) in press.

[4]   See Proc. of the STM Workshop in Oberlech, Austria, IBM Europe Institute (1985), published in: (1986) IBM J. Rev. Develop. 30, (4) and (5).

[5]   Eigler, D.M. and Schweizer, E.K. (1990) Nature 344, 524.

[6]   Awschalom, D.D., McCord, M.A. and Grinstein, G. (1990) Phys. Rev. Lett. 65, 783.

[7]   Berndt, R., Gimzewski, J.K. and Johansson, P. (1991) Phys. Rev. Lett. 67, 3796.

[8]   Otto, A., Mrozek, I., Grabhorn, H. and Akemann, W. (1992) J. Phys.: Condens. Matter 4, 1143.

[9]   Lang, N.D. (1985) Phys. Rev. Lett. 55, 230; (1986) Phys. Rev. B 34, 5947.

[10]   Berndt, R., Gimzewski, J.K. and Schlittler, R.R. to be published

[11]   Bertel, E., Stockbauer, R. and Madey, T.E. (1984) Surf. Sci. 141, 355; Biwer B.W. and Bernasek, S.L. (1986) Surf. Sci. 167, 207; Konishi, R., Ikeda, S., Osaki, T. and Sasakura, H. (1990) Jpn. J. Appl. Phys. 29, 1805.

[12]   Johansson, P., Monreal, R. and Apell, P. (1990) Phys. Rev. B 42, 9210.

[13]   Berndt, R., Baratoff, A. and Gimzewski, J.K. in Ref. 1, pp. 269-280.

[14]   Persson, B.N.J. and Baratoff, A. (1992) Phys. Rev. Lett. 68, 3324.

[15]   Berndt, R. and Gimzewski, J.K. (1992) Physica Status Solidi A 131, 31.

[16]   Berndt, R., Gimzewski, J.K. and Schlittler, R.R. (1991) in Proc. Int'l Conf. on Scanning Tunneling Microscopy "STM 91," Interlaken, Switzerland, Aug. 12-16, 1991, Ultramicroscopy, in press.

[17]   Berndt, R., Gimzewski, J.K. and Johansson, P., to be published.

[18]   Berndt, R. (1992) Ph.D. thesis, University of Basel, Switzerland.
[19]   Many different recipes exist for preparing tips. In our case freshly etched W tips were annealed in UHV to ~1300 K, then sputter-cleaned using Ne ions, and run in the field emission mode prior to mounting in the STM. The procedure and the materials used would indicate that W, O and C are the most likely elements at the tip apex. During tunneling an additional coating with sample material is also possible.
[20]   Berndt, R. and Gimzewski, J.K. (1992) Surf. Sci. 269/270, 556.

INDEX

0D-0D tunneling 231,232,233
1-D chains 26
2-D (two-dimensional) structures/overlayers 28,30-34,228,272
23x 3 Au(111) reconstruction 17,304,305,312
3-layer resist 236
3-terminal device 241
a-Si:H films 50
A-type atomic step 42
activation barrier 79,89-95
activation energy 81,84
adaptive alignment 169,170
adsorption of Au 310
AES (Auger Electron Spectroscopy) 303,304
AFM data storage 149-151,200
AFM images, calculation of 247-257,296
Ag particles 331
Ag(111) surface 271,283,331,332
Al(001) surface 114-116
alkali metals 25-34, 213,223,224
alkane lamellae, cooperative tilt flips of 268
alkane lamellae, roughening of 268,269
alkane thiols 282-286
alkanes 267,268,272
alkyl-derivatives 267,268,278
alkylsiloxane 275-277,279,281,282,286,288,289
aluminum 135
amorphous silicon 49
anodic oxidation 271
atom, lateral motion of 116-119
atom switch 5-7
atom transfer 91,93,94,97,103,104,107,108,115,116,133
atom transfer, directionality in 14,94,96,106,115
atom transfer, rate of 103
atomic beam 215,217,218
atomic force microscope (AFM) 149,247
atomic gold wire 247
atomic manipulation/modification 1,3,7-9,11,15-17,25,38,67,133,139,166
atomic rearrangements, large scale 18
atomic structure 111
atomic wire 247,248,253,258,259
atoms, displacement of 45
Au(110) surface 331,332
Au(111) surface 17,18,22,23,282-287,303-305,307,309,310,312
Au atomic wire 247
autocompensation model 55
azobenzenes 273

ballistic electron emission spectroscopy (BEEM) 153-163
ballistic regime 118
band-bending 31,34,60
band gap 32,33
band structure 33
band-tail states 55,61
barrier height measurements 305,309,312
BEEM 153
bias-dependent imaging 327
bilayers 279
biocatalytic activity 321

337